Probability and Mathematical Statistics (Continued)

SCHEFFE • The Analysis of Variance
SEBER • Linear Regression Analysis
SEN • Sequential Nonparametrics: Invariance Principles and Statistical Inference
SERFLING • Approximation Theorems of Mathematical Statistics
TJUR • Probability Based on Radon Measures
WILLIAMS • Diffusions, Markov Processes, and Martingales, Volume I: Foundations
ZACKS • Theory of Statistical Inference

Applied Probability and Statistics

ABRAHAM and LEDOLTER • Statistical Methods for Forecasting
AICKIN • Linear Statistical Analysis of Discrete Data
ANDERSON, AUQUIER, HAUCK, OAKES, VANDAELE, and WEISBERG • Statistical Methods for Comparative Studies
ARTHANARI and DODGE • Mathematical Programming in Statistics
BAILEY • The Elements of Stochastic Processes with Applications to the Natural Sciences
BAILEY • Mathematics, Statistics and Systems for Health
BARNETT • Interpreting Multivariate Data
BARNETT and LEWIS • Outliers in Statistical Data
BARTHOLOMEW • Stochastic Models for Social Processes, *Third Edition*
BARTHOLOMEW and FORBES • Statistical Techniques for Manpower Planning
BECK and ARNOLD • Parameter Estimation in Engineering and Science
BELSLEY, KUH, and WELSCH • Regression Diagnostics: Identifying Influential Data and Sources of Collinearity
BHAT • Elements of Applied Stochastic Processes
BLOOMFIELD • Fourier Analysis of Time Series: An Introduction
BOX • R. A. Fisher, The Life of a Scientist
BOX and DRAPER • Evolutionary Operation: A Statistical Method for Process Improvement
BOX, HUNTER, and HUNTER • Statistics for Experimenters: An Introduction to Design, Data Analysis, and Model Building
BROWN and HOLLANDER • Statistics: A Biomedical Introduction
BROWNLEE • Statistical Theory and Methodology in Science and Engineering, *Second Edition*
BURY • Statistical Models in Applied Science
CHAMBERS • Computational Methods for Data Analysis
CHATTERJEE and PRICE • Regression Analysis by Example
CHOW • Analysis and Control of Dynamic Economic Systems
CHOW • Econometric Analysis by Control Methods
COCHRAN • Sampling Techniques, *Third Edition*
COCHRAN and COX • Experimental Designs, *Second Edition*
CONOVER • Practical Nonparametric Statistics, *Second Edition*
CONOVER and IMAN • Introduction to Modern Business Statistics
CORNELL • Experiments with Mixtures: Designs, Models and The Analysis of Mixture Data
COX • Planning of Experiments
DANIEL • Biostatistics: A Foundation for Analysis in the Health Sciences, *Third Edition*
DANIEL • Applications of Statistics to Industrial Experimentation
DANIEL and WOOD • Fitting Equations to Data: Computer Analysis of Multifactor Data, *Second Edition*
DAVID • Order Statistics, *Second Edition*
DAVISON • Multidimensional Scaling

continued on back

Linear Statistical Analysis of Discrete Data

Linear Statistical Analysis of Discrete Data

MIKEL AICKIN

John Wiley & Sons

New York · Chichester · Brisbane · Toronto · Singapore

Library of Congress Cataloging in Publication Data:

Aickin, Mikel.
    Linear statistical analysis of discrete data.

    (Wiley series in probability and mathematical statistics, ISSN 0271-6356. Applied probability and statistics)
    Bibliography: p.
    Includes index.
    1. Contingency tables. I. Title. II. Series.

QA277.A5 1983 519.5′6 83-10609
ISBN 0-471-09774-8

Printed in the United States of America

10 9 8 7 6 5 4 3 2 1

To Mary

Preface

Contingency tables are the oldest form of data display, but most of the important work on their analysis has been done in about the past two decades. My aim in this book is to present a view of the most useful results of this modern development and to introduce one or two new ideas.

Contingency-table analysis means the fitting of models to the data in a table. Typical sorts of models specify the independence of some factors from others, the conditional independence of some factors given the values of others, lack of high-order interaction among variables, and the expression of probabilities or functions of probabilities in terms of fixed background factors (covariates). The material presented here includes log-linear models for cross-classifications, models specified in terms of linear constraints on probabilities, and logistic regression and its more complicated relatives.

What distinguishes this book most completely from other books and journal articles on the topic is its integration of three themes: application, computation, and theory. I view these three aspects of statistics as being mutually reinforcing, and I have tried to organize the material so that anyone interested in only one facet is naturally drawn to consider the others.

There are numerous examples of analyses of data that arose in actual experiments or surveys. There are no fabricated data sets. I regard some of the analyses to be essentially complete as presented, while others are partial, some only tentative. I refrain from divulging which analyses I see as complete and which partial, in the hope that you will try to improve on all analyses you find here. Anyone who feels timid in this endeavor should recall the $2K$ rule: if you can see K plausible ways of analyzing a particular set of data, then you will be able to find $2K$ statisticians, each of whom insists on an analysis different from all the others.

You will also encounter a computer program, LAMDA, whose use greatly facilitates the analysis of discrete data. There are other programs oriented towards contingency tables, but LAMDA is the only one tailored to the methods presented here. For information about obtaining LAMDA, write me at Statistical Consulting Services, 517 East Lodge Drive, Tempe, Arizona 85283.

All of the theory supporting the usual practice of discrete statistics is given here. I have tried to reduce the number of required mathematical tools to an absolute minimum, while giving essentially complete proofs of all results specifically pertaining to contingency-table analysis.

I recognize that various readers will approach this book with more interest in some topics than others. Thus I offer three plans for reading the book. Important sections are marked *, very important sections * *. (In some cases it may be necessary to trace definitions through the general index, or symbol usage through the special-symbol index.)

Section	Application	Computation	Theory
I.A	* *	* *	* *
I.B	* *		
II.A			* *
II.B.1–3			* *
II.B.4	*		*
II.B.5	* *	*	
II.B.6		* *	
II.C			* *
II.D	* *		
III.A.1–3			* *
III.A.4	* *		
III.A.5		* *	
III.A.6			*
III.B	* *	*	
IV.A.1			* *
IV.A.2	* *		*
IV.A.3	*		*
IV.A.4–5	*		
IV.B	* *	*	
V.A.1	*	*	*
V.A.2			*
V.A.3	*		*
V.A.4		*	

Section	Application	Computation	Theory
V.B	* *		
VI.A.1		* *	
VI.A.2	*	*	
VI.B–D	*		*
VI.E		*	*
Appendix A	*	* *	

Paradoxically, discrete data are harder to analyze than continuous data. Although most of us can *look* at a table of counts, few of us can *see* what the pattern of the frequencies means without the help of statistically sophisticated methods. For the most part, continuous data are analyzed by relating the means of various Normal distributions to underlying factors, and in many frequently occurring situations one can in fact visualize this relationship diagrammatically or geometrically. The key to the paradox is that the Normal distribution is very simple, whereas most of the distributions encountered in discrete statistics are more complex.

However, discrete data analysis is not as difficult as one might deduce from the literature. There is much overlap in published work, with different authors preferring various notations and sundry ways of describing what is essentially the same body of information. I am not so deluded as to believe that the present book will solve the redundancy problem—this has been tried before—but I think and hope that the successful reader will easily recognize the common threads running through superficially different accounts of topics in discrete statistics.

I have divided the literature references into three categories. Entries in Group I are both important and readable. Group II contains items which have at least one of these properties. Group III is a far more comprehensive listing of sources relevant to discrete data analysis. I have done this with the intention of facilitating the reader's entry into the literature, because the published work on contingency-table analysis is rather redundant and of greatly varying quality, and so it seems to me unfair to simply list it without any guideposts. Moreover, as is often true in a rapidly expanding field, newer expositions tend to subsume earlier work. I realize that there is a certain risk in organizing the bibliography in this manner, but I offer it only as being representative of my personal view; I am sure others would demand some rearrangement.

Appropriate preparation for this book consists of a course in statistics beyond the elementary level, some calculus, and a course in linear algebra. Readers with this minimal preparation will benefit most from the assistance of a tutor or instructor. To the extent that you are comfortable with multivariate calculus, abstract linear algebra, or advanced statistical theory, you will feel more at home with the material. There are substantive analogies between discrete data analysis on the one hand and the general linear model (regression, analysis of variance, analysis of covariance) and likelihood-ratio tests on the other, and readers familiar with at least one of these should be able to avoid mystification handily.

It is a great satisfaction for me to acknowledge my debt to Norman Breslow, who introduced me to contingency tables and provided me with much needed help and inspiration.

I also owe thanks to several of my colleagues who endured a course given from an earlier version of the book: Michael F. Driscoll, Dennis L. Young, Richard K. Burdick, Daniel F. Brooks, Jose A. Cobas, and Ronald Jacobowitz. I thank Andrew Olshan for help in gathering information on computer programs and other literature.

MIKEL AICKIN

Tempe, Arizona
August 1983

Contents

Symbols, xv

I. FREQUENCIES AND PROBABILITIES 1

A. Contingency Tables, 1
1. Discrete Data, 1
2. Probability Distributions, 2
B. Examples, 7
1. A Table of Type 2, 7
2. A Table of Type 7, 9
3. An Ordered Table, 11
4. A 2^2 Table, 11
5. A 4^2 Table, 13
6. A $2^2 \times 9$ Table, 14
7. A Triangular Table, 16
8. A Four-Way Table, 17
9. Covariate Models, 19
10. A Table Without Factors, 20
11. Conclusion, 20
Problems, 21
Remarks, 22

II. INFERENCE FOR A SINGLE DISTRIBUTION 25

A. Function Spaces, 26
1. Moments, 26
2. Standardization, 28

B. The Empirical Distribution, 31
1. Consistency, 31
2. Asymptotic Theory, 32
3. Pearson's Chi-Square, 37
4. Computation of P-Values, 39
5. Examples, 40
6. LAMDA Notes, 46
C. Discrepancies, 47
1. Definition and Asymptotic Properties, 47
2. Convex Discrepancies, 48
3. The Discrepancies $\mathbb{D}_\omega$, 50
4. Asymptotic Behavior Under Contiguous Sequences, 52
D. Summary, 55
1. Discrepancies, 55
2. Test of $\pi = q$, 56
3. Power of the Test, 56
4. Normality of the Empiricial Distribution, 56
5. Maximal Standard Functions, 57
Problems, 58
Remarks, 65

III. INFERENCE FOR A SINGLE MODEL **68**

A. Asymptotic Theory for Models, 69
1. Linear and Exponential Models, 69
2. Rapid Convergence, 76
3. Minimum-Discrepancy Estimators, 80
4. The ANALYSIS Table, 92
5. LAMDA Notes, 96
6. Existence of ML and MDI Estimates, 98
B. Data Examples, 100
1. One-Way Tables, 100
2. Multi-Way Tables, 109
Problems, 125
Remarks, 131

IV. INFERENCE FOR MODEL CHAINS **133**

A. Nested Models, 134
1. Asymptotic Behavior of Discrepancies, 134
2. The General ANALYSIS Table, 136
3. Classical Log-linear Analysis, 141

4. Residuals, 151
5. Multiple Confidence Intervals for Means, 153
B. Data Examples, 155
1. 2^3 Table, 155
2. $2 \times 3 \times 4$ Table, 158
3. Residuals for Model Fitting, 161
4. 2×3^2 with Linear Effects, 162
5. A Flow Model, 166
6. A Four-Way Table, 169
Problems, 174
Remarks, 180
Interference, 180
Other Topics, 181

V. COVARIATE INFERENCE **183**

A. Bilinear Models, 183
1. Responses and Covariates, 183
2. ML Inference, 188
3. Conditional Inference, 190
4. LAMDA Notes, 196
B. Data Examples, 198
1. Logistic Regression: Indicators of a Single Factor, 198
2. Logistic Regression: Two Linear Factors, 200
3. Logistic Regression: Two Quadratic Factors, 204
4. Logistic Regression: Four Binary Factors, 208
5. $2^2|2^2$, 210
6. Poisson Regression, 212
7. Triangular Table, 214
Problems, 217
Remarks, 225
Computation, 225
BMDP, 225
ECTA, 226
GLIM, 227
SAS, 227

VI. FURTHER TOPICS **229**

A. Computational Algorithm, 229
1. The Newton–Raphson Procedure, 229
2. Convergence Problems, 232

B. Markov Models, 234
C. Conditional and Reversed Inference, 240
D. Homogeneity Tests, 246
1. Full Homogeneity, 246
2. Partial Homogeneity, 250
3. Example of Collapsibility, 252
4. Homogeneity Data Example, 253
5. Collapsibility Data Example, 254
E. Local Approximation by Exponential Models, 256
Problems, 259
Remarks, 263

APPENDIX A: USING LAMDA 265

Instructions, 265
1. Introduction to LAMDA, 265
2. Entering Data, 267
3. Modifying Data, 269
4. Outputting Data, 274
5. Active and Passive Commands, 275
6. Screening Data, 276
7. Creating, Sorting, and Ordering Data, 278
8. Standardization and Fitting of Variables, 280
Examples, 282
1. A Contingency Table, 282
2. Orthogonal Polynomials, 285
3. Binomial Probabilities, 288
Problems, 291

APPENDIX B: EXAMPLE LXM OUTPUT 292

APPENDIX C: LAMDA RUNS 295

APPENDIX D: LAMDA DICTIONARY 335

REFERENCES 337

INDEX 355

Symbols

Symbol	Description	Page
$\rightarrow$	Hypothesis test	139
$\rightarrow_p$	Convergence in probability	31
$\rightarrow_d$	Convergence in distribution	32
$\uparrow_F$	Conditioning	148
$\downarrow_F$	Marginalizing	149
$\bar{\alpha}$	Serial level	137
[ ]	Hierarchical model notation	110
$\langle\ \rangle$	Encloses P-value	38
C	Covariance	26
δ	Normalizing factor	71
$\mathbb{D}$	Discrepancy	47
$\mathbb{D}_0$	Information discrepancy	51
$\mathbb{D}_1$	Likelihood discrepancy	51
$\mathbb{D}_2$	Pearson's discrepancy	31
$\mathcal{E}$	Exponential model	71
F_0	Standard normal cdf	33
F_ν	$\chi^2(\nu)$ cdf	37
$F_{\nu,\lambda}$	$\chi^2(\nu, \lambda)$ cdf	52
$\mathbf{I}$	Identity matrix	—
$\mathcal{L}$	Linear model	69
M	Mean	26
$\mathcal{M}$	Model	91
N	Normal distribution	3
$N(y)$	Number of occurrences of y	6

Symbol	Description	Page
n	Sample size	6
$\mathbf{0}$	Zero vector	—
p, q	Probability distributions	3
$\mathcal{P}_0$ ($\mathcal{P}$)	Set of (strictly positive) probability distributions	3
$\overline{P}$	Serial P-value	137
$\hat{p}$	Empirical distribution	7
π	True distribution	7
$\hat{q}, \hat{q}_n$	MDE in a model	81
$\mathbb{R}$	Real numbers	26
$\mathcal{S}$	Function space	26
U	Unlikelihood	188
V	Variance	27
$\chi^2(\nu)$	Chi-square distribution, ν df	37
$\chi^2(\nu, \lambda)$	Chi-square distribution, ν df, noncentrality parameter λ	52
$\mathcal{Y}$	Contingency table	2
$\otimes$	Tensor or Kronecker product	189
$\#A$	Number of elements in set A	26
$\#\mathbf{u}$	Number of components in $\mathbf{u}$	29

Linear Statistical Analysis of Discrete Data

CHAPTER I

Frequencies and Probabilities

A. CONTINGENCY TABLES

1. Discrete Data

Counting how many things fall into various categories is surely the oldest kind of data analysis. Indeed, part of the reason for inventing the positive integers in the first place was to provide a means for recording numbers of objects in mercantile transactions, a practice which goes back many centuries.

It is a relatively recent idea that one can infer facts about nature by looking at the frequencies with which various events occur. John Graunt's *Natural and Political Observations Mentioned in a following Index and made upon the Bills of Mortality*, published in London in 1662, is usually given credit for ushering in this modern idea. Graunt collected the frequencies of deaths from several causes, as well as frequencies of births. His most noted discovery was that there were more male than female children born. From this he drew support for a natural law against polygamy, and while his reasoning might be regarded as bizarre by modern standards, the fundamental process of inferring conclusions from frequencies is the same as that used today.

The greatest difference between Graunt and us is that we have a more sophisticated technology for studying frequencies. Most important is the notion of probability and the idea that different natural laws lead to different probabilities of various outcomes. Statistical inference consists of reversing this chain; different patterns of frequencies help us to form different conclusions about the underlying probabilities, and thus to draw inferences about natural laws based on the frequencies we observe.

The purpose of this book is to describe a fairly wide class of procedures for taking the first of these steps, drawing conclusions about probabilities on the basis of frequencies. We will be concerned about the theory which justifies these procedures, the means by which solutions to actual data problems can be computed, and the art of using probabilistic models. These three topics constitute discrete data analysis.

The "discrete" part of discrete data refers to the fact that in any given situation there are only finitely many different categories into which a given individual might fall. Technically, in mathematical statistics "discrete" means "countably infinite," but probability models which permit infinitely many outcomes (like the Poisson) are inherently unrealistic on philosophical grounds, and serve primarily as approximations to realistic finite models. For us, discrete means finite.

A finite collection of non-overlapping categories is called a *contingency table*. Naturally, all possible categories should be allowed for in specifying the contingency table for an experiment or survey. The word "cell" is used interchangeably with "category" here. At the end of data collection there will be frequencies or counts in the cells telling how many individuals fell into each category.

The fundamental assumption behind all discrete data analysis is that each subject who enters the study sorts himself into one of the categories according to a probability distribution. This distribution is unknown, but under various hypotheses about the process which does the sorting, one can specify conditions which it must satisfy. One usually takes it for granted that different subjects sort themselves independently of each other, and we will adopt this assumption without comment. (In Chapter VI, however, we will see some models under which this independence requirement is modified.) This means, of course, that we usually assume the subjects are sampled at random from some population, and despite its importance for our methods, we will not concern ourselves here with means for accomplishing this. In dealing with data collected by others, or data not collected according to a formal random sampling scheme, we will generally make a leap of faith and assume that the random-sampling assumption is a reasonable approximation to the actual sampling plan. We do this not because it is good statistical practice (it isn't), but because survey sampling and experimental design are not the topics of the book.

We now turn to the notation for probabilities that will be used from here on.

2. Probability Distributions

Our most general notation for a contingency table is $\mathcal{Y}$, and a typical cell or category in the table is denoted y. We will also have some specialized notation for particular tables, but many results hold for general tables and then the specialized notation just gets in the way.

A probability distribution over the table $\mathcal{Y}$ is defined to be a function p defined on $\mathcal{Y}$, taking values in the interval [0, 1], and satisfying $\Sigma_y p(y) = 1$. We interpret $p(y)$ as the probability that a given subject will fall into cell y. If $A \subset \mathcal{Y}$ is a collection of cells, then the probability that the individual falls into one of the cells in A is $\Sigma_{y \in A} p(y)$. The set of all possible probability distributions on a table $\mathcal{Y}$ is denoted $\mathcal{P}_0(\mathcal{Y})$. The collection of all strictly positive probability distributions [that is, $p(y) > 0$ for all y] is denoted $\mathcal{P}(\mathcal{Y})$. When $\mathcal{Y}$ is clear from the context, we will write only $\mathcal{P}_0$ and $\mathcal{P}$.

Specialized notation for contingency tables is required because in practice most tables have some natural structure. This structure is determined by *factors* having certain numbers of *levels*. Since the probability models for tables are usually described in terms of the factors, it is important to become familiar with them.

The simplest table, in terms of structure, is a one-dimensional table, that is, a table described by a single factor. Let A stand for the name of that factor. If there are m cells in the table, then we denote them by $A0, A1, \ldots, A(m-1)$. This convention of beginning our numbering at 0 (rather than 1) is done for technical reasons related to parameter identifiability that will be explained in Chapter III. Now we can write $\{Ai : 0 \leqq i < m\}$ in our specialized notation, and $p(Ai)$ denotes the probability that an individual will fall into cell Ai. Keeping in mind that this is a one-factor table, we can and will abbreviate $p(Ai)$ to $p(i)$ whenever there is no chance of confusion.

Another convention we will use throughout is that replacement of a factor level by "+" means summation over all levels of the factor. Thus $p(+) = p(A+) = 1$ is true by definition of a probability distribution.

The next most complicated table is a two-dimensional table, one having two factors. Let A and B denote the names of the two factors. Suppose that A has levels $0, 1, \ldots, r-1$ and B has levels $0, 1, \ldots, c-1$. Then a typical cell of the table is $AiBj$, meaning the category in which A is at level i and B is at level j. If all possible combinations of factor levels occur in the table, then we can write $\{AiBj : 0 \leqq i < r, 0 \leqq j < c\}$ as the specialized notation for the table. This is an example of a *cross-classification*; each category consists of subjects at one level for each of the factors, and all factor-level combinations are possible. As before, $p(AiBj)$ denotes the probability of an individual falling into cell $AiBj$. If we agree that the order of the factors is fixed as A, B, then we can and will write $p(ij)$ in place of $p(AiBj)$ whenever this does not cause confusion. It may be worthwhile to point out that $p(121)$ would be ambiguous, since it could stand for either $p(A12B1)$ or $p(A1B21)$. We can always use the more elaborate notation when it is necessary for clarity.

It is natural to visualize the two-dimensional cross-classification as a rectangle, with rows determined by the levels of A and columns determined by the levels of B. By summing probabilities across the columns (within rows) we obtain the *marginal distribution* of the A factor. Following the convention stipulated above,

$$p(Ai) = p(AiB+) = \sum_j p(AiBj)$$

or more simply

$$p(Ai) = p(i+) = \sum_j p(ij)$$

In the same way, summing down the rows (within columns), we have the marginal distribution of B,

$$p(Bj) = p(+j) = \sum_i p(ij)$$

If we make observations in the A, B-table, but then ignore the levels of B, we are *collapsing* the table to the A-margin, or *collapsing across* B. If $p(AiBj)$ are the probabilities for the full table, then $p(AiB+)$ are the implied probabilities for the one-dimensional A-marginal table, that is, the probabilities of A ignoring B.

At the other extreme, suppose we record the level of B for each subject, and then look to the probabilities of various levels for A. We are now looking at c one-dimensional tables, one for each level of B. The appropriate probability distribution of A within the table determined by Bj is

$$p(Ai|Bj) = \frac{p(AiBj)}{p(A+Bj)}$$

the *conditional distribution* of A given Bj. Note also that we can write

$$p(i|j) = \frac{p(ij)}{p(+j)},$$

although this simplified notation depends on our remembering to associate i with A and j with B.

It is important to be clear about the fact that looking at the A marginal distribution amounts to ignoring B, while looking at the conditional distributions of A given B amounts to taking B into account to the greatest extent possible. Thus, marginalization and conditionalization are opposite ideas.

However, when they happen to coincide, we say that A and B are *independent*. Thus, the condition for independence is

$$p(Ai|Bj) = p(Ai)$$

or

$$p(i|j) = p(i+)$$

or again

$$p(ij) = p(i+)p(+j)$$

The independence condition has been fundamental for the development of discrete data analysis, and so we will take some time to examine it in more detail. We observe that the distribution p is determined by three pieces of information: (1) the A-marginal, $p(i+)$; (2) the B-marginal, $p(+j)$; and (3) the fact that A and B are independent. If we specify only (3), then we have specified a collection of distributions on the table. Such a collection is called a *model*, and here in the case of the independence model there is one distribution in the model for each possible pair of A- and B-marginal distributions. By saying that A and B are independent, we mean that the distribution which sorts individuals into the table has unknown marginals, but is determined from its marginals by multiplication.

We gain further insight into independence by taking logarithms:

$$\ln p(ij) = \ln p(i+) + \ln p(+j)$$

This suggests that we might be able to describe the independence model by writing

$$\ln p(ij) = \theta_{Ai} + \theta_{Bj}$$

where θ_{Ai} and θ_{Bj} are unknown parameters. We can verify this by noting

$$p(i+) = e^{\theta_{Ai}} \sum_j e^{\theta_{Bj}}$$

$$p(+j) = \sum_i e^{\theta_{Ai}} e^{\theta_{Bj}}$$

$$p(++) = \sum_i e^{\theta_{Ai}} \sum_j e^{\theta_{Bj}} = 1$$

so that

$$p(ij) = p(i+)p(+j)$$

Thus we see that independence is equivalent to the condition that $\ln p(ij) = \theta_{Ai} + \theta_{Bj}$ for some parameters θ_{Ai} and θ_{Bj}. The ability to express a model in terms of parameters is a great help in the analysis.

For example, with the parametric representation of independence, we can easily see that independence is equivalent to the condition

$$\ln p(ij) - \ln p(i'j) - \ln p(ij') + \ln p(i'j') = 0$$

for any choice of i, j, i', and j'.

It can be shown that any distribution in $\mathcal{P}$ can be written as

$$\ln p(ij) = \theta_{Ai} + \theta_{Bj} + \theta_{AiBj}$$

with certain side conditions on the θ-parameters. It is then true that setting $\theta_{AiBj} = 0$ for all i, j is equivalent to specifying the independence model. This is a simple example of a *log-linear parametrization*, and illustrates the general idea that setting some log-linear parameters equal to zero results in an interesting model.

It should go without saying that the next most complex kind of cross-classification has three factors, A, B, and C. A typical cell is $AiBjCk$ with probability $p(ijk)$. The marginal A, B-distribution is defined on the table collapsed across C by $p(ij+)$. Likewise, the A, C-marginal is $p(i+j)$, collapsed across B. The A-marginal is gotten by collapsing across both B and C: $p(i++)$.

There are now several interesting conditional distributions. For instance, the A, B-conditional distribution given Ck is

$$p(AiBj|Ck) = \frac{p(AiBjCk)}{p(A+B+Ck)}$$

If we have independence of A and B within each Ck table,

$$p(AiBj|Ck) = p(Ai|Ck)p(Bj|Ck) \qquad (\text{all } k)$$

then we say that A and B are *conditionally independent* given C. This is not the same as saying that A and B are independent (ignoring C). Indeed, independence on the marginal A, B-table and conditional independence within each table determined by a level of C are unrelated concepts. Another concept yet is *total independence* of A, B, and C, expressed by

$$p(AiBjCk) = p(AiB+C+)p(A+BjC+)p(A+B+Ck)$$

or

$$p(ijk) = p(i++)p(+j+)p(++k)$$

Extension of these ideas to higher-way tables is straightforward. The list of factors grows longer, and there are more and more marginal tables and various conditional distributions, but they are all direct extensions of the ideas in the three-way table.

We conclude this section by introducing the notation $N(y)$ to stand for the number of individuals who fell in cell y of the table. Thus $N(+) = n$ is the total sample size. In our specialized notation for cross-classifications, $N(AiBj)$ is the frequency in the $AiBj$ cell, while $N(AiB+)$ is the frequency of individuals at level i of factor A, that is, the sum across the ith row of the table.

Under our ubiquitous random-sampling assumption, a natural estimate of the distribution on the table is the *empirical distribution*, $\hat{p}(y) = N(y)/N(+)$. Although $\hat{p}$ always forms the basis for an analysis, it is of great interest that it is often not the best estimator. This fact motivates most of the analytical techniques to be presented here.

Our final notational convention will be to use π to stand for the true distribution according to which the individuals sorted themselves into the table. The letters p and q, with subscripts and superscripts, will be reserved for other distributions on the table.

B. EXAMPLES

1. A Table of Type 2

The simplest possible experiment results in one of two possible outcomes. A sample drawn at random from a population of voters might result in Table I.1. We will describe this table by saying that it has a single *factor*, P = Political Affiliation, with two *levels*, $P0$ = Republican and $P1$ = Democrat. The frequency of $P0$-observations will be denoted by $N(P0)$ or $N(0)$, and the frequency of $P1$-observations by $N(P1)$ or $N(1)$; thus, $N(0) = 220$ and $N(1) = 275$.

In Table I.1 the problem would be to do some inference about $\pi(0)$, the proportion of Republicans in the population. This would, of course, be the same thing as inference about $\pi(1) = 1 - \pi(0)$, the proportion of Democrats. The only reasonable estimator of $\pi(0)$ is the empirical proportion, $\hat{p}(0)$, which is $220/495 = .4444$. We would probably be willing to assume that the number of Republicans in the sample was Binomial(495, $\pi(0)$), and so we use $\sqrt{\pi(0)\pi(1)/495}$ for the standard deviation of $\hat{p}(0)$. In the context of a test of the hypothesis that $\pi(0) = .5$, we refer $\sqrt{495}\,(\hat{p}(0) - .5)/.5 = -2.47$ to the standard Normal distribution for

Table I.1

A Single Factor with Two Levels

Republican	Democrat
220	275

significance. Since the interval $[-1.96, 1.96]$ captures the middle 95% of the standard Normal probability distribution, this observation would cause us to reject the hypothesis that $\pi(0) = .5$. If there were no hypothesis about $\pi(0)$, it would be reasonable to estimate it with the .95 confidence interval $\hat{p}(0) \pm 1.96\sqrt{\hat{p}(0)\hat{p}(1)/495}$, yielding the interval [.4007, .4882].

Although this is the traditional way for statisticians to approach Table I.1, one could choose to look instead at a measure such as the *odds* favoring Republicans in the population, $\pi(0)/\pi(1)$. Bookmakers prefer to use the terminology of odds, and recently epidemiologists have tended to go along, or go even further by introducing the *ln odds*, $\ln[\pi(0)/\pi(1)]$. The obvious estimator of the ln odds is $\ln[\hat{p}(0)/\hat{p}(1)] = -.233$. An expression for its approximate standard deviation is $1/\sqrt{n\pi(0)\pi(1)}$. Thus, to test the hypothesis $\pi(0) = .5$ we refer

$$.5\sqrt{495}\ln\frac{\hat{p}(0)}{\hat{p}(1)} = -2.48$$

again to the standard Normal for significance. A .95 confidence interval for $\ln[\pi(0)/\pi(1)]$ is then

$$\ln\frac{\hat{p}(0)}{\hat{p}(1)} \pm \frac{1.96}{\sqrt{495\hat{p}(0)\hat{p}(1)}}$$

or $-.233 \pm .177$. One can very easily convert the confidence interval for the ln odds into one for the odds itself:

$$[e^{-.233-.177}, e^{-.233+.177}] = [.66, .95]$$

Since $d = \pi(0)/\pi(1)$ is equivalent to $\pi(0) = d/(1 + d)$, we can again convert the confidence interval so that it is appropriate for $\pi(0)$:

$$\left[\frac{.66}{1 + .66}, \frac{.95}{1 + .95}\right] = [.40, .49]$$

All of these slightly different ways of approaching the table give about the same answer. This is to be expected, because 495 observations is a large number to have for the estimation of a single parameter, and so any good inferential procedure should give about the same result. In smaller samples one can get disagreement between different methods. Here, we prefer to use the Normal approximation to $\ln[\hat{p}(0)/\hat{p}(1)]$, because this transformation straightens out some of the skewness in the distribution of $\hat{p}(0)$.

A consequence of this choice is that we prefer to focus on the parameter $\theta = \ln[p(1)/p(0)]$. This is called a *log-linear parametrization*, and leads to the parametric equations

$$p^{\theta}(1) = \frac{e^{\theta}}{1 + e^{\theta}}, \qquad p^{\theta}(0) = \frac{1}{1 + e^{\theta}}$$

This is the simplest case of a kind of parametrization we will see often in the sequel.

2. A Table of Type 7

In Table I.2 we see the frequencies of breast-cancer cases found in various cities in a medical study. There is also a single factor in this table, C = City, but whereas there were only two levels in the preceding case, now there are seven, $C0$ = Boston to $C6$ = Tokyo. An interesting question about these data might have to do with whether there are more cases than one should expect in some cities, although this raises complex questions about the methods of detecting breast cancer.

What we need for an analysis is an assessment of what the distribution of breast-cancer cases would be if there were no differential among the cities. An obvious assumption here is that breast-cancer incidence should be proportional to population, and this assumption gave rise to the distribution q which appears below. There may be serious difficulties with this notion, because the means of collecting cases may not sample the region which geographers use to estimate population size, and moreover, the effort to detect breast cancer may vary widely in the cities.

Nonetheless, we can proceed with a preliminary analysis by computing a standardized residual, $\sqrt{n}(\hat{p} - q)/\sqrt{q(1 - q)}$, for each cell. The resulting

Table I.2

A Single Factor Table with Seven Levels

Boston	577
Glamorgan	607
Athens	795
Slovenia	754
Sao Paulo	532
Taipei	211
Tokyo	847

numbers, shown below, are so large as to cast doubt on the reasonableness of our approach altogether. However, we can note that the two largest cities have the negative residuals, so that some of the problems mentioned earlier may be related to the size of the city. We will now decompose the analysis into one part for the large cities, and one for the small. This gives the second set of residuals below. These are still rather large, but at least in the case of Athens the divergence between the observed and the expected proportion is not statistically significant.

Although this analysis is clearly exploratory, we may well want to summarize our results with confidence intervals. Since there are several residuals to be compared, we require a multiple-confidence-interval approach. To obtain a .95 multiple confidence interval for residuals in the five small cities, each individual confidence interval will require confidence coefficient .99. Similarly, for a .95 confidence interval for the two largest cities, a .975 confidence interval will be required for each. The displayed results are gotten by using

$$\ln\frac{\hat{p}}{q} \pm 2.58\sqrt{\frac{1-\hat{p}}{n\hat{p}}}$$

for a confidence interval for $\ln[\pi/q]$, and then converting to a confidence interval for π/q. The value 2.24 was used in place of 2.58 for the larger cities. Thus, each of the two subcollections of cities has multiple .95 confidence intervals for its odds. With this kind of multiple inference, we also find Slovenia showing incidence in line with its population.

	Bo	Gl	At	Sl	Sa	Ta	To
$\hat{p}$	.1335	.1404	.1839	.1744	.1231	.0488	.1959
q	.0478	.0771	.0880	.0931	.2180	.0319	.4442
St. res.	26.41	15.60	22.26	18.40	−15.11	6.32	−32.86

	Bo	Gl	At	Sl	Ta	Sa	To
$\hat{p}$	.1960	.2062	.2700	.2561	.0717	.3858	.6147
q	.1415	.2283	.2604	.2755	.0943	.3292	.6708
St. res.	8.48	−2.86	1.19	−2.36	−4.20	4.47	−4.43
CI	1.26–1.54	0.83–0.99	0.96–1.13	0.86–1.01	0.64–0.90	1.08–1.23	0.87–0.96

Table I.3

A Table with a Single Ordered Factor

Age	1	1–4	5–9	10–14	15–19	20–29	30–39	40–59	50 +
Frequency	2	32	53	55	23	18	7	5	1

3. An Ordered Table

In Table I.3 we see the number of cases of diphtheria during an outbreak in 1970, classified according to the age group of the patient. Again we have a table with a single factor, A = Age, with levels $A0$ = "1 year" to $A8$ = "50 + years." The form of this table differs from that of the preceding one in that the age factor defines an ordering among the cells. One could associate with each cell the midpoint of the age class (with some suitable convention for the oldest class), and we will see in Chapter III, Section B.1 how this quantity can be used to help us analyze the data.

4. A 2^2 Table

In the political survey of Table I.1, each individual might have been classified according to socio-economic status in addition to political affiliation. This gives us Table I.4, with factors P = Political Affiliation and C = Socio-economic Class ($C0$ = Working Class and $C1$ = Middle Class). The cell containing working-class Republicans is $C0P0$, while working-class Democrats are in $C0P1$. The frequency of observation of $CiPj$ will be denoted $N(CiPj)$ or $N(ij)$. With the presence of two factors it becomes possible to ask whether there is an association between them, and how it might be measured and estimated.

A common approach to the analysis of Table I.4 is to ask whether political affiliation is independent of socio-economic status. If the two

Table I.4

A Table with Two Factors, Each Having Two Levels

	Republican	Democrat	
Working Class	60	181	241
Middle Class	160	94	254
	220	275	

factors are not independent, then one wants an estimate of their degree of association.

The condition of independence can be imposed on the table by requiring $p(ij) = p(i+)p(+j)$. Since this model puts no restrictions on the marginal distributions $p(i+)$ and $p(+j)$, a reasonable procedure would be to estimate $\pi(i+)$ by $\hat{p}(i+)$ and $\pi(+j)$ by $\hat{p}(+j)$. This leads to the estimate of $\pi(ij)$ as $\hat{q}(ij) = \hat{p}(i+)\hat{p}(+j)$. Note that $\hat{q}(i+) = \hat{p}(i+)$ and $\hat{q}(+j) = \hat{p}(+j)$, so that $\hat{q}(ij) = \hat{q}(i+)\hat{q}(+j)$. This gives us the following distributions:

$\hat{p}$

.1212	.3657	.4869
.3232	.1899	.5131
.4444	.5556	

$\hat{q}$

.2164	.2705	.4869
.2280	.2851	.5131
.4444	.5556	

Visual inspection suggests that $\hat{p}$ and $\hat{q}$ are not very close, and clearly the differences between them can be used to test the hypothesis of independence. The statistic $\mathbb{D}_2(\hat{p}, \hat{q})$ is formed by summing the terms

$$\frac{[\hat{p}(ij) - \hat{q}(ij)]^2}{\hat{q}(ij)}$$

and referring $n\mathbb{D}_2(\hat{p}, \hat{q}) = 72.72$ to the chi-square (1 df) distribution for significance. Since large values of chi-square lead to rejection, it is clear that the independence hypothesis is untenable. Another chi-square statistic which we will use extensively is

$$\mathbb{D}_1(\hat{p}, \hat{q}) = \sum_{ij} \hat{p}(ij) \ln \frac{\hat{p}(ij)}{\hat{q}(ij)}$$

In large samples, the difference between $n\mathbb{D}_1(\hat{p}, \hat{q})$ and $n\mathbb{D}_2(\hat{p}, \hat{q})$ is negligible.

Using the log-linear approach, a measure of the row–column association is the ln odds ratio,

$$\ln p(00) - \ln p(01) - \ln p(10) + \ln p(11)$$

since this will vanish only when there is independence. The name derives from the fact that it can be written

$$\ln\left[\frac{p(11)}{p(10)}\bigg/\frac{p(01)}{p(00)}\right]$$

and the term in brackets is the ratio of the odds favoring Democrats in the Middle Class to the odds favoring Democrats in the Lower Class.

The estimate from Table I.4 is -1.636. An appropriate expression for the standard deviation is

$$\sqrt{\frac{1}{n}\left(\frac{1}{\pi(00)}+\frac{1}{\pi(10)}+\frac{1}{\pi(01)}+\frac{1}{\pi(11)}\right)}$$

which would be estimated by replacing π by $\hat{p}$. This gives the value .039, and leads to the .95 confidence interval -1.636 ± 0.077. We are led to the conclusion that there is a large negative association between being Democrat and being Middle Class. We have also provided a rather precise estimate of that association, which might be useful in subsequent studies on political trends.

5. A 4^2 Table

Table I.5 shows the frequencies of grandfather–grandson pairs classified according to the occupational status of each. We thus have two factors, F = Grandfather's Status, with levels $F0$ = Lower Blue Collar to $F3$ = Upper White Collar, and S = Grandson's Status, with similar levels $S0$ to $S3$. The experimental unit in this survey is the *pair* of individuals, whereas each *person* was an experimental unit in the preceding cases. A problem for data like this is to try to develop a process which describes social mobility and can explain the frequencies in the table.

Table I.5

A Social-Mobility Table

	Grandson's Status			
Grandfather's Status	LB	UB	LW	UW
Lower Blue Collar	11	18	6	5
Upper Blue Collar	19	62	33	40
Lower White Collar	1	8	6	9
Upper White Collar	10	28	31	66

Data like those in Table I.5 are often studied by transition models. The idea is that characteristics of the grandfather's class partially determine the likelihood of the grandson being in various classes. We can think of the family having passed from one state (the grandfather's class) to another state (the grandson's class). If the intervening family member (the grandson's father) had been included, we could have studied two transitions for each family.

One among many models for such data is

$$\ln p(FiSj) = \theta_{Fi} + \theta_{Sj} - \theta_D \sqrt{|i-j|}$$

If the θ_F and θ_S parameters alone had been present, this would have been the independence model. Of course, one does not expect grandfather's status to be independent of grandson's, and so the θ_D parameter reflects a decreasing rate of transitions between classes that are farther apart.

The fitted distribution under the model has been obtained by the LAMDA program. The empirical distribution $\hat{p}$ and the fitted distribution $\hat{q}$ are shown below. The chi-square (8 df) statistic is $n\mathbb{D}_1(\hat{p}, \hat{q}) = 9.34$, indicating that the model fits the data well. The estimate of θ_D is $\hat{\theta}_D = .4916$; when divided by an estimate of its standard deviation, this result yields 5.59, indicating that it is highly significantly different from zero.

$\hat{p}$

	S0	S1	S2	S3
F0	.0312	.0510	.0170	.0142
F1	.0538	.1756	.0935	.1133
F2	.0028	.0227	.0170	.0255
F3	.0283	.0793	.0878	.1870

$\hat{q}$

	S0	S1	S2	S3
F0	.1262	.0362	.0228	.0280
F1	.0515	.1900	.0898	.1050
F2	.0066	.0182	.0230	.0202
F3	.0319	.0841	.0797	.1868

6. A $2^2 \times 9$ Table

In Table I.6 we see a sample of British coal miners classified according to their responses to two respiratory questions, and their ages. Thus we have B = Breathlessness with levels $B0$ = Yes, $B1$ = No; W = Wheeze with $W0$ = Yes, $W1$ = No; and A = Age with $A0$ = 20–24 years to $A8$ = 60–64 years. A typical cell in this table would be denoted $BiWjAk$, and would contain $N(BiWjAk)$ or $N(ijk)$ observations. The (B, W) marginal table would contain entries $N(ij+)$ obtained from the (B, W, A) table by summing over levels of age. Similarly, the (B, A) marginal table would have frequencies $N(i+k)$, the summation having been performed over levels of W.

When several factors are present, they can be included in more complex log-linear parametrizations. For example, the data in Table I.6 concern three factors, B, W, and A. A log-linear parametrization for the class $\mathcal{P}$ of all strictly positive distributions on the table can be given as

$$\ln p(BiWjAk) = \theta_{Bi} + \theta_{Wj} + \theta_{Ak} + \theta_{BiWj} + \theta_{BiAk} + \theta_{WjAk} + \theta_{BiWjAk}$$

Setting various parameters equal to zero results in interesting submodels.

As an illustration, the model in which Breathlessness and Age are conditionally independent, given Wheeze, satisfies

$$p(BiAk|Wj) = p(Bi|Wj)\,p(Ak|Wj)$$

Under this condition, any association between Breathlessness and Age would be due to their common association with Wheeze. Multiplying both sides of the above equation by $p(Wj)$, we get

$$p(BiWjAk) = \frac{p(BiWj)\,p(WjAk)}{p(Wj)}$$

This can be written in log-linear form as

$$\ln p(BiWjAk) = \ln p(BiWj) + \ln p(WjAk) - \ln p(Wj)$$

Table I.6

A Three-Factor Table

		Breathlessness			
		Yes		No	
	Wheeze:	Yes	No	Yes	No
Age	20–24	9	7	95	1841
	25–29	23	9	105	1654
	30–34	54	19	177	1863
	35–39	121	48	247	2357
	40–44	169	54	273	1778
	45–49	269	88	324	1712
	50–54	404	117	245	1324
	55–59	406	152	225	967
	60–64	372	106	132	526

which suggests that the log-linear model specification is

$$\ln p(BiWjAk) = \theta_{Bi} + \theta_{Wj} + \theta_{Ak} + \theta_{BiWj} + \theta_{WjAk}$$

—a speculation which turns out to be correct.

A reasonable estimate of π under these model conditions is

$$\hat{q}(ijk) = \frac{\hat{p}(ij+)\hat{p}(+jk)}{\hat{p}(+j+)}$$

since the model places no restrictions on the (B, W) and (W, A) margins. Again, there is a chi-square (18 df) statistic, $n\mathbb{D}_1(\hat{p}, \hat{q}) = 173.84$, and this enormous value decisively rejects the hypothesis of conditional independence.

There are several other models available for the three-way table, each of which involves setting some of the θ-parameters to zero, or imposing linear constraints on them. A full analysis requires the determination of fitted distributions, like $\hat{q}$ above, and the measuring of lack of fit with statistics like $\mathbb{D}_1(\hat{p}, \hat{q})$. For some models it is possible to compute $\hat{q}$ by hand, as it was in the conditional-independence case, but in general an iterative computer routine is necessary. The program LAMDA, mentioned in the Preface, will be used for all analyses in this book. An introduction to the basic LAMDA commands may be found in Appendix A.

7. A Triangular Table

There is an unfortunate tendency to put discrete data in the form of a cross-classification regardless of whether this fulfills the purpose for which the data were collected. For example, suppose that of each individual in a medical study it is asked whether the subject had ever smoked, and if so whether he had quit, and if not how much he smoked currently. The factors are E = Ever Smoked, Q = Ever Quit, and S = Amount Smoked. Clearly these three factors do not form a cross-classification, since if one never smoked it does not make sense to ask how much. An appropriate contingency table looks like

$E0$	$E1Q1$	$E1Q0S0$
	$E1Q2$	$E1Q0S1$
		$E1Q0S2$

The factor E has levels $E0$ = Never Smoked and $E1$ = Smoked, and the factor Q has levels $Q0$ = Never Quit, $Q1$ = Quit a Long Time Ago, and

$Q2$ = Quit Recently. We say that Q is *nested* in $E1$, because classification on the basis of Q is only possible within the $E1$-cells. Likewise, S has levels $S0$ = Less Than One Pack, $S1$ = One Pack and $S2$ = More Than One Pack, and we say that S is nested in $E1Q0$. The other factors which were recorded in this study were the age (A), sex (X), and disease status (D) of the subjects. There is one of the above triangular tables for each of the possible combinations of (A, X, D). Thus the data involve both *crossed* factors (E, A, X, D) and *nested* factors (Q, S). This structure is used to analyze some data in Chapter V, Section B.7.

8. A Four-Way Table

Part of the results of an animal study is shown in Table I.7. B represents the presence or absence of a certain experimental condition, X represents experimental or control animal, T denotes the trial number for a given animal, and C denotes a successful or unsuccessful response. The (B, X, C) factors are crossed, but not all T-values appear in conjunction with all (B, X, C) combinations. Some cells are automatically empty; for example, there are no T15 observations for $B0X0$, and so these cells contain no

Table I.7

An Incomplete Table

	$B0X0$		$B1X0$		$B0X1$		$B1X1$	
	$C1$	$C0$	$C1$	$C0$	$C1$	$C0$	$C1$	$C0$
$T0$	1	19	1	19	0	20	3	17
$T1$	1	19	6	14	3	17	4	16
$T2$	0	20	15	5	2	18	4	16
$T3$	0	20	19	1	0	20	6	14
$T4$	4	16	17	3	0	20	8	12
$T5$	4	16	19	1	0	20	15	5
$T6$	5	15	17	3	3	17	17	3
$T7$	7	13	19	1	2	18	15	5
$T8$	13	7	19	1	8	12	14	6
$T9$	17	3	20	0	12	8	16	4
$T10$	15	5	—	—	15	5	19	1
$T11$	15	5	—	—	18	2	18	2
$T12$	19	1	—	—	17	3	19	1
$T13$	18	2	—	—	18	2	20	0
$T14$	20	0	—	—	18	2	—	—
$T15$	—	—	—	—	20	0	—	—

observations as a result of the design of the experiment, rather than being empty by chance like $B0X1C1T0$, where $N(0110) = 0$. We say that this table is incomplete because there are some meaningful combinations of factors which are known ahead of time not to contain any observations. Cells which are empty for this reason are called *fixed zero* cells, while those which are empty by chance are called *random zero* cells. (There is an associated problem with these data in that it is clear the experiment was continued for each animal until 20 successes were achieved.)

The data in Table I.7 can also be treated from the point of view of a single probability. We will concentrate on the first experimental animal,

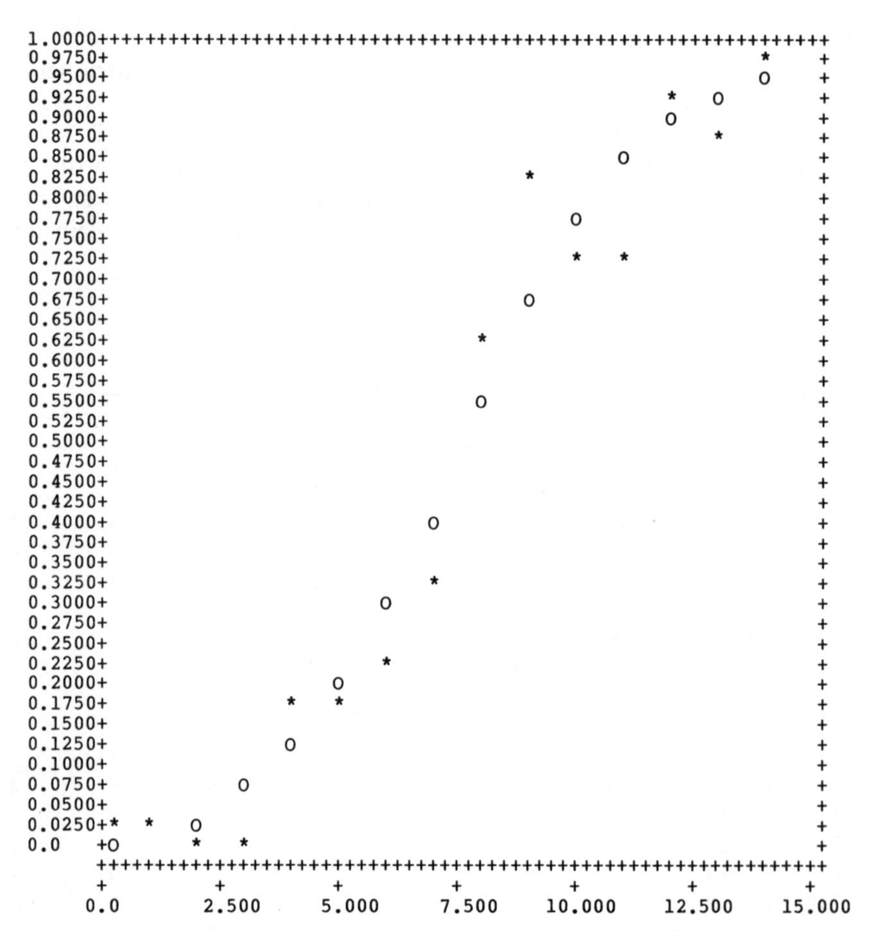

Fig. I.1 A plot of the proportion of correct responses (∗) and the fitted logistic proportion (O), against the number of the trial, for the first animal in Table I.7.

whose results appear in the first two columns of the table. In the initial trial ($T0$), the animal got one ($C1$) correct response and 19 ($C0$) incorrect responses. This situation is like the example in Section B.1 in that there is a single probability involved. However, as we move from the first to later trials, it is unreasonable to assume that the animal never learns, and so we should permit our single probability to depend on the number of the trial. We can do this quite simply by permitting the ln odds favoring a correct response to be linearly related to the trial number:

$$\ln \frac{p(C1|Ti)}{p(C0|Ti)} = \beta_0 + \beta_1 i$$

This can be expressed as

$$p(C1|Ti) = \frac{e^{\beta_0 + \beta_1 i}}{1 + e^{\beta_0 + \beta_1 i}}$$

which is a special case of the *logistic regression* model.

The estimated parameters from the data in the first two columns of Table I.7 are $\hat{\beta}_0 = -.40617$ and $\hat{\beta}_1 = .5365$. The logistic regression model appears to fit quite well, as judged by the plot of observed (*) and fitted (0) proportions of correct responses, plotted against trial number in Figure I.1. Both the estimation and the plot were done by the LAMDA program.

The importance of being able to fit the data with the logistic regression model is that a fairly complex pattern of data can be summarized by only two parameter estimates. Eventually one wants to compare curves for different animals, and as we shall see, comparisons based on a small number of parameters are more powerful than comparisons involving many parameters.

9. Covariate Models

We have seen that it is possible to build a very simple model (for a two-cell table, for instance), and then relate the parameter in that model to background factors, or *covariates*. In the animal experiment discussed above, the trial number played the role of a covariate. In the data on coal miners from Table I.6, we can let Age be the covariate. The ln odds ratio between Breathlessness and Wheeze measures their association, and we might want to relate this measure to Age. For this purpose it is appropriate to compute the conditional ln odds ratios within levels of Age,

$$\ln p(B0W0|Ak) - \ln p(B1W0|Ak) - \ln p(B0W1|Ak)$$
$$+ \ln p(B1W1|Ak)$$

Table I.8

A-level	Conditional ln Odds	SD
$A0$	3.216	.2650
$A1$	3.695	.1649
$A2$	3.398	.0773
$A3$	3.180	.0336
$A4$	3.015	.0287
$A5$	2.782	.0188
$A6$	2.926	.0159
$A7$	2.441	.0145
$A8$	2.638	.0216

with estimated standard deviations

$$\sqrt{\frac{1}{N(00k)}+\frac{1}{N(10k)}+\frac{1}{N(01k)}+\frac{1}{N(11k)}}$$

These are given in Table I.8.

These results suggest that the association between B and W is high, but declines with Age, perhaps in a nearly linear fashion. This may seem a contradiction, but it is easily explained by the hypothesis that the $B0W0$ cell tends to become depleted as miners tend to acquire one or the other of the symptoms, and the $B1W1$ cell tends to become depleted by deaths due to lung disease.

This idea of writing log-linear parameters as simple functions of covariates is very important in the analysis of complex data.

10. A Table Without Factors

The final example of this section appears in Table I.9. Here the experimental unit is one family, classified according to its pattern of children. For instance, a pattern *ADSA* would indicate that the first child was adopted (A), the second died at birth (D), the third survived (S), and the fourth was adopted (A). It is clear that the notion of a factor, as we have applied it previously, is not really appropriate here. What one wants is an analysis based on a model which accounts for how the process of child-bearing and adoption operates. A possible analysis will be suggested in Problem VI.10.

11. Conclusion

These brief, preliminary analyses should give some idea of the sorts of results and approaches that are possible in discrete data analysis. The

Table I.9

Some Data on Childbearing and Adoption

Pattern	Frequency	Pattern	Frequency
A	10	*SSSA*	1
AA	15	*SSSSA*	1
AAA	2	*DAA*	1
AAS	1	*DDA*	1
AS	4	*DASS*	1
ASS	1	*DASSSS*	1
ASSS	3	*DSA*	1
ADS	1	*DSSSA*	1
SA	9	*SDDAAS*	1
SAS	1	*SDDDADADA*	1
SSA	2		

purpose of the remainder of the book is to explain why these steps are justified, and to create a feeling for how analyses can be formulated, for standard models as well as variations of them. Most analyses also require a great deal of computing, and so we will provide the necessary support with the LAMDA program.

PROBLEMS

I.1. Construct two probability distributions over a three-way table with factors A, B, and C, such that under the first distribution A and B are independent, but A and B are not conditionally independent given C, and under the second distribution A and B are conditionally independent given C, but they are not independent.

I.2. Construct a probability distribution on a 2^3 table, having factors A, B and C, so that

$$(1)\quad \ln \frac{p(A1B1|Ci)\,p(A0B0|Ci)}{p(A1B0|Ci)\,p(A0B1|Ci)} > 0 \qquad \text{for} \quad i = 0, 1$$

but

$$(2)\quad \ln \frac{p(A1B1)\,p(A0B0)}{p(A1B0)\,p(A0B1)} = 0$$

Construct another distribution so that both the conditional ln odds ratios in (1) are zero, but the marginal ln odds ratio in (2) is nonzero. What implication does this have for using the ln odds ratio as a measure of association?

I.3. Show that

$$p(AiBj|CkDm) = \frac{p(AiBjCk|Dm)}{p(Ck|Dm)}$$

and note that this means that one can condition on (C, D) by first conditioning on D, then conditioning on C.

I.4. 1. Consider an (A, B)-cross-classification. Let (ij) denote $AiBj$, and abuse the notation by writing $p_i(j) = p(Bj|Ai)$ and $p_j(i) = p(Ai|Bj)$. Prove that

$$p(i+) = \frac{1}{\sum_j \frac{p_i(j)}{p_j(i)}}$$

2. Using the result of Problem I.3, prove that in the ABC-table with typical cell $ijk = AiBjCk$, we have

$$p_i(j+) = \frac{1}{\sum_k \frac{p_{ij}(k)}{p_{ik}(j)}}$$

and thus conclude that the conditional distributions $p_{ij}(k)$, $p_{ik}(j)$, and $p_{jk}(i)$ determine p (provided they are all strictly positive).

I.5. The following quotation is from a research report: "If A and B are dependent, and B and C are dependent, then it is impossible that A and C should be independent." Either verify this assertion or provide a counter-example.

REMARKS

It is important to visualize the filling of a contingency table as a process in which an experiment is repeated a number of times, and each repetition adds one count to the table. This focuses attention where it belongs, on the probability distribution governing the placing of each count in the table. It is very tempting to concentrate on the counts themselves, and to ignore the underlying probabilities. This can easily lead to confusion and miscomputation [see Lewis and Burke (1949), and Problem II.17], and so it is always good practice to keep in mind this stochastic procedure by which experimental units are sorted into the cells of the table.

In very general terms, the purpose of analyzing a contingency table is to relate aspects of the probability distribution to various factors which might influence how the experimental units are placed in the cells. To a certain extent this idea is implicit in the use of the contingency-table structure itself. The cells of the table are not displayed in just any order—they are arranged so that the position of each cell with respect to the others indicates what kind of relationships are of interest. This is well and good, but one should realize that in more sophisticated analyses the number of important interrelationships increases greatly, and so it eventually becomes extremely difficult to visualize the composition of the table geometrically or diagrammatically. Even in a table as simple as Table I.7 one can ask whether there is an interaction of the B- and X-factors in their influence on the (T, C)-interrelationship. To try to display this it would be necessary to rearrange the entire table, and as soon as this was done other comparative features of the table would be lost.

The way out of this problem is to give up a reliance on contingency-table displays, and to substitute algebraic analytic methods. What is lost by this is ease and intuitive appeal, but it is more than recovered in power and flexibility. A simple example is the ln odds ratio, introduced in Section A.2. In order to assess whether two factors are independent of each other, one can either scrutinize the two-way table for row (or column) proportionality, or command that certain ln odds ratios be computed. The two approaches are equally reasonable in such a simple case, but one can generalize the analytic approach, so that certain sums and differences of ln odds ratios can be used to answer the question about a (B, X)-interaction on the (T, C)-interaction. Moreover, this can be done without sacrificing analytical information about other aspects of Table I.7.

The general plan of the book from this point onwards is to develop the machinery for asking and answering increasingly complex questions about the data in a contingency table. In Chapter II we consider how one might decide whether a particular distribution could have generated the data, and if not, in what way the true distribution probably differs from the hypothesized distribution. In Chapter III we treat the fundamental unit of discrete inference, the test of the hypothesis that the true distribution lies in a particular model. Since it is more common to have many interesting models, rather than only one, in Chapter IV we give a complete treatment of the problem of doing simultaneous inference for many interrelated hypotheses. In Chapter V we present the discrete analogues of regression and analysis of variance, important techniques used throughout applied statistics. Finally, Chapter VI contains several topics which might be considered to be of a more advanced nature, including more general models and some very useful analytic techniques.

Among the problems in this chapter, Problem I.2 is particularly interesting, because historically it has been a source of difficulty in developing the notion of association between factors [see Simpson (1951), and more generally, any of the references whose titles contain the word "interaction" or "association"].

CHAPTER II

Inference for a Single Distribution

The main purpose of this chapter is to present a method for doing inference about a single distribution, but an important additional aim is to introduce several concepts which will be required throughout the book. Thus this chapter is rather more complex than is strictly necessary to derive the statistical assertions, but it serves to bring in some tools which will ease the establishment of more difficult results in subsequent chapters. The most important of these tools is the idea of a *function space*, because it ties together the theoretical asymptotic results, the strategy for model fitting, and the computations. A technical device that is extraordinarily useful in this regard is the *standardization* procedure, which is carried out by the LAMDA instruction **STD**.

It is somewhat surprising that even the very rapid computations made possible by modern electronic computers are not rapid enough to obtain the exact sampling distributions of many of the statistics used in discrete data analysis. Thus we will need the asymptotic theory presented here for inference based on large samples, although it is problematic what "large" means in this context. The *Normal*, *Chi-square*, and *Noncentral Chi-square* distributions play a considerable role, and we will give approximations to these distributions which make tables unnecessary in routine applications.

Another important notion is that of measuring the distance between two distributions. The most natural approach to making a decision whether $\pi = p$ is to reason that the empirical distribution $\hat{p}$ ought to approximate π and then to see how far $\hat{p}$ is from p. For this purpose there is a broad class of reasonable ways of measuring distances, which we will call *discrepancies*. We will see that in the context of formal hypothesis testing the power of a discrepancy-based test to detect a departure from the hypothesis $\pi = p$ can also be expressed in terms of discrepancies.

This chapter contains an amount of technical detail which may be heavy going for some readers, and so a simplified summary of the important consequences is given in Section D.

A. FUNCTION SPACES

We now turn to the basic facts about functions defined on a contingency table, their means, variances, and covariances.

1. Moments

Let $\mathcal{Y}$ be the finite set of possible outcomes of an experiment, and let $\#\mathcal{Y}$ stand for the number of elements in $\mathcal{Y}$. Let $p \in \mathcal{P}_0$, and for any real-valued function $u: \mathcal{Y} \to \mathbb{R}$, define the *mean* of u with respect to p by

$$M[u: p] = \sum_y u(y) p(y)$$

If $\mathbf{u}$ is a vector or matrix-valued function on $\mathcal{Y}$, then $M[\mathbf{u}: p]$ is also defined by the above expression, the summation on the right then representing addition of vectors or matrices. In particular, if u_i is the ith component of the vector-valued function $\mathbf{u}$, then $M[u_i: p]$ is the ith component of $M[\mathbf{u}: p]$.

If $\mathbf{u}: \mathcal{Y} \to \mathbb{R}^k$, then for any vector $\boldsymbol{\alpha}' = (\alpha_1, \alpha_2, \ldots, \alpha_k)$ we have the function $\boldsymbol{\alpha}'\mathbf{u}$ defined by

$$\boldsymbol{\alpha}'\mathbf{u} = \sum_{i=1}^{k} \alpha_i u_i$$

The collection of all functions of this form is denoted by $\mathbb{S}(\mathbf{u})$, and is a linear space in the sense that whenever v_1 and v_2 are in $\mathbb{S}(\mathbf{u})$, then so is $\alpha v_1 + \beta v_2$ for any numbers α and β. The means of all functions in $\mathbb{S}(\mathbf{u})$ are determined by $M[\mathbf{u}: p]$, since $M[\boldsymbol{\alpha}'\mathbf{u}: p] = \boldsymbol{\alpha}' M[\mathbf{u}: p]$. We will occasionally abuse the notation by writing $\mathbb{S}(\mathbf{u}, \mathbf{v})$ in place of $\mathbb{S}\begin{pmatrix}\mathbf{u}\\ \mathbf{v}\end{pmatrix}$.

If $\mathbf{u}$ and $\mathbf{v}$ are two vector-valued functions on $\mathcal{Y}$, then $\mathbf{u}\mathbf{v}'$ is the matrix-valued function whose ijth component is $u_i v_j$. Moreover,

$$M[\mathbf{u}\mathbf{v}': p] = \sum_y \mathbf{u}(y)\mathbf{v}(y)' p(y)$$

We define the *covariance matrix* of $\mathbf{u}$ and $\mathbf{v}$ with respect to p by

$$C[\mathbf{u}, \mathbf{v}: p] = M[\mathbf{u}\mathbf{v}': p] - M[\mathbf{u}: p] M[\mathbf{v}: p]'$$

a matrix whose ijth element is the *covariance* between u_i and v_j,

$$M[u_i v_j : p] - M[u_i : p] M[v_j : p]$$

The covariance of a function u with itself is its *variance*,

$$V[u : p] = M[u^2 : p] - M[u : p]^2$$

The covariance matrix of $\mathbf{u}$ with itself will be abbreviated $C[\mathbf{u} : p]$.

The variances and covariances of all functions in $\mathbb{S}(\mathbf{u})$ are determined by $C[\mathbf{u} : p]$ according to the equation $C[\boldsymbol{\alpha}'\mathbf{u}, \boldsymbol{\beta}'\mathbf{v} : p] = \boldsymbol{\alpha}' C[\mathbf{u}, \mathbf{v} : p]\boldsymbol{\beta}$. Since $V[u : p] = M[(u - M[u : p])^2 : p] \geqq 0$ with equality only when u is constant, we have

$$V[\boldsymbol{\alpha}'\mathbf{u} : p] = \boldsymbol{\alpha}' C[\mathbf{u} : p]\boldsymbol{\alpha} \geqq 0$$

with equality only when $\boldsymbol{\alpha}'\mathbf{u}$ is constant. The vector-valued function $\mathbf{u}$ has *affinely linearly independent components* if and only if $\boldsymbol{\alpha} = \mathbf{0}$ is the only vector for which $\boldsymbol{\alpha}'\mathbf{u}$ is constant. This is evidently equivalent to $C[\mathbf{u} : p]$ being strictly positive definite, that is, $\boldsymbol{\alpha}' C[\mathbf{u} : p]\boldsymbol{\alpha} \geqq 0$ with equality only when $\boldsymbol{\alpha} = \mathbf{0}$.

Proposition 1. If the components of $\mathbf{u}$ are affinely linearly independent, then every $v \in \mathbb{S}(1, \mathbf{u})$ has exactly one representation

$$v = \alpha_0 + \boldsymbol{\alpha}'\mathbf{u}.$$

Proof. If also $v = \beta_0 + \boldsymbol{\beta}'\mathbf{u}$, then $0 = \alpha_0 + \boldsymbol{\alpha}'\mathbf{u} - \beta_0 - \boldsymbol{\beta}'\mathbf{u} = (\alpha_0 - \beta_0) + (\boldsymbol{\alpha} - \boldsymbol{\beta})'\mathbf{u}$ implies that $\boldsymbol{\alpha} = \boldsymbol{\beta}$, whence $\alpha_0 = \beta_0$. □

On account of the properties established above it will be very convenient in most circumstances to deal with spaces $\mathbb{S}(1, \mathbf{u})$ under the assumption that $\mathbf{u}$ has affinely linearly independent components.

Factorial Indicators. Although we will use many different functions for the construction and analysis of statistical models, a class which is frequently important consists of *indicators*, that is, functions which assume only the values 0 and 1. In a contingency table determined by several factors $A, B, C, \ldots$, we will define a (*first-order*) *factorial indicator* by

$$u_{Ai}(y) = \begin{cases} 1 & \text{if cell } y \text{ is at level } i \text{ of factor } A \\ 0 & \text{otherwise} \end{cases}$$

and similarly for u_{Bj}, u_{Ck}, and so on. An *mth-order factorial indicator* is a nonzero product of m first-order factorial indicators. For instance, $u_{AiBjCk} = u_{Ai}u_{Bj}u_{Ck}$ is a third-order factorial indicator. The means of factorial indicators are the marginal probabilities of the table. For example, $M[u_{Ai}: p] = p(Ai)$, and $M[u_{AiBjCk}: p] = p(AiBjCk)$.

2. Standardization

Properties related to $M[\mathbf{u}: p]$ and $\mathcal{S}(1, \mathbf{u})$ are much easier to describe and prove when an additional condition is imposed on $\mathbf{u}$, a condition we will call standardization. The purpose of this subsection is to outline the process of standardization and show that in a sense there is no lack of generality in passing from arbitrary $\mathbf{u}$ to standardized $\mathbf{u}$.

A vector-valued function $\mathbf{u}: \mathcal{Y} \to \mathbb{R}^k$ is *p-standard* if and only if $M[\mathbf{u}: p] = \mathbf{0}$ and $C[\mathbf{u}: p] = \mathbf{I}$, where $\mathbf{I}$ denotes an identity matrix of the correct dimension. Since $\mathbf{I}$ is strictly positive definite, the reasoning of the preceding subsection shows that every p-standard $\mathbf{u}$ has affinely linearly independent components. The converse of this statement is not true, but it is always possible to construct a p-standard function from any vector-valued function.

The procedure for doing this is called *p-standardization*, and is based on the Gram–Schmidt steps, which we now define. To begin with, let $\mathbf{v}$ be a vector-valued function with affinely linearly independent components. The first Gram–Schmidt step is to define

$$w_1 = v_1 - M[v_1 : p]$$

and then set $u_1 = w_1 / \sqrt{M[w_1^2 : p]}$. The remainder of the Gram–Schmidt steps are defined inductively by

$$w_i = v_i - M[v_i : p] - \sum_{j=1}^{i-1} M[v_i u_j : p] u_j$$

$$u_i = \frac{w_i}{\sqrt{M[w_i^2 : p]}}$$

Since by induction $M[u_j : p] = 0$ ($j < i$), we have $M[u_i : p] = 0$. For any $m < i$, again by induction

$$\begin{aligned} M[w_i u_m : p] &= M[v_i u_m : p] - M[v_i : p] M[u_m : p] \\ &\quad - \sum_{j=1}^{i-1} M[v_i u_j : p] M[u_j u_m : p] \\ &= M[v_i u_m : p] - M[v_i u_m : p] = 0 \end{aligned}$$

so that $M[u_i u_m : p] = 0$. Clearly $V[u_i : p] = 1$. This shows that at step i the vector $(u_1, u_2, \ldots, u_i)$ is p-standard and moreover is defined as a linear combination of $(v_1, v_2, \ldots, v_i)$.

Note that it was tacitly assumed at each stage that $M[w_i^2 : p] > 0$. This follows from the assumptions, since $M[w_i^2 : p] = 0$ implies that $v_i \in \mathbb{S}(1, u_1, \ldots, u_{i-1}) \subset \mathbb{S}(1, v_1, \ldots, v_{i-1})$, an occurrence which is ruled out by the assumption that $\mathbf{v}$ has affinely linearly independent components. Finally, we observe

$$v_i = u_i \sqrt{M\left[w_i^2 : p\right]} + M[v_i : p] + \sum_{j=1}^{i-1} M[v_i u_j : p] u_j$$

so that $v_i \in \mathbb{S}(1, u_1, \ldots, u_i)$.

We will call $\mathbf{u}$, obtained from $\mathbf{v}$ by applying Gram–Schmidt steps, the *p-standardization* of $\mathbf{v}$. We have now established the following result.

Proposition 2. Suppose that $\mathbf{v}$ has affinely linearly independent components, and $\mathbf{u}$ is the p-standardization of $\mathbf{v}$. Then each $(u_1, \ldots, u_i)$ is p-standard, and $\mathbb{S}(1, u_1, \ldots, u_i) = \mathbb{S}(1, v_1, \ldots, v_i)$.

If $\mathbf{u}$ is p-standard, then its components are affinely linearly independent, and so each $w \in \mathbb{S}(1, \mathbf{u})$ has a unique representation $w = \alpha_0 + \boldsymbol{\alpha}'\mathbf{u}$. We may compute $M[w : p] = \alpha_0$ and $M[w u_i : p] = \alpha_i$ by using the definition of a p-standard function, so that the coefficients in the unique representation of w are easily computed. This is one of the features of p-standard functions which makes them so convenient.

Now let $\mathbf{v}\colon \mathcal{Y} \to \mathbb{R}^k$ be arbitrary, not necessarily having affinely linearly independent components. If we apply the Gram–Schmidt steps to this function, and if at some stage $v_i \in \mathbb{S}(1, v_1, \ldots, v_{i-1})$, then

$$w_i = v_i - M[v_i : p] - \sum_{j=1}^{i-1} M[v_i u_j : p] u_j = 0$$

by the observations made in the preceding paragraph, and so the Gram–Schmidt procedure breaks down. In this situation we simply discard v_i and proceed to v_{i+1}. Each time we come to a function v_i which is in $\mathbb{S}(1, v_1, \ldots, v_{i-1})$, we discard it. We will still call $\mathbf{u}$ the p-standardization of $\mathbf{v}$, even though the number of components of $\mathbf{u}$, denoted $\#\mathbf{u}$, will be smaller than $\#\mathbf{v}$. The important point about this procedure is that if $(u_1, \ldots, u_i)$ is the p-standardization of $(v_1, \ldots, v_k)$, then $\mathbb{S}(1, u_1, \ldots, u_i) = \mathbb{S}(1, v_1, \ldots, v_k)$ even if $i \neq k$.

Maximal p-Standard Functions. Let $\mathcal{Y} = \{y_1, y_2, \ldots, y_m\}$, and define $u_i(y)$ to be 1 if $y = y_i$ and 0 otherwise. By deleting one of the u_i's, say u_m, it is easy to see that $(u_1, \ldots, u_{m-1})$ has affinely linearly independent components. Further, if w is any function, then

$$w(y) = w(y_m) + \sum_{i=1}^{m-1} [w(y_i) - w(y_m)] u_i(y)$$

so that $\mathbb{S}(1, u_1, \ldots, u_{m-1})$ is the collection of all functions defined on $\mathcal{Y}$. As a consequence there is no vector-valued function which has affinely linearly independent components and whose components properly include $u_1, \ldots, u_{m-1}$. A moment's reflection shows that the p-standardization of $(u_1, \ldots, u_{m-1})$ also cannot be enlarged without destroying the affine linear independence of its components. We will say that a p-standard $\mathbf{u}$ is *maximal p-standard* when there is no p-standard function whose components properly include those of $\mathbf{u}$. We have in effect shown that maximal p-standard functions always exist. To summarize the properties of a maximal p-standard $\mathbf{u}$:

1. $M[\mathbf{u} : p] = \mathbf{0}$.
2. $C[\mathbf{u} : p] = \mathbf{I}$.
3. For any $\mathbf{w} : \mathcal{Y} \to \mathbb{R}^k$ we have

$$\mathbf{w} = M[\mathbf{w} : p] + M[\mathbf{w}\mathbf{u}' : p]\mathbf{u}$$

and the coefficients in this representation are unique.

The following result is the primary reason that maximal standard functions are so important for discrete data analysis.

Proposition 3. Let $q \in \mathcal{P}$, and let $\mathbf{u}$ be maximal q-standard. Then

$$p = (1 + \boldsymbol{\alpha}'\mathbf{u})q, \qquad \boldsymbol{\alpha} = M[\mathbf{u} : p]$$

establishes a continuous one-to-one correspondence between $\mathcal{P}_0$ and a closed subset of $\boldsymbol{\alpha} \in \mathbb{R}^{\#\mathbf{u}}$.

Proof. Since p/q is a function on $\mathcal{Y}$, we have

$$\frac{p}{q} = M\left[\frac{p}{q} : q\right] + \sum_{i=1}^{\#\mathbf{u}} M\left[\frac{u_i p}{q} : q\right] u_i = 1 + M[\mathbf{u} : p]'\mathbf{u}$$

and this representation is unique. □

This result will be much used later on, but to hint at the next step in the development, suppose we define

$$\mathbb{D}_2(p, q) = \sum_y \frac{[p(y) - q(y)]^2}{q(y)}$$

which will be recognized as being related to Pearson's chi-square statistic. Letting $\boldsymbol{\alpha} = M[\mathbf{u} : p]$, by algebra

$$\begin{aligned}
\mathbb{D}_2(p, q) &= \sum_y \frac{p(y)^2 - 2p(y)q(y) + q(y)^2}{q(y)} \\
&= \sum_y \frac{p(y)}{q(y)} p(y) - 1 \\
&= \sum_y [1 + \boldsymbol{\alpha}'\mathbf{u}(y)] p(y) - 1 \\
&= \boldsymbol{\alpha}' \sum_y \mathbf{u}(y) p(y) = \boldsymbol{\alpha}'\boldsymbol{\alpha} \\
&= \|\boldsymbol{\alpha}\|^2 = \|M[\mathbf{u} : p]\|^2
\end{aligned}$$

This relation is of fundamental importance in establishing the large-sample behavior of a wide variety of customary chi-square statistics. The terms $M[u_i : p]^2$ form a decomposition of $\mathbb{D}_2(p, q)$, and are useful for increasing the precision of statistical assertions about discrete data.

B. THE EMPIRICAL DISTRIBUTION

1. Consistency

If X_n is a sequence of random variables, then we will write $X_n \to_p x$ and say that X_n *converges in probability* to x if and only if for every $\varepsilon > 0$,

$$P[|X_n - x| > \varepsilon] \to 0 \qquad \text{as} \quad n \to \infty$$

where P denotes probability with respect to the system in which the X_n are observed. Letting M denote the mean of a random variable with respect to P, from Chebyschev's inequality

$$P[|X_n - x| > \varepsilon] \leqq \frac{M\left[(X_n - x)^2\right]}{\varepsilon^2}$$

we have that $M[(X_n - x)^2] \to 0$ implies $X_n \to_p x$.

In the discrete situation, if n independent observations have been made, then let $N_n(y)$ be the number of times y was observed. Writing $\mathcal{Y} = \{y_1, \ldots, y_m\}$, it can be shown that

$$P[N_n(y_i) = k_i (1 \leqq i \leqq m)]$$

$$= \frac{n!}{k_1! k_2! \cdots k_m!} \pi(y_1)^{k_1} \pi(y_2)^{k_2} \cdots \pi(y_m)^{k_m}$$

which is the Multinomial distribution with probabilities determined by the true distribution π. As a consequence

$$M[N_n(y)] = n\pi(y)$$

$$C[N_n(y_i), N_n(y_j)] = \begin{cases} n\pi(y_i)[1 - \pi(y_i)] & \text{if } i = j \\ -n\pi(y_i)\pi(y_j) & \text{if } i \neq j \end{cases}$$

This implies that for the empirical distribution, $\hat{p}_n(y) = N_n(y)/n$, we have $M[\hat{p}_n(y)] = \pi(y)$ and $V[\hat{p}_n(y)] = \pi(y)[1 - \pi(y)]/n$. Thus $V[\hat{p}_n(y)] \to 0$ as $n \to \infty$, which establishes that $\hat{p}_n(y)$ is *consistent*, that is, $\hat{p}_n(y) \to_p \pi(y)$ for all $y \in \mathcal{Y}$.

This is, of course, an extremely important property of $\hat{p}_n$, since it says that the values $\hat{p}_n(y)$ will be close to the true values $\pi(y)$ in a probabilistic sense as the sample size increases, and in this sense $\hat{p}_n$ is an estimator of π.

Note that if h is a continuous function, then $h(\hat{p}_n(y)) \to_p h(\pi(y))$, and moreover it can be shown that if $h(x_1, x_2, \ldots, x_m)$ is continuous in all its arguments, then $h(\hat{p}_n(y_1), \ldots, \hat{p}_n(y_m)) \to_p h(\pi(y_1), \ldots, \pi(y_m))$. This shows, for example, that $M[\mathbf{u} : \hat{p}_n] \to_p M[\mathbf{u} : \pi]$ for any vector-valued function $\mathbf{u}$.

2. Asymptotic Theory

Although it is theoretically possible to make probability statements about $\hat{p}_n$ using the Multinomial distribution, in practice the computation involved is excessive, and so one generally relies on approximate probability statements whose validity improves as the sample size increases. The main tool of asymptotic theory is the notion of convergence in distribution, and the cornerstones are the multivariate central limit theorem (MCLT) and the Mann–Wald theorem (MWT).

We begin by considering the general case of a sequence $\{X_n\}$ of random variables. If X is a random variable having cumulative distribution function (cdf) F defined by $P[X \leq x] = F(x)$, and if F is continuous, then we say that X_n *converges in distribution* to X and write $X_n \to_d X$ to mean that

$P[X_n \leqq x] = F_n(x) \to F(x)$ for all x. Notice that $X_n \to_d X$ is really only a statement about the cdf's of the X_n converging to the cdf of X. The random variable X plays no role except to carry the distribution F. We could be more precise by always referring to cdf's rather than random variables, but this would lead to annoyingly repetitive prose.

The most important case of convergence in distribution occurs when F is the *standard Normal* distribution,

$$F_0(x) = \int_{-\infty}^{x} \frac{1}{\sqrt{2\pi}} e^{-t^2/2}\, dt$$

(The π in this expression is pi = 3.14159265..., not the true distribution.) If X has this distribution, we say that X is $N(0, 1)$, since it can be shown that $M[X] = 0$ and $V[X] = 1$. Moreover, we write $X_n \to_d N(0, 1)$ to mean that X_n converges in distribution to such a random variable.

If $(X - \mu)/\sigma$ is $N(0, 1)$ for a choice of μ and σ, then we say that X is $N(\mu, \sigma^2)$, and clearly $M[X] = \mu$ and $V[X] = \sigma^2$. If $\mathbf{X}$ is a random vector, then we say that $\mathbf{X}$ is $N(\mathbf{0}, \mathbf{I})$ to mean that for any vector $\boldsymbol{\alpha} \neq \mathbf{0}$, $\boldsymbol{\alpha}'\mathbf{X}$ is $N(0, \|\boldsymbol{\alpha}\|^2)$. This is the *standard multivariate Normal distribution*. For any vector $\boldsymbol{\mu}$ and matrix $\mathbf{C}$ we say that $\mathbf{X}$ is $N(\boldsymbol{\mu}, \mathbf{C})$ to mean that for any vector $\boldsymbol{\alpha}$ either $\boldsymbol{\alpha}'X$ is constant or it is $N(\boldsymbol{\alpha}'\boldsymbol{\mu}, \boldsymbol{\alpha}'\mathbf{C}\boldsymbol{\alpha})$. Clearly the matrix $\mathbf{C}$ must be positive definite, and if $\mathbf{C}$ is strictly positive definite, then $\boldsymbol{\alpha}'\mathbf{X} = c$ (a constant) can only happen with $\boldsymbol{\alpha} = \mathbf{0}$, in which case we speak of a *nonsingular* Normal distribution. It is obvious that $M[X] = \boldsymbol{\mu}$ and $C[\mathbf{X}] = \mathbf{C}$.

Consequently, if $\mathbf{X}_n$ is a sequence of random variables, $\mathbf{X}_n \to_d N(\boldsymbol{\mu}, \mathbf{C})$ means that for every $\boldsymbol{\alpha}$ either $\boldsymbol{\alpha}'\mathbf{X}_n \to_p c$ (a constant) or $\boldsymbol{\alpha}'\mathbf{X}_n \to_d N(\boldsymbol{\alpha}'\boldsymbol{\mu}, \boldsymbol{\alpha}'\mathbf{C}\boldsymbol{\alpha})$. The following result is extremely important for the remainder of the theory, and is proved in several textbooks and monographs.

Proposition 4 (MCLT). If $\{\mathbf{X}_i : 1 \leqq i < \infty\}$ are independent random vectors, each having the same distribution, with mean $\boldsymbol{\mu}$ and covariance matrix $\mathbf{C}$, and if

$$\mathbf{Y}_n = \frac{1}{n} \sum_{i=1}^{n} \mathbf{X}_i$$

then $\sqrt{n}\,(\mathbf{Y}_n - \boldsymbol{\mu}) \to_d N(\mathbf{0}, \mathbf{C})$.

We may apply the MCLT to $\hat{p}_n$ in order to gain a considerable amount of information about its asymptotic behavior.

Proposition 5. If the observations are independent with distribution π, and if $\mathbf{u}$ is π-standard, then $\sqrt{n}\,M[\mathbf{u}:\hat{p}_n] \to_d N(\mathbf{0},\mathbf{I})$.

Proof. Let $\mathcal{Y} = \{y_1, y_2, \ldots, y_m\}$, define

$$N_{in} = \begin{cases} 1 & \text{if the } n\text{th observation is } y_i \\ 0 & \text{otherwise} \end{cases}$$

and let $\mathbf{N}_n$ be the vector with ith component N_{in}. The vectors $\mathbf{N}_n$ are independent, all having the same distribution, and consequently the random vectors

$$\mathbf{U}_n = \sum_{i=1}^{m} \mathbf{u}(y_i) N_{in}$$

are also independent with a common distribution. On account of

$$M[\mathbf{N}_n] = \begin{pmatrix} \pi(y_1) \\ \pi(y_2) \\ \vdots \\ \pi(y_m) \end{pmatrix}$$

and

$$C[\mathbf{N}_n] = \begin{bmatrix} \pi(y_1) & 0 & 0 & \cdots & 0 \\ 0 & \pi(y_2) & 0 & \cdots & 0 \\ \vdots & \vdots & & & \vdots \\ 0 & 0 & \cdots & 0 & \pi(y_m) \end{bmatrix}$$

$$- \begin{pmatrix} \pi(y_1) \\ \pi(y_2) \\ \vdots \\ \pi(y_m) \end{pmatrix} \begin{pmatrix} \pi(y_1) \\ \pi(y_2) \\ \vdots \\ \pi(y_m) \end{pmatrix}'$$

we have

$$M[\mathbf{U}_n] = M[\mathbf{u}:\pi] = \mathbf{0}$$

$$C[\mathbf{U}_n] = \sum_i \mathbf{u}(y_i)\mathbf{u}(y_i)'\pi(y_i) = \mathbf{I}$$

and so the observation that

$$M[\mathbf{u}: \hat{p}_n] = \frac{1}{n} \sum_{i=1}^{n} \mathbf{U}_i$$

and the MCLT complete the proof. □

Another very important result which is often used in conjunction with the MCLT is the Mann–Wald theorem:

Proposition 6 (MWT). If $X_n \to_d X$ and $Y_n \to_p y$, and if h is continuous, then

$$h(X_n, Y_n) \to_d h(X, y)$$

Applying the MWT with $h(x, y) = x + y$, we can say that $X_n \to_d X$ and $X_n - Y_n \to_p 0$ implies $Y_n = h(X_n, Y_n - X_n) \to_d h(X, 0) = X$. We will frequently use the MWT in this form.

For instance, suppose that $\tilde{p}_n$ is a random sequence of distributions satisfying $\sqrt{n}(\hat{p}_n - \tilde{p}_n) \to_p 0$. Then we have $\sqrt{n}\, M[\mathbf{u}: \tilde{p}_n] \to_d N(\mathbf{0}, \mathbf{I})$ if $\mathbf{u}$ is π-standard, which shows precisely how close $\tilde{p}_n$ needs to be to $\hat{p}_n$ in order to have the same asymptotic behavior.

It is worth remarking that the MWT remains valid if both X_n and Y_n are replaced by random vectors.

A random sequence $\{X_n\}$ is *bounded in probability* whenever for each $\varepsilon > 0$ there are values b and n_0 for which $P[|X_n| > b] < \varepsilon$ for all $n \geqq n_0$. Thus, although the sequence X_n may not itself be bounded, the probabilities of each of its terms being bounded can be made arbitrarily near 1. An important companion to the MWT can now be stated.

Proposition 7.

1. If $X_n \to_d X$, then X_n is bounded in probability.
2. If X_n is bounded in probability and $Y_n \to_p 0$, then also $X_n Y_n \to_p 0$.

Another useful consequence of this development is an extension of the MCLT which is often called the "δ-method."

Proposition 8. Suppose that $\sqrt{n}(\mathbf{X}_n - \boldsymbol{\mu}) \to_d N(\mathbf{0}, \mathbf{C})$, where $\mathbf{X}_n \in \mathbb{R}^k$. Assume that $\mathbf{f}: \mathbb{R}^k \to \mathbb{R}^m$ has continuous second partial derivatives. Then

$$\sqrt{n}\,[\mathbf{f}(\mathbf{X}_n) - \mathbf{f}(\boldsymbol{\mu})] \to_d N(\mathbf{0}, \mathbf{FCF}')$$

where $\mathbf{F}$ is the matrix of partial derivatives of $\mathbf{f}$, evaluated at $\boldsymbol{\mu}$; that is, F_{ij} is $(\partial/\partial x_j)f_i(\boldsymbol{\mu})$.

Proof. We consider only the one-dimensional case, in which $\sqrt{n}(X_n - \mu) \to_d N(0, \sigma^2)$ and we need to show that $\sqrt{n}[f(X_n) - f(\mu)] \to_d N(0, f'(\mu)^2\sigma^2)$.

By Taylor's theorem with remainder,

$$f(x) = f(\mu) + f'(\mu)(x - \mu) + (x - \mu)^2 \int_0^1 (1 - t) f''(\mu + t(x - \mu))\, dt$$

so that

$$\sqrt{n}[f(X_n) - f(\mu)] = f'(\mu)\sqrt{n}(X_n - \mu) + \sqrt{n}(X_n - \mu)(X_n - \mu) \times \int_0^1 (1 - t) f''(\mu + t(X_n - \mu))\, dt$$

It is easy to see that $f'(\mu)\sqrt{n}(X_n - \mu) \to_d N(0, f'(\mu)^2\sigma^2)$, and so it is only necessary to show that the remaining term $\to_p 0$, by the MWT. Since $\sqrt{n}(X_n - \mu)$ converges in distribution, it is bounded in probability. Then $X_n - \mu \to_p 0$ and the integral is bounded in probability, so that the product of these last three terms $\to_p 0$ according to Proposition 7.2. □

Linear Combinations of Log Ratios. As an application of these results, let $p \in \mathcal{P}$ and $\mathbf{w} : \mathcal{Y} \to \mathbb{R}^k$ be arbitrary. Define

$$\mathbf{W}_n = \sum_y \mathbf{w}(y) \ln \frac{\hat{p}_n(y)}{p(y)}$$

From Proposition 3 we can write $\hat{p}_n = (1 + \hat{\boldsymbol{\alpha}}_n'\mathbf{u})\pi$, where $\mathbf{u}$ is an arbitrarily chosen maximal π-standard function and $\hat{\boldsymbol{\alpha}}_n = M[\mathbf{u} : \hat{p}_n]$. Define

$$\mathbf{f}(\boldsymbol{\alpha}) = \sum_y \mathbf{w}(y) \ln \frac{(1 + \boldsymbol{\alpha}'\mathbf{u})\pi(y)}{p(y)}$$

and compute

$$\frac{\partial}{\partial \boldsymbol{\alpha}'}\mathbf{f} = \sum_y \frac{\mathbf{w}(y)\mathbf{u}(y)'}{1 + \boldsymbol{\alpha}'\mathbf{u}(y)} = \sum_y \mathbf{w}(y)\mathbf{u}(y)'$$

the second equation holding when $\boldsymbol{\alpha} = \mathbf{0}$. Now apply the δ-method to conclude

$$\sqrt{n}\left(\mathbf{W}_n - \sum_y \mathbf{w}(y)\ln\frac{\pi(y)}{p(y)}\right) \to_d N\left(\mathbf{0}, \sum_y \mathbf{w}(y)\mathbf{u}(y)'\sum_y \mathbf{u}(y)\mathbf{w}(y)'\right)$$

The apparent dependence of the covariance matrix on the arbitrarily chosen **u** is dispelled by using the representation

$$\frac{1}{\pi}\mathbf{w} = \sum_y \mathbf{w}(y) + \left(\sum_y \mathbf{w}(y)\mathbf{u}(y)'\right)\mathbf{u}$$

so that

$$\sum_y \frac{1}{\pi(y)}\mathbf{w}(y)\mathbf{w}(y)' = \sum_y \mathbf{w}(y)\sum_y \mathbf{w}(y)'$$

$$+ \left(\sum_y \mathbf{w}(y)\mathbf{u}(y)'\right)\left(\sum_y \mathbf{u}(y)\mathbf{w}(y)'\right)$$

and so the covariance matrix given above can be written

$$\sum_y \frac{1}{\pi(y)}\mathbf{w}(y)\mathbf{w}(y)' - \left(\sum_y \mathbf{w}(y)\right)\left(\sum_y \mathbf{w}(y)\right)'$$

3. Pearson's Chi-Square

As we observed before, if **u** is maximal π-standard, then we have the decomposition

$$\mathbb{D}_2(p, \pi) = \|M[\mathbf{u} : p]\|^2$$

Since $\sqrt{n}\, M[\mathbf{u} : \hat{p}_n] \to_d N(\mathbf{0}, \mathbf{I})$, from the MWT it follows that $n\|M[\mathbf{u} : \hat{p}_n]\|^2$ converges in distribution to a random variable which has the distribution of a sum of squared independent standard Normals.

We formally define the Chi-square distribution with ν degrees of freedom (df), denoted $\chi^2(\nu)$, as the distribution of $\|\mathbf{Z}\|^2$ when $\mathbf{Z}$ is $N(\mathbf{0}, \mathbf{I})$ and $\#\mathbf{Z} = \nu$. The cdf of $\chi^2(\nu)$ can be computed to be

$$F_\nu(x) = \int_0^x \frac{\left(\frac{1}{2}\right)^{\nu/2} t^{(\nu/2)-1} e^{-t/2}}{\Gamma(\nu/2)}\,dt \qquad (x \geqq 0)$$

and we have

Proposition 9. If $\mathbf{u}$ is maximal π-standard, then

$$n\mathbb{D}_2(\hat{p}_n, \pi) = n\|M[\mathbf{u}: \hat{p}_n]\|^2 \rightarrow_d \chi^2(\#\mathcal{Y} - 1)$$

For historical reasons $n\mathbb{D}_2(\hat{p}_n, \pi)$ is called Pearson's chi-square statistic.

Test of a Simple Hypothesis. Proposition 9 shows that the following hypothesis test tends to become valid with increasing sample size:

Reject the hypothesis $\pi = p$ whenever

$$1 - F_m(n\mathbb{D}_2(\hat{p}_n, p)) \leqq \alpha,$$

where $m = \#\mathcal{Y} - 1$, and otherwise *confirm* the hypothesis. The probability of rejecting $\pi = p$ when it is in fact true (the *significance level* of the test) is approximately α.

The random variable $1 - F_m(n\mathbb{D}_2(\hat{p}_n, p))$ is called the *P-value* of the test, and in fact when $\pi = p$ it has (asymptotically) a uniform distribution on $[0, 1]$. We will follow the convention of enclosing P-values in angular brackets $\langle \ \rangle$ to distinguish them.

Since $\mathbb{D}_2(\hat{p}_n, p)$ is fairly obviously a kind of measure of distance between $\hat{p}_n$ and p, it seems clear that $\pi = p$ should be rejected only for large values of $\mathbb{D}_2(\hat{p}_n, p)$. Of course the procedure given above is identical to rejecting whenever $n\mathbb{D}_2(\hat{p}_n, p) \geqq \xi$, where $F_m(\xi) = 1 - \alpha$.

Empty Cells. In small samples it frequently happens that several values of $\hat{p}_n(y)$ will be zero, and it is felt that the approximation involved in the above testing procedure is then likely to be rather poor. Thus $\hat{p}_n$ is often replaced by any one of a number of pseudo-empirical distributions which eliminate the empty cells. One popular choice is to set

$$N_n^*(y) = \begin{cases} N_n(y) & \text{if } N_n(y) > 0 \\ .5 & \text{otherwise} \end{cases}$$

and then define $p_n^*(y) = N_n^*(y)/N_n^*(+)$. Another method uses $N_n^*(y) = N_n(y) + .5$ for all y. In either case $\sqrt{n}\,[p_n^*(y) - \hat{p}_n(y)] \rightarrow_p 0$, a reasonable property in the light of the comments following Proposition 6. For any such adjustment, replacing $\hat{p}_n$ by $p_n^* \in \mathcal{P}$, we have the following:

Proposition 10. If $\sqrt{n}(p_n^* - \hat{p}_n) \to_p 0$, then

$$n\mathbb{D}_2(p_n^*, \pi) - n\mathbb{D}_2(\hat{p}_n, \pi) \to_p 0.$$

Proof. Compute

$$n\mathbb{D}_2(p_n^*, \pi) = n\sum_y \frac{\{[p_n^*(y) - \hat{p}_n(y)] - [\hat{p}_n(y) - \pi(y)]\}^2}{\pi(y)}$$

$$= n\mathbb{D}_2(\hat{p}_n, \pi) + 2n\sum_y \frac{[p_n^*(y) - \hat{p}_n(y)][\hat{p}_n(y) - \pi(y)]}{\pi(y)}$$

$$+ n\sum_y \frac{[\hat{p}_n(y) - p_n^*(y)]^2}{\pi(y)}$$

The last term on the right $\to_p 0$ immediately from the assumptions, and the middle term also $\to_p 0$ because it consists of terms

$$\frac{\sqrt{n}[p_n^*(y) - \hat{p}_n(y)]\sqrt{n}[\hat{p}_n(y) - \pi(y)]}{\pi(y)}$$

each of which $\to_p 0$ according to Proposition 7. □

Thus in terms of asymptotically valid tests based on Pearson's Chi-square it is immaterial whether one uses $\hat{p}_n$ or p_n^*.

4. Computation of *P*-Values

In order to produce the *P*-values for various hypothesis tests we will often need the values of the standard Normal cdf

$$F_0(x) = \int_{-\infty}^{x} \frac{1}{\sqrt{2\pi}} e^{-t^2/2}\, dt$$

and the Chi-square cdf with ν df,

$$F_\nu(x) = \int_0^x \frac{\left(\frac{1}{2}\right)^{\nu/2} t^{(\nu/2)-1} e^{-t/2}}{\Gamma(\nu/2)}\, dt$$

Although tables of these distributions are available, for most practical purposes we may approximate F_0 by

$$F_0(x) = \frac{1}{1 + \exp[-1.5976x(1 + 0.04417x^2)]}$$

the error usually being in the fourth decimal over the range $-4 \leqq x \leqq 4$.

An approximation to F_ν can then be made by a modification of the Wilson–Hilferty transformation:

If $\nu = 1$, then $F_\nu(x) = F_0(\sqrt{x}) - F_0(-\sqrt{x})$.
If $\nu = 2$, then $F_\nu(x) = 1 - \exp(-x/2)$.
If $\nu > 2$, then

$$F_\nu(x) = F_0\left(\frac{(x/\nu)^{1/3} - 1 + 2/(9\nu)}{\sqrt{2/(9\nu)}}\right)$$

The first two of these cases are exact, while the latter amounts to saying that $(X/\nu)^{1/3}$ is $N(1 - 2/(9\nu), 2/(9\nu))$ when X is $\chi^2(\nu)$. This procedure is least satisfactory for $\nu = 3, 4$, and as ν increases, so does the accuracy of the approximation.

All of the P-values we compute will be based on the above method, unless specific mention to the contrary is made. LAMDA programs for carrying out the computations are given in Section B.6. Since the upper tail probabilities are most frequently required, we note for convenience that these are

$$1 - F_0(x) = \frac{1}{1 + \exp[1.5976x(1 + 0.04417x^2)]}$$

$$1 - F_\nu(x) = \begin{cases} 2[1 - F_0(\sqrt{x})] & (\nu = 1) \\ \exp(-x/2) & (\nu = 2) \\ 1 - F_0\left(\dfrac{(x/\nu)^{1/3} - 1 + 2/(9\nu)}{\sqrt{2/(9\nu)}}\right) & (\nu > 2) \end{cases}$$

5. Examples

In this section we will consider two examples which illustrate the ideas of standardization, Pearson's chi-square $\mathbb{D}_2$, and decompositions of $\mathbb{D}_2$.

2² Tables. Let $\mathcal{Y} = \{ij : i = 0, 1; j = 0, 1\}$ be a 2^2 contingency table, and let q and $\mathbf{u}$ be defined as in Table II.1. A few seconds of computation shows

Table II.1

Maximal q-standard Functions on a 2^2 Table

y	u_1	u_2	u_3	q
00	1	1	1	.25
01	1	−1	−1	.25
10	−1	1	−1	.25
11	−1	−1	1	.25

that **u** is maximal q-standard, and so for any p,

$$\mathbb{D}_2(p,q) = M[u_1:p]^2 + M[u_2:p]^2 + M[u_3:p]^2$$

$$= [p(0+) - p(1+)]^2 + [p(+0) - p(+1)]^2$$

$$+ [p(00) - p(01) - p(10) + p(11)]^2$$

Substituting $\hat{p}_n$ for p and multiplying by n, we have $n\mathbb{D}_2(\hat{p}_n, q)$ decomposed into three terms, each of which has a limiting $\chi^2(1)$ distribution if $\pi = q$. Just looking at the form of the statistics, it seems reasonable to imagine that $n[\hat{p}_n(0+) - \hat{p}_n(1+)]^2$ reflects the degree to which $\pi(0+) - \pi(1+)$ differs from zero, while $n[\hat{p}_n(+0) - \hat{p}_n(+1)]^2$ likewise measures the evidence against $\pi(+0) - \pi(+1) = 0$. Moreover, Proposition 5 shows that

$$nM[u_1:\hat{p}_n]^2 + nM[u_2:\hat{p}_n]^2 = n[\hat{p}_n(0+) - \hat{p}_n(1+)]^2$$

$$+ n[\hat{p}_n(+0) - \hat{p}_n(+1)]^2$$

could be considered a $\chi^2(2)$ statistic for testing the hypothesis that $\pi(0+) = \pi(1+)$ and $\pi(+0) = \pi(+1)$. If this were confirmed, then it seems reasonable to use $nM[u_3:\hat{p}_n]^2$ as a $\chi^2(1)$ statistic for testing $\pi = q$.

This is a fairly crude procedure, depending as it does on the assumption that $\pi = q$, and we will greatly improve it later on, but for the moment the important point is that the computation of $n\mathbb{D}_2(\hat{p}_n, q)$ is the beginning, not the end, of the analysis. When this statistic is so large as to cause rejection of $\pi = q$, one wants to know why (that is, in what particular ways $\hat{p}_n$ provides evidence against q), and this is the purpose of partitioning $\mathbb{D}_2$.

To illustrate an application of this method, we consider the data in Table II.2, which pertain to the occurrence of two types of stone tools found at two levels of an archeological site. Compute the components of $\mathbb{D}_2$ and

Table II.2

Archeological Data

	Level I	Level II	Total
Flakes	70	93	163
Scrapers	63	74	137
Total	133	167	300

P-values (in brackets) as follows:

$$nM[u_1 : \hat{p}_n]^2 = 2.253 \ \langle .134 \rangle$$

$$nM[u_2 : \hat{p}_n]^2 = 3.853 \ \langle .049 \rangle$$

$$nM[u_3 : \hat{p}_n]^2 = 0.480 \ \langle .488 \rangle$$

$$n\mathbb{D}_2(\hat{p}_n, q) = 6.586 \ \langle .085 \rangle$$

Although $n\mathbb{D}_2(\hat{p}_n, q)$ is not below the traditional .05 level, it does cast some doubt on the hypothesis that $\pi = q$, and it appears from the partition that the direction of the departure of π from q is in the marginal Level distribution.

In Table II.3 we have some data pertaining to a certain subsample of mathematics faculty members in 1979–1980. A plausible value for q was obtained from a larger similar sample, and appears in Table II.4. Also in this table is the q-standardization of the u-functions, here renamed v. Compute the components of Pearson's Chi-square:

$$nM[v_1 : \hat{p}_n]^2 = 0.486 \ \langle .485 \rangle$$

$$nM[v_2 : \hat{p}_n]^2 = 15.628 \ \langle .000 \rangle$$

$$nM[v_3 : \hat{p}_n]^2 = 2.901 \ \langle .089 \rangle$$

$$n\mathbb{D}_2(\hat{p}_n, q) = 19.015 \ \langle .000 \rangle$$

One would clearly reject $\pi = q$ here, and the point is that by decomposing $\mathbb{D}_2$ we have found that the main source of difference seems to be in the tenured–untenured marginal distribution. See Section B.6 for the LAMDA programs for this and the preceding analysis.

Table II.3

Data Relating the Possession of a Doctorate

	Untenured	Tenured
Without Doctorate	18	50
With Doctorate	78	373

Table II.4

Maximal q-standard Functions on a 2^2 Table

y	v_1	v_2	v_3	q
00	2.4611	1.6818	4.1310	.0370
01	2.4611	−0.5946	−1.4594	.1047
10	−0.4063	1.6818	−0.6934	.2242
11	−0.4063	−0.5946	0.2453	.6341

A One-Way Periodic Table. The data in Table II.5 pertain to a source of infant mortality which has been called the "sudden infant death syndrome" (SIDS). The second column of Table II.5 may be considered to show the population frequencies of birth months, while the third column reflects the random observations of the birth months of SIDS cases. In the next two columns these frequencies have been converted to probabilities.

The hypothesis of interest is whether the distribution of birth months of SIDS infants is the same as the population distribution, that is, $\pi = q$, where π is the true distribution for SIDS cases. If π were to depart from q, one might expect one or more systematic cyclic trends, and so it seems reasonable to choose functions having periodic components. Define

$$u_i(y) = \sin\left(\frac{i\pi}{6}(y-1)\right), \qquad v_i(y) = \cos\left(\frac{i\pi}{6}(y-1)\right)$$

for $1 \leqq i \leqq 6$ and with the months y numbered consecutively beginning with January = 1. Thus the cycle is taken to begin in the middle of winter. (The π in these expressions is the usual pi = 3.14159265, not the true distribution.)

Since $u_6 = 0$, we see that the eleven functions u_i $(1 \leqq i \leqq 5)$ and v_i $(1 \leqq i \leqq 6)$ will serve the purpose of decomposing Pearson's chi-square. The next step should be to compute the q-standardization of these functions and then compute their $\hat{p}$-means. It is important to see from the definition of

Table II.5

Population Distribution of Birth Months and Distribution of Birth Months of SIDS Infants

Month	Average Number of Births	SIDS Cases	q	$\hat{p}$
January	1460	29	.0770	.0898
February	1446	22	.0763	.0681
March	1635	14	.0863	.0433
April	1547	19	.0816	.0588
May	1619	18	.0854	.0557
June	1586	23	.0837	.0712
July	1630	28	.0860	.0867
August	1625	36	.0858	.1115
September	1611	27	.0850	.0836
October	1603	39	.0846	.1207
November	1529	38	.0807	.1176
December	1659	30	.0875	.0929

standardization that the order in which the functions are standardized is significant, different orderings leading to different standardized functions, and thus to different $\hat{p}$-means and different decompositions of $\mathbb{D}_2$.

If we do not standardize the entire set of functions, but instead only normalize each one separately, we obtain the $\hat{p}$-means and P-values of Table II.6. Each P-value in this table is proper only for the case in which the function on the same row is the first to be standardized. We now order these functions according to their P-values, and standardize the entire set in the resulting order. This is done in Table II.7 and the resulting P-values are computed. We note that these new P-values are highly related to the old ones, but we must remember that the functions in Table II.7 represent the standardizations of the functions in Table II.6. Notice that this procedure violates the strict hypothesis-testing discipline, since the order of standardization is determined by the data. Thus the P-values in Table II.7 are not strictly justified, but they do give some evidence in a descriptive sense that some of the trend functions come much closer than others to fitting the data.

Since the period of u_i and v_i is $12/i$, we see that the two most important components have period 12 and 3, respectively, while the remainder of the functions do not appear to fall in any particular pattern or to make very significant contributions to $\mathbb{D}_2$. In fact, the sum of the last nine terms $nM[u_i : \hat{p}]^2$ is only 3.527, whereas that of the first two is 20.794. Thus

Table II.6

Empirical Means and P-values of the Normalized Sines and Cosines

u	$M[u:\hat{p}]$	$\langle P \rangle$
u_1	−.2373	⟨.000⟩
u_2	−.0334	⟨.551⟩
u_3	−.0075	⟨.893⟩
u_4	.0941	⟨.093⟩
u_5	.0301	⟨.591⟩
v_1	.0151	⟨.787⟩
v_2	.0519	⟨.354⟩
v_3	−.0188	⟨.746⟩
v_4	.0397	⟨.478⟩
v_5	.0553	⟨.323⟩
v_6	−.0354	⟨.527⟩

Table II.7

Empirical Means and P-values of the Standardized Sines and Cosines

u	$M[u:\hat{p}]$	$\langle P \rangle$
u_1	−.2373	⟨.000⟩
u_4	.0942	⟨.093⟩
v_5	.0568	⟨.310⟩
v_2	.0514	⟨.359⟩
v_4	.0390	⟨.486⟩
v_6	−.0289	⟨.605⟩
u_2	−.0356	⟨.525⟩
u_5	.0289	⟨.605⟩
v_3	.0202	⟨.718⟩
v_1	.0179	⟨.749⟩
u_3	.0068	⟨.903⟩

$n\mathbb{D}_2(\hat{p}, q) = 24.321$ $\langle .007 \rangle$ leads to rejection of $\pi = q$, and one is led to suspect that there is an intra-year trend reflected in the u_1 and u_4 components. In the next subsection we will see how this analysis may be done by LAMDA.

6. LAMDA Notes

Although the process of standardizing functions and computing means can, in theory, be done with a pocket calculator, one maintains a more cheerful countenance when a computer does the work.

The first analysis of the preceding section can be carried out by Run II.B.1 (see Appendix C). At line 3 the *C*-file is dimensioned to have four rows and columns, and on lines 6–9 the functions and frequencies are entered. Line 11 declares variable 4 as the weight variable, and then at line 13 summary statistics on variables 1, 2, and 3 are requested. The output will contain the means and standard deviations of these three variables with respect to the probability distribution having probabilities proportional to the values of variable 4. The output will also display the covariance matrix of the three variables, and the inverse of the covariance matrix, all computations using variable 4 as the weight variable. In order to compute the components of Pearson's Chi-square, simply square each of the means and multiply by sample size.

The analysis of Table II.3 is carried out by Run II.B.2. The data are entered in the same way as in the previous run. The functions are in variables 2, 3, and 4, and at line 13 these are standardized using the weight variable in column 1. Note that the distribution q has been placed in variable 1. The output from the STD command includes (1) the means and standard deviations, (2) the covariance and inverse covariance matrices of the normalizations of the functions, and (3) the coefficients of the weighted least-squares fit of each normalized variable in terms of its predecessors. At line 18 the observed frequencies are declared as the weights, and then the summary statistics are requested at line 19. Again, the means here would be squared and multiplied by sample size to produce the components of Pearson's Chi-square.

The analysis of Table II.5 is rather more complex. In Run II.B.3, the values of the sines and cosines have been previously stored on file 8. As they are read in at line 5, two blank variables are created in columns 1 and 2, according to the SEL command at line 3. Columns 1 and 2 are then filled with the population and observed frequencies. The sines and cosines are normalized with respect to the population frequencies at line 15, and then the summary statistics are computed with respect to the observed frequencies at line 20. At line 22 the sines and cosines are standardized in the order determined by Table II.6, as described in the preceding subsection. Again the summary statistics are computed with respect to the observed frequencies at line 26. Squaring the means and multiplying by sample size gives the components of Pearson's Chi-square.

Programs for P-values. LAMDA programs to compute approximate Normal and Chi-square P-values appear in Run II.B.4 and Run II.B.5. In both of these runs the number of rows to be entered is put in place of X in row 3 of the program, and then the values of the Normal statistics, or the Chi-square statistics and degrees of freedom, are put in after the **ENT** command. The first program computes F_0, and the second F_m for $m \geqq 3$.

C. DISCREPANCIES

The procedure for testing $\pi = q$ has been given in terms of rejecting $\pi = q$ when $\mathbb{D}_2(\hat{p}, q)$ is too large. The idea is that since $\hat{p}_n \to_p \pi$, it follows from the MWT that $\mathbb{D}_2(\hat{p}_n, q) \to_p \mathbb{D}_2(\pi, q)$, and this latter quantity will be zero only when $\pi = q$. This procedure seems logical and is justified by the asymptotic theory of Section B.3, but what is perhaps not clear is why $\mathbb{D}_2$ is chosen to measure the difference between two probability distributions. In this section we will see that there is a very large collection of other ways of measuring differences.

1. Definition and Asymptotic Properties

A function $\mathbb{D} : \mathcal{P} \times \mathcal{P} \to [0, \infty[$ is a *discrepancy* if and only if for all $p, q \in \mathcal{P}$ we have $\mathbb{D}(p, q) \geqq 0$ with equality only when $p = q$. $\mathbb{D}$ is *continuous* if and only if whenever $p_n \to q \in \mathcal{P}$ we have $\mathbb{D}(p_n, q) \to 0$. Clearly $\mathbb{D}_2$ is a continuous discrepancy.

Recall the representation given in Proposition 3,

$$p = (1 + \boldsymbol{\alpha}'\mathbf{u})q, \qquad \boldsymbol{\alpha} = M[\mathbf{u} : p]$$

in which $q \in \mathcal{P}$ and $\mathbf{u}$ (maximal q-standard) are fixed, establishing a continuous one-to-one correspondence between $\mathcal{P}$ and an open subset of $\mathbb{R}^k$. For a given discrepancy $\mathbb{D}$ we thus have an associated function d defined by $d(\boldsymbol{\alpha}) = \mathbb{D}((1 + \boldsymbol{\alpha}'\mathbf{u})q, q)$. Obviously $\mathbb{D}$ is continuous if and only if d is continuous. Moreover, we will call $\mathbb{D}$ *smooth* if and only if the second-derivative matrix

$$\frac{\partial^2}{\partial \boldsymbol{\alpha}\, \partial \boldsymbol{\alpha}'} d$$

whose ijth element is $(\partial^2/\partial\alpha_i\, \partial\alpha_j)d$, is a continuous function of α.

Proposition 11. If $\mathbb{D}$ is a smooth discrepancy, if

$$\frac{\partial^2}{\partial \boldsymbol{\alpha}\, \partial \boldsymbol{\alpha}'} d = 2\mathbf{I}$$

when evaluated at $\boldsymbol{\alpha} = \mathbf{0}$, and if $\pi = q$, then

1. $n\mathbb{D}(\hat{p}_n, q) - n\mathbb{D}_2(\hat{p}_n, q) \to_p 0$,
2. $n\mathbb{D}(\hat{p}_n, q) \to_d \chi^2(\#\mathcal{Y} - 1)$.

Proof. From a Taylor expansion with remainder about the point $\boldsymbol{\alpha} = \mathbf{0}$,

$$\mathbb{D}(p, q) = \boldsymbol{\alpha}' \int_0^1 (1 - t) \frac{\partial^2}{\partial \boldsymbol{\alpha}\, \partial \boldsymbol{\alpha}'} d(t\boldsymbol{\alpha})\, dt\, \boldsymbol{\alpha}, \qquad \boldsymbol{\alpha} = M[\mathbf{u}: p]$$

and since $\hat{p}_n \to_p q$, by the MWT we have

$$n\mathbb{D}(\hat{p}_n, q) - \tfrac{1}{2} nM[\mathbf{u}: \hat{p}_n]' \frac{\partial^2}{\partial \boldsymbol{\alpha}\, \partial \boldsymbol{\alpha}'} d(\mathbf{0}) M[\mathbf{u}: \hat{p}_n] \to_p 0$$

and so the assumption of the proposition gives

$$n\mathbb{D}(\hat{p}_n, q) - n\|M[\mathbf{u}: \hat{p}_n]\|^2 \to_p 0$$

which proves part 1. Now part 2 follows from the discussion after Proposition 6. □

If $\mathbb{D}$ satisfies the conditions of the above proposition, we will say that $\mathbb{D}$ is *equivalent* to $\mathbb{D}_2$. Notice that if $\tilde{p}_n$ is equivalent to $\hat{p}_n$ in the sense that $\sqrt{n}(\tilde{p}_n - \hat{p}_n) \to_p 0$, then the above result remains true with $\hat{p}_n$ replaced throughout by $\tilde{p}_n$.

We will frequently be able to define discrepancies over the wider classes $\mathcal{P}_0 \times \mathcal{P}$, $\mathcal{P} \times \mathcal{P}_0$, or even $\mathcal{P}_0 \times \mathcal{P}_0$, though we always assume $\pi \in \mathcal{P}$. For instance, $\mathbb{D}_2$ is in fact a discrepancy on $\mathcal{P}_0 \times \mathcal{P}$.

2. Convex Discrepancies

We now outline a broad class of discrepancies which might be used in place of $\mathbb{D}_2$. Recall that a function $f: [0, \infty[\to] - \infty, \infty]$ is *convex* if and only if for every $u: \mathcal{Y} \to [0, \infty[$ and every $q \in \mathcal{P}_0$ it is true that $M[f(u): q] \geq f(M[u: q])$. Further, f is *strictly convex* if and only if the above inequality is an equality only when $f(u(y)) = f(M[u: q])$ for every y having $q(y) > 0$. The only convex functions we will deal with in practice are twice continuously differentiable, and so for us f is convex if and only if $f'' \geqq 0$, and strictly convex if and only if $f'' > 0$. A function $g : [0, \infty[\rightarrow]-\infty, \infty]$ is *x-convex* if and only if the function $x \cdot g(x)$ is strictly convex.

Proposition 12. Assume that g is x-convex, $g(1) = 0$, and the solutions of $g(x) = 0$ lie in $\{0, 1\}$. Then $\mathbb{D}$ defined by

$$\mathbb{D}(p, q) = M\left[g\left(\frac{p}{q}\right) : p\right]$$

is a discrepancy on $\mathcal{P}_0 \times \mathcal{P}$.

Proof. $\mathbb{D}(p, q) = M[(p/q)g(p/q) : q] \geqq M[p/q : q] \cdot g(M[(p/q) : q]) = g(1) = 0$, so that $\mathbb{D}(p, q) = 0$ if and only if $(p/q)g(p/q) = 0$, which is equivalent to p/q assuming values in $\{0, 1\}$, which forces $p = q$. $\square$

Note that if we assume only $g :]0, \infty] \rightarrow]-\infty, \infty]$ (that is, $g(0)$ is not defined), then the above proof goes through and we conclude that $\mathbb{D}$ is a discrepancy on $\mathcal{P} \times \mathcal{P}$.

We will call any discrepancy obtained in the above fashion a *convex discrepancy*. We now establish the conditions g must satisfy in order for the associated discrepancy to be equivalent to $\mathbb{D}_2$.

Proposition 13. Let $\mathbb{D}$ be the convex discrepancy based on g. Then

$$n\{\mathbb{D}(\hat{p}_n, \pi) - [g'(1) + \tfrac{1}{2}g''(1)]\mathbb{D}_2(\hat{p}_n, \pi)\} \rightarrow_p 0$$

Thus $g'(1) + \frac{1}{2}g''(1) = 1$ is necessary and sufficient for $n\mathbb{D}(\hat{p}_n, \pi) \rightarrow_d \chi^2(\#\mathcal{Y} - 1)$.

Proof. Again let $q = (1 + \boldsymbol{\alpha}'\mathbf{u})\pi$ for maximal π-standard $\mathbf{u}$. Compute

$$\frac{\partial}{\partial \boldsymbol{\alpha}}\left[\frac{q}{\pi}g\left(\frac{q}{\pi}\right)\right] = \frac{\partial}{\partial \boldsymbol{\alpha}}\left[\frac{q}{\pi}\right] \cdot g\left(\frac{q}{\pi}\right) + \frac{q}{\pi}\frac{\partial}{\partial \boldsymbol{\alpha}}g\left(\frac{q}{\pi}\right)$$

$$= g\left(\frac{q}{\pi}\right)\mathbf{u} + \frac{q}{\pi}g'\left(\frac{q}{\pi}\right)\mathbf{u}$$

$$\frac{\partial^2}{\partial \alpha\, \partial \alpha'}\left[\frac{q}{\pi}g\left(\frac{q}{\pi}\right)\right] = \left[g'\left(\frac{q}{\pi}\right)\mathbf{u} + g'\left(\frac{q}{\pi}\right)\mathbf{u} + \frac{q}{\pi} \cdot g''\left(\frac{q}{\pi}\right)\mathbf{u}\right]\mathbf{u}'$$

$$= \left[2g'\left(\frac{q}{\pi}\right) + \frac{q}{\pi}g''\left(\frac{q}{\pi}\right)\right]\mathbf{u}\mathbf{u}'$$

Thus for $d(\boldsymbol{\alpha}) = \mathbb{D}((1 + \boldsymbol{\alpha}'\mathbf{u})\pi, \pi)$ we have

$$\frac{\partial^2}{\partial\boldsymbol{\alpha}\,\partial\boldsymbol{\alpha}'} d(\mathbf{0}) = [2g'(1) + g''(1)]M[\mathbf{uu}' : \pi]$$

which completes the proof by Proposition 11. □

3. The Discrepancies $\mathbb{D}_\omega$

For any ω other than 0 or 1, define

$$g_\omega(x) = 2 \cdot \frac{1 - x^{\omega-1}}{\omega(1-\omega)} \qquad (x > 0)$$

For $\omega > 1$ we can define $g_\omega(0) = 2/\omega(1-\omega)$. Compute

$$x \cdot g_\omega(x) = 2 \cdot \frac{x - x^\omega}{\omega(1-\omega)}$$

$$\frac{\partial}{\partial x}[x \cdot g_\omega(x)] = 2 \cdot \frac{1 - \omega x^{\omega-1}}{\omega(1-\omega)}$$

$$\frac{\partial^2}{\partial x^2}[x \cdot g_\omega(x)] = 2x^{\omega-2} > 0 \qquad \text{for all} \quad x > 0$$

Thus g_ω is x-convex, and since $g_\omega(x) = 0$ has unique solution $x = 1$, by Proposition 12 we can define $\mathbb{D}_\omega$ to be the convex discrepancy based on g_ω,

$$\mathbb{D}_\omega(p, q) = M\left[g_\omega\left(\frac{p}{q}\right) : p\right]$$

Since $g'_\omega(x) = 2x^{\omega-2}/\omega$, $g''_\omega(x) = 2x^{\omega-3}(\omega-2)/\omega$, we have $g'_\omega(1) + \frac{1}{2}g''_\omega(1) = 2/\omega + (\omega-2)/\omega = 1$, and so each $\mathbb{D}_\omega$ is equivalent to Pearson's Chi-square by Proposition 13.

Note that

$$\mathbb{D}_\omega(p, q) = 2 \cdot \frac{1 - \Sigma q(y)^{1-\omega} p(y)^\omega}{\omega(1-\omega)}$$

which immediately yields the equation $\mathbb{D}_\omega(p, q) = \mathbb{D}_{1-\omega}(q, p)$. Observe also that the notation we have been using for Pearson's Chi-square is consistent with the $\mathbb{D}_\omega$ notation:

$$\mathbb{D}_2(p,q) = M\left[g_2\left(\frac{p}{q}\right) : p\right] = 2\sum_y \frac{1 - p(y)/q(y)}{2(-1)} p(y)$$

$$= \sum_y \frac{p(y)^2}{q(y)} - 1 = \sum_y \frac{[p(y) - q(y)]^2}{q(y)}$$

Another well-known Chi-square statistic is the case $\omega = -1$: $n\mathbb{D}_{-1}(\hat{p}_n, q) = n\mathbb{D}_2(q, \hat{p}_n)$ is *Neyman's Chi-square*. $n\mathbb{D}_{1/2}(\hat{p}_n, q)$ is a Chi-square associated with *Matusita*, and has the distinction of being the only $\mathbb{D}_\omega$-based Chi-square that is symmetric in $\hat{p}_n$ and q.

To extend the definition of $\mathbb{D}_\omega$ to the case $\omega = 1$, set $\xi = \omega - 1$ and let $\xi \to 0$ to obtain

$$g_\omega(x) = \frac{2}{\omega} \frac{x^\xi - 1}{\xi} \to 2\ln x$$

and so we define $g_1(x) = 2\ln x$ and compute

$$\frac{\partial}{\partial x}[2x\ln x] = 2 + 2\ln x$$

$$\frac{\partial^2}{\partial x^2}[2x\ln x] = \frac{2}{x} > 0$$

so that

$$\mathbb{D}_1(p,q) = 2\sum p(y)\ln\frac{p(y)}{q(y)}$$

is a discrepancy, and $g_1'(x) = 2/x$, $g_1''(x) = -2$ shows that $\mathbb{D}_1$ is equivalent to $\mathbb{D}_2$. $n\mathbb{D}_1(\hat{p}_n, q)$ may be called the *Likelihood Chi-square statistic.*

By symmetry $\mathbb{D}_0$ is defined as

$$\mathbb{D}_0(p,q) = \lim_{\omega\to 0}\mathbb{D}_\omega(p,q) = \lim_{\omega\to 0}\mathbb{D}_{1-\omega}(q,p)$$

$$= \mathbb{D}_1(q,p)$$

so that

$$\mathbb{D}_0(p,q) = 2\sum q(y)\ln\frac{q(y)}{p(y)}$$

Likewise, we can compute that $g_0(x) = (2/x)\ln(1/x)$ and check that $\mathbb{D}_0$ is a convex discrepancy equivalent to $\mathbb{D}_2$. $n\mathbb{D}_0(\hat{p}_n, q)$ is called either the *Kullback–Leibler* or the *Information Chi-square statistic.*

We note that $\mathbb{D}_\omega$ is now defined on

$$\mathcal{P}_0 \times \mathcal{P} \text{ for } \omega \geqq 1,$$
$$\mathcal{P}_0 \times \mathcal{P}_0 \text{ for } 0 < \omega < 1,$$
$$\mathcal{P} \times \mathcal{P}_0 \text{ for } \omega \leqq 0,$$

and that the class $\{\mathbb{D}_\omega\}$ contains most of the commonly used Chi-square statistics.

4. Asymptotic Behavior Under Contiguous Sequences

Suppose that we intend to reject the hypothesis $\pi = q$ whenever $n\mathbb{D}_2(\hat{p}_n, q) \geqq \xi$, where $F_m(\xi) = 1 - \alpha$. We know that $\mathbb{D}_2(\hat{p}_n, q) \to_p \mathbb{D}_2(\pi, q)$, so that $n\mathbb{D}_2(\hat{p}_n, q)$ diverges to ∞ if $\pi \neq q$. Consequently, if $\pi \neq q$, the above hypothesis-testing procedure will reject $\pi = q$ with larger and larger probability as $n \to \infty$. We would like to make some estimate of this probability, and clearly we cannot do this for any fixed $q \neq \pi$, so instead we consider a sequence for which $q_n \to \pi$. We will say that $\{q_n\}$ is a *contiguous sequence* if and only if $\sqrt{n}\,(q_n - \pi)$ converges to a finite function on $\mathcal{Y}$.

For the next proposition we will need to recall that the *Noncentral Chi-square* distribution with ν df and noncentrality parameter λ, denoted $\chi^2(\nu, \lambda)$, is that of $\|\mathbf{Z}\|^2$ when $\mathbf{Z}$ is $N(\boldsymbol{\mu}, \mathbf{I})$, where $\nu = \#\boldsymbol{\mu}$ and $\lambda = \|\boldsymbol{\mu}\|^2$. (Some authors use $\|\boldsymbol{\mu}\|^2/2$ for the noncentrality parameter.)

Proposition 14. If $\{p_n\}$ is a contiguous sequence, then

$$n\mathbb{D}_2(\hat{p}_n, p_n) \to_d \chi^2(\#\mathcal{Y} - 1, \lim n\mathbb{D}_2(p_n, \pi))$$

Proof. Let $\mathbf{u}$ be maximal π-standard, and let

$$\frac{\hat{p}_n - p_n}{\pi} = \hat{\beta}_0 + \hat{\boldsymbol{\beta}}_n'\mathbf{u}$$

so that

$$\hat{\beta}_0 = M\left[\frac{\hat{p}_n - p_n}{\pi} : \pi\right] = 0$$

$$\hat{\boldsymbol{\beta}}_n = M\left[\mathbf{u}\frac{\hat{p}_n - p_n}{\pi} : \pi\right] = M[\mathbf{u} : \hat{p}_n] - M[\mathbf{u} : p_n]$$

By assumption, $\sqrt{n}\,M[\mathbf{u}: p_n]$ converges, and so

$$\sqrt{n}\,\hat{\boldsymbol{\beta}}_n \rightarrow_d N\left(-\lim\sqrt{n}\,M[\mathbf{u}: p_n], \mathbf{I}\right)$$

Now

$$n\|\hat{\boldsymbol{\beta}}_n\|^2 = \frac{n\Sigma[\hat{p}_n(y) - p_n(y)]^2}{\pi(y)}$$

$$= n\sum_y \frac{[\hat{p}_n(y) - p_n(y)]^2}{p_n(y)}$$

$$+ n\sum_y [\hat{p}_n(y) - p_n(y)]^2\left[\frac{1}{\pi(y)} - \frac{1}{p_n(y)}\right]$$

and we have that the left-hand side $\rightarrow_d \chi^2(\#\mathcal{Y} - 1, \lim n\mathbb{D}_2(p_n, \pi))$ by the MWT. Now each term

$$\sqrt{n}\,[\hat{p}_n(y) - p_n(y)] = \sqrt{n}\,[\hat{p}_n(y) - \pi(y)] + \sqrt{n}\,[\pi(y) - p_n(y)]$$

converges in distribution, and thus is bounded in probability. Since $1/\pi(y) - 1/p_n(y) \rightarrow 0$, we have from Proposition 6 that

$$n\|\hat{\boldsymbol{\beta}}_n\|^2 - n\mathbb{D}_2(\hat{p}_n, p_n) \rightarrow_p 0$$

which completes the proof. □

Proposition 15. Let $\mathbb{D}$ be a smooth discrepancy equivalent to $\mathbb{D}_2$, and let $\mathbf{u}$ be maximal π-standard. Let $\mathbf{u}_p$ be the p-standardization of $\mathbf{u}$. Define

$$d(\boldsymbol{\alpha}, p) = \mathbb{D}\left((1 + \boldsymbol{\alpha}'\mathbf{u}_p)p, p\right)$$

and assume that the second-derivative matrix of d is continuous in both $\boldsymbol{\alpha}$ and p. Then whenever $\{p_n\}$ is a contiguous sequence, we have

$$n\mathbb{D}(\hat{p}_n, p_n) - n\mathbb{D}_2(\hat{p}_n, p_n) \rightarrow_p 0$$

Proof. Again from the equation

$$d(\boldsymbol{\alpha}, p) = \boldsymbol{\alpha}'\int_0^1 (1 - t)\frac{\partial^2}{\partial\boldsymbol{\alpha}\,\partial\boldsymbol{\alpha}'}d(t\boldsymbol{\alpha}, p)\,dt\,\boldsymbol{\alpha}$$

we have

$$n\mathbb{D}(\hat{p}_n, p_n) = \sqrt{n}\,\hat{\boldsymbol{\alpha}}_n' \int_0^1 (1 - t)\frac{\partial^2}{\partial \boldsymbol{\alpha}\, \partial \boldsymbol{\alpha}'} d(t\hat{\boldsymbol{\alpha}}_n, p_n)\, dt \sqrt{n}\,\hat{\boldsymbol{\alpha}}_n$$

where $\hat{\boldsymbol{\alpha}}_n = M[\mathbf{u}_n, \hat{p}_n]$ and $\mathbf{u}_n$ is the p_n-standardization of $\mathbf{u}$. Since the integral term above converges to $\mathbf{I}$, the result follows from $\|\hat{\boldsymbol{\alpha}}_n\|^2 = \mathbb{D}_2(\hat{p}_n, p_n)$. □

Asymptotic Power. As a consequence of these last two propositions we have for a wide class of discrepancies, including the $\mathbb{D}_\omega$, that

$$n\mathbb{D}(\hat{p}_n, p_n) \to_d \chi^2(\#\mathscr{Y} - 1, \lim n\mathbb{D}(p_n, \pi))$$

the form of the noncentrality parameter being obtained by the same argument as in the last proposition. In fact,

$$\lim n\mathbb{D}(p_n, \pi) = \lim n\mathbb{D}_2(p_n, \pi)$$

To return to the hypothesis test which began this section, we may compute the power of the tests of the contiguous sequence p_n by computing the sequence of probabilities $P[n\mathbb{D}(\hat{p}_n, p_n) \geqq \xi]$, and this is done by treating $n\mathbb{D}(\hat{p}_n, p_n)$ as if it were $\chi^2(\#\mathscr{Y} - 1, \lim n\mathbb{D}(p_n, \pi))$. Since the critical value ξ and the sample size are in actuality fixed, only a single power computation will be made. It should be kept in mind, however, that the approximation is only good for p_n near π, and so it should not be used for large values of $\mathbb{D}(p_n, \pi)$.

Since the power of the test at $q \neq p_n$ is the probability that the test of $\pi = p_n$ will correctly lead to rejection when $\pi = q$, we see that high power is a desirable property of a test, and ought to play a role in choosing one hypothesis-testing procedure over another. We also see, however, that there is a very broad class of discrepancies whose asymptotic behavior is identical to that of $\mathbb{D}_2$, with regard to both the hypothesis $\pi = q$ and the power along contiguous sequences. Thus, considerations of power do not lead to a preference of some discrepancies over others.

If we imagine ourselves in the situation in which either $\pi = p$ or $\pi = q$ (so that no other possibilities are under consideration), then when we observe $n\mathbb{D}(\hat{p}_n, p) \geqq \xi$ we reject $\pi = p$. Because α was chosen before the test, we can say that the probability of our rejection being in error is (approximately) α. However, if we observe $n\mathbb{D}(\hat{p}_n, p) < \xi$, then we confirm $\pi = p$, but what is the probability that this conclusion is in error? By the results above, if $\pi = q$, then $n\mathbb{D}(\hat{p}_n, p)$ is approximately $\chi^2(\#\mathscr{Y} -$

$1, n\mathbb{D}(p, q))$, so that the probability of error is approximately $F_{\nu,\lambda}(\xi)$, where $F_{\nu,\lambda}$ is the cdf of $\chi^2(\nu, \lambda)$, $\nu = \#\mathcal{Y} - 1$, $\lambda = n\mathbb{D}(p, q)$. The power of the test at q is just $1 - F_{\nu,\lambda}(\xi)$.

In order to carry out these computations it is necessary to be able to compute $F_{\nu,\lambda}$. The Noncentral Chi-square distribution may be approximated by the (central) Chi-square distribution according to

$$F_{\nu,\lambda}(x) = F_{\nu'}(cx)$$

which is the same as saying that "X is $\chi^2(\nu, \lambda)$" is approximately equivalent to "cX is $\chi^2(\nu')$." By matching the first two moments of these distributions we obtain

$$\nu' = c(\nu + \lambda), \qquad \nu' = c^2(\nu + 2\lambda)$$

so that

$$c = \frac{\nu + \lambda}{\nu + 2\lambda}, \qquad \nu' = \frac{(\nu + \lambda)^2}{\nu + 2\lambda}$$

We may summarize the computational procedure as follows:

1. Test $\pi = q$ by rejecting when $n\mathbb{D}(\hat{p}_n, q) \geqq \xi$, where $F_m(\xi) = 1 - \alpha$, $m = \#\mathcal{Y} - 1$.
2. Compute the power of the test at p as $1 - F_{m,\lambda}(\xi)$, where $\lambda = n\mathbb{D}(q, p)$, by using the approximation $1 - F_{m,\lambda}(\xi) = 1 - F_{m'}(c\xi)$ with $c = (m + \lambda)/(m + 2\lambda)$, $m' = (m + \lambda)^2/(m + 2\lambda)$.

D. SUMMARY

The important consequences of this chapter are presented here, stripped of the mathematical details. The reader is referred to the body of the chapter for more precise statements and for proofs.

1. Discrepancies

The three most important ways of measuring the distance between two probability distributions are

Information: $\mathbb{D}_0(p, q) = 2\Sigma q(y)\ln[q(y)/p(y)]$;
Likelihood: $\mathbb{D}_1(p, q) = 2\Sigma p(y)\ln[p(y)/q(y)]$;
Pearson: $\mathbb{D}_2(p, q) = \Sigma[p(y) - q(y)]^2/q(y)$.

In each case the summation is over all $y \in \mathcal{Y}$. We will use $\mathbb{D}_0$ and $\mathbb{D}_1$ most frequently in following chapters. They all have the property that

$$\mathbb{D}_i(p, q) \geqq 0 \text{ with equality if and only if } p = q$$

for $i = 0$, 1, or 2.

2. Test of $\pi = q$

Any of the discrepancies can be used to test the hypothesis that $\pi = p$. This is done by using the fact that $n\mathbb{D}_i(\hat{p}_n, p)$ will have (nearly) a chi-square distribution with $\#\mathcal{Y} - 1$ df if in fact $\pi = p$. This means

$$P\left[n\mathbb{D}_i(\hat{p}_n, p) \leqq x\right] = F_m(x) \qquad \text{(approximately)}$$

where $m = \#\mathcal{Y} - 1$, and instructions for computing F_m are given in Section B.4. The testing procedure can be phrased in either of two equivalent forms:

1. Choose the level of the test, α. Determine ξ so that $F_m(\xi) = 1 - \alpha$. Reject $\pi = p$ if $n\mathbb{D}_i(\hat{p}_n, p) \geqq \xi$, and otherwise confirm $\pi = p$.
2. Compute the P-value, $1 - F_m(n\mathbb{D}_i(\hat{p}_n, p))$. If this quantity falls below α, then reject $\pi = p$ at level α; otherwise, confirm $\pi = p$.

The value α has the interpretation of the probability of rejecting $\pi = p$ when it is in fact true.

3. Power of the Test

In testing $\pi = p$ by the above procedure, if in fact $\pi \neq p$, then

$$P\left[n\mathbb{D}_i(\hat{p}_n, p) \leqq x\right] = F_{m,\lambda}(x) \qquad \text{(approximately)}$$

where $F_{m,\lambda}$ is the noncentral chi-square cdf, $m = \#\mathcal{Y} - 1$, and $\lambda = n\mathbb{D}_i(p, \pi)$. Instructions for computing $F_{m,\lambda}$ are given in Section C.4. Letting ξ be as in form 1 above, the power of the test is $1 - F_{m,\lambda}(\xi)$, the probability of rejecting $\pi = p$ when in fact $\pi \neq p$.

4. Normality of the Empirical Distribution

If u is any real-valued function defined on the contingency table $\mathcal{Y}$, then for any distribution $p \in \mathcal{P}_0$ we define

$$M[u : p] = \sum u(y) p(y)$$

$$V[u : p] = M\left[u^2 : p\right] - M[u : p]^2$$

The statistics $\sqrt{n}\,(M[u : \hat{p}_n] - M[u : \pi])/\sqrt{V[u : \pi]}$ will tend to have a

Normal distribution, which means

$$P\left[\frac{\sqrt{n}\,(M[u:\hat{p}_n] - M[u:\pi])}{\sqrt{V[u:\pi]}} \leqq x\right] = F_0(x) \qquad \text{(approximately)}$$

where instructions for computing F_0 are given in Section B.4. A test at level α of the hypothesis $M[u:\pi] = \mu$ against the alternative $M[u:\pi] \neq \mu$ would be to reject if the P-value

$$2\left\{1 - F_0\left(\frac{\sqrt{n}\,|M[u:\hat{p}_n] - \mu|}{\sqrt{V[u:\hat{p}_n]}}\right)\right\}$$

fell below α. A $1 - \alpha$ confidence interval for $M[u:\pi]$ would be $M[u:\hat{p}_n] \pm \sqrt{V[u:\hat{p}_n]} \cdot \xi$, where $F_0(\xi) = 1 - \alpha/2$. A one-sided test of $M[u:\pi] = \mu$ against the alternative $M[u:\pi] > \mu$ would be based on the P-value

$$1 - F_0\left(\frac{\sqrt{n}\,(M[u:\hat{p}_n] - \mu)}{\sqrt{V[u:\hat{p}_n]}}\right)$$

5. Maximal Standard Functions

Let $\mathcal{Y}$ be a contingency table, let $m = \#\mathcal{Y} - 1$, and let $\mathbf{u}$ be a vector-valued function defined on $\mathcal{Y}$ and assuming values in $\mathbb{R}^k$ for some $k \leqq m$. For any $p \in \mathcal{P}_0$ the mean vector and covariance matrix of $\mathbf{u}$ with respect to p are defined by

$$M[\mathbf{u}:p] = \sum \mathbf{u}(y)p(y)$$

$$C[\mathbf{u}:p] = \sum \mathbf{u}(y)\mathbf{u}(y)'p(y) - M[\mathbf{u}:p]M[\mathbf{u}:p]'$$

$\mathbf{u}$ is called maximal p-standard if and only if

1. $\mathbf{u}$ takes values in $\mathbb{R}^m$,
2. $M[\mathbf{u}:p] = \mathbf{0}$ (the zero vector),
3. $C[\mathbf{u}:p] = \mathbf{I}$ (the identity matrix).

If $\mathbf{u}$ is maximal π-standard, then any function $w: \mathcal{Y} \to \mathbb{R}$ can be written in one and only one way as $w = \alpha_0 + \boldsymbol{\alpha}'\mathbf{u}$, and moreover we have $\alpha_0 =$

$M[w:\pi]$, $\boldsymbol{\alpha} = M[w\mathbf{u}:\pi]$. For any $p \in \mathcal{P}_0$, p/π is such a function, and so $p = (1 + \boldsymbol{\alpha}'\mathbf{u})\pi$ with $\boldsymbol{\alpha} = M[\mathbf{u}:p]$, and this representation is unique. Consequently, any function defined on $\mathcal{P}_0$ can be defined in terms of $\boldsymbol{\alpha}$. That is, if we have a function $f\colon \mathcal{P}_0 \to \mathbb{R}$, then there is a function φ such that $f(p) = \varphi(\boldsymbol{\alpha})$ where p and $\boldsymbol{\alpha}$ are linked as above. Noting that $\boldsymbol{\alpha} = \mathbf{0}$ corresponds to $p = \pi$, and letting $\hat{\boldsymbol{\alpha}}_n = M[\mathbf{u}:\hat{p}_n]$ correspond to $\hat{p}_n$, we can state that

$$\frac{\sqrt{n}\,[f(\hat{p}_n) - f(\pi)]}{\left\|\dfrac{\partial}{\partial\boldsymbol{\alpha}}\varphi(\mathbf{0})\right\|}$$

is (nearly) Normal. Thus, a test at level α of $f(\pi) = f_0$ is to reject whenever

$$\frac{\sqrt{n}\,|f(\hat{p}_n) - f_0|}{\left\|\dfrac{\partial}{\partial\alpha}\varphi(\hat{\alpha}_n)\right\|} \geqq \xi$$

where $F_0(\xi) = 1 - \alpha/2$. A confidence interval for $f(\pi)$ would be $f(\hat{p}_n) \pm \|(\partial/\partial\boldsymbol{\alpha})\varphi(\hat{\boldsymbol{\alpha}}_n)\| \cdot \xi$.

Pearson's Chi-square can be decomposed by a maximal standard function. Specifically, if $\mathbf{u}$ is maximal π-standard, then

$$n\mathbb{D}_2(\hat{p}_n, \pi) = n\|M[\mathbf{u}:\hat{p}_n]\|^2 = \sum_{i=1}^{m} nM[u_i:\hat{p}_n]^2$$

and the statistics $nM[u_i:\hat{p}_n]^2$ are (nearly) independent Chi-squares with one df each. This permits one to ask why the test of $\pi = p$ rejects, by observing which values $nM[u_i:\hat{p}_n]^2$ are too large.

PROBLEMS

II.1. Let $p \in \mathcal{P}$, and let $\binom{\mathbf{u}}{\mathbf{v}}$ be maximal p-standard. Establish for an arbitrary function w the equation

$$M\left[(w - \alpha_0 - \boldsymbol{\alpha}'\mathbf{u})^2 : p\right] = (M[w:p] - \alpha_0)^2 + \|M[w\mathbf{u}:p] - \boldsymbol{\alpha}\|^2 + \|M[w\mathbf{v}:p]\|^2$$

and conclude that $w^* = M[w:p] + M[w\mathbf{u}:p]'\mathbf{u}$ minimizes $M[(w - w^*)^2 : p]$ among all $w^* \in \mathbb{S}(1, \mathbf{u})$. w^* is called the *p-projection* of w on $\mathbb{S}(1, \mathbf{u})$.

Using the idea of a projection, establish that

1. $w \in \mathbb{S}(1, \mathbf{u})$ if and only if $w = w^*$;
2. $w \in \mathbb{S}(1, \mathbf{u})$ if and only if $M[w\mathbf{v} : p] = \mathbf{0}$;
3. for a vector-valued function $\mathbf{w}$ with ith component w_i, letting $\mathbf{w}^*$ have ith component w_i^*, we have $(\mathbf{a}'\mathbf{w})^* = \mathbf{a}'\mathbf{w}^*$;
4. for any $\mathbf{w}$, $\mathbb{S}(1, \mathbf{u}, \mathbf{w}) = \mathbb{S}(1, \mathbf{u}, \mathbf{w} - \mathbf{w}^*)$, and we also have $M[(\mathbf{w} - \mathbf{w}^*)\mathbf{u}' : p] = 0$.

II.2. Given functions $\mathbf{u}$ and $\mathbf{v}$, explain how one could determine a function $\mathbf{w}$ such that $\mathbb{S}(1, \mathbf{w}) = \mathbb{S}(1, \mathbf{u}) \cap \mathbb{S}(1, \mathbf{v})$. (Hint: Problem II.1, point 2.)

II.3. Show that if $\mathbf{u}$ is both maximal p-standard and maximal q-standard for some p and q in $\mathcal{P}_0$, then in fact $p = q$.

II.4. Show that for w: $\mathcal{Y} \to \mathbb{R}$ we have

$$\sqrt{n}\,(M[w : \hat{p}_n] - M[w : \pi]) \to_d N(0, V[w : \pi]).$$

Generalize to $\mathbf{w} : \mathcal{Y} \to \mathbb{R}^k$.

II.5. If $\pi \in \mathcal{P}$ and $\mathbf{u}$ is maximal π-standard, then determine the limiting behavior of

$$\sum_y \mathbf{u}(y)\pi(y)\ln \frac{\hat{p}_n(y)}{\pi(y)}$$

II.6. Suppose $\pi \in \mathcal{P}$. Show that if $\mathbf{u}$ and $\mathbf{v}$ are maximal π-standard, then $\mathbf{v} = \mathbf{A}\mathbf{u}$ for an orthogonal matrix $\mathbf{A}$ (that is, $\mathbf{A}'\mathbf{A} = \mathbf{A}\mathbf{A}' = \mathbf{I}$). Thus, show that for any w: $\mathcal{Y} \to \mathbb{R}^k$ the matrix $M[\mathbf{w}\mathbf{u}' : \pi]M[\mathbf{u}\mathbf{w}' : \pi]$ does not in fact depend on which maximal π-standard $\mathbf{u}$ is chosen.

II.7. Letting the variance of $M[u : \hat{p}_n]$ (under the independent-identically-distributed assumption) be denoted $V[M[u : \hat{p}_n]]$, show that $V[M[u : \hat{p}_n]] = V[u : \pi]/n$. (Hint: use Problem II.6.)

II.8. Let $\mathcal{Y} = \{ij : i = 0, 1, j = 0, 1\}$ be a 2^2 contingency table. Determine the asymptotic behavior of the ln odds ratio

$$\Lambda = \ln \frac{\hat{p}(00)\hat{p}(11)}{\hat{p}(01)\hat{p}(10)}$$

and use this to determine the asymptotic behavior of Yule's Q,

$$Q = \frac{\hat{p}(00)\hat{p}(11) - \hat{p}(10)\hat{p}(01)}{\hat{p}(00)\hat{p}(11) + \hat{p}(10)\hat{p}(01)}$$

II.9. Suppose that $\mathbf{u}_1' = (u_{11}, u_{12}, \ldots, u_{1m_1})$ is maximal π_1-standard on $\mathcal{Y}_1$, and that $\mathbf{u}_2' = (u_{21}, u_{22}, \ldots, u_{2m_2})$ is maximal π_2-standard on $\mathcal{Y}_2$.

Define for every $(y_1, y_2) \in \mathcal{Y}_1 \times \mathcal{Y}_2$

$$\pi_{12}(y_1, y_2) = \pi_1(y_1)\pi_2(y_2)$$

$$\mathbf{u}_1(y_1, y_2) = \mathbf{u}_1(y_1)$$

$$\mathbf{u}_2(y_1, y_2) = \mathbf{u}_2(y_2)$$

and define $\mathbf{u}_1 \otimes \mathbf{u}_2$ on $\mathcal{Y}_1 \times \mathcal{Y}_2$ by

$$\mathbf{u}_1 \otimes \mathbf{u}_2 = \begin{pmatrix} u_{11}\mathbf{u}_2 \\ u_{12}\mathbf{u}_2 \\ \vdots \\ u_{1m_1}\mathbf{u}_2 \end{pmatrix}$$

Show that the function

$$\begin{pmatrix} \mathbf{u}_1 \\ \mathbf{u}_2 \\ \mathbf{u}_1 \otimes \mathbf{u}_2 \end{pmatrix}$$

is maximal π_{12}-standard.

II.10. Using the definition of $\chi^2(\nu)$, argue that if $X_1, X_2, \ldots, X_k$ are independent with X_i being $\chi^2(\nu_i)$, then ΣX_i is $\chi^2(\Sigma \nu_i)$.

II.11. Let $\mathbf{u}$ be p-standard, and let $\mathbf{u}_n$ be the p_n-standardization of $\mathbf{u}$. Show

1. $p_n \to p$ implies $\mathbf{u}_n \to \mathbf{u}$;
2. $\sqrt{n}(p_n - p) \to 0$ implies $\sqrt{n}(\mathbf{u}_n - \mathbf{u}) \to 0$.

II.12. Let $\mathcal{Y}$ be a contingency table of dimension $r_1 \times r_2 \times \cdots \times r_k$. Let $\mathbf{u}_i$ be the indicators of all but one of the levels of factor i. Determine the maximal affinely linearly independent functions built up of products of indicators. Determine the number of factorial indicators of each order.

II.13. Let $\mathcal{Y} = \{y_0, y_1, \ldots, y_m\}$. Define

$$\Pi = \begin{bmatrix} \pi(y_1) & 0 & 0 & \cdots & 0 \\ 0 & \pi(y_2) & 0 & \cdots & 0 \\ \vdots & & & & \vdots \\ 0 & \cdots & 0 & 0 & \pi(y_m) \end{bmatrix}, \qquad \boldsymbol{\pi} = \begin{pmatrix} \pi(y_1) \\ \pi(y_2) \\ \vdots \\ \pi(y_m) \end{pmatrix}$$

1. Show that $[\Pi - \boldsymbol{\pi}\boldsymbol{\pi}']^{-1} = \Pi^{-1} + \dfrac{1}{\pi(y_0)}\mathbf{1}\mathbf{1}'$

2. Prove the conclusion of Proposition 9, that $n\mathbb{D}_2(\hat{p}_n, \pi) \to_d \chi^2(\#\mathcal{Y} - 1)$, by using

$$C\left[\begin{pmatrix} p(y_1) \\ p(y_2) \\ \vdots \\ p(y_m) \end{pmatrix}\right] = \frac{1}{n}(\Pi - \boldsymbol{\pi}\boldsymbol{\pi}')$$

(Hint: you will need the fact that every symmetric positive definite matrix $\mathbf{A}$ has a square root, that is, a matrix $\mathbf{C}$ satisfying $\mathbf{CC}' = \mathbf{A}$.)

II.14. Let $\mathbf{u}$ be maximal π-standard, and let $\hat{\mathbf{u}}_n$ be the $\hat{p}_n$-standardization of $\mathbf{u}$. Prove that

$$\sqrt{n}\, M[\hat{\mathbf{u}}_n : \pi] \to_d N(\mathbf{0}, \mathbf{I})$$

and thus that $n\mathbb{D}_2(\pi, \hat{p}_n) \to_d \chi^2(\#\mathcal{Y} - 1)$.

II.15. Show that if $\mathbf{u}$ has affinely linearly independent components, then so does the function $\mathbf{v}(y) = \mathbf{u}(y) - \mathbf{a}$ for any fixed vector $\mathbf{a}$, and moreover $\mathbb{S}(1, \mathbf{v}) = \mathbb{S}(1, \mathbf{u})$.

II.16. Suppose there are two independent experiments, involving n_1 and n_2 individuals, and suppose that the outcome for each is the contingency table $\mathcal{Y} = \{0, 1\}$. Letting π_1 and π_2 denote the true distributions, and $\hat{p}_1$ and $\hat{p}_2$ denote the empirical distributions, determine the asymptotic distributions of the following:

1. $\ln[\hat{p}_1(1)/\hat{p}_2(1)]$;
2. $\dfrac{\hat{p}_1(1) - \hat{p}_2(1)}{\hat{p}_1(1) + \hat{p}_2(1)}$;
3. $\sqrt{\hat{p}_1(1)\hat{p}_2(1)} - \sqrt{\hat{p}_1(0)\hat{p}_2(0)}$;
4. $\ln \dfrac{\hat{p}_1(1)\hat{p}_2(0)}{\hat{p}_1(0)\hat{p}_2(1)}$;
5. $\arcsin\sqrt{\hat{p}_1(1)} - \arcsin\sqrt{\hat{p}_2(1)}$ (note: $(d/dx)\arcsin\sqrt{x} = 1/[2\sqrt{x(1-x)}]$).

II.17. The famous Buffon's needle problem goes as follows. Take a sheet of paper and rule it with lines one inch apart. Drop a needle of length one inch at random on the sheet. What is the probability that it crosses a line? The answer is $2/\pi$ [see Feller (1966)]. An instructor devised the following experiment to acquaint his students with Pearson's Chi-square and its use in hypothesis testing. The sheet was ruled as described above and a handful of needles was dropped, the number crossing lines being counted. This was repeated twelve times. Each time, the instructor recorded the number of needles n_i which landed on the sheet, and the number r_i which crossed lines. The traditional formula for Pearson's Chi-square is $\Sigma_i(O_i - E_i)^2/E_i$, where O_i stands for an observed frequency and E_i for the expected frequency. The instructor put in r_i for O_i and $n_i 2/\pi$ for E_i, and summed over $i = 1$ to 12. Comment on the usefulness of this educational demonstration. Derive the actual distribution of the above "Chi-square." (You should be able to add more objections to your list after reading Chapter III.)

II.18. The data in Table II.8 consist of the frequencies of SIDS cases by birth weight. Consider the functions in Table II.9 for analyzing the data. Using q as the distribution of birth weights, q-standardize u_1,

Table II.8

Birth-Weight Category	SIDS Cases	Total Births
0	1	279
1	3	413
2	12	904
3	12	2,838
4	37	11,509
5	52	24,941
6	34	21,832
7	13	8,408
8	2	2,061

Table II.9

y	u_1	u_2	u_3
0	0	0	0
1	0	0	1
2	0	1	0
3	0	1	1
4	0.5	0	0
5	1	0	0
6	1	0	1
7	1	1	0
8	1	1	1

u_2, and u_3 (in that order) and determine

1. Pearson's Chi-square for testing $\pi = q$,
2. the components of Pearson's Chi-square corresponding to (the standardization of) u_1, u_2, and u_3,
3. the sum of the components associated with all other functions (besides u_1, u_2, and u_3).

Compute the P-values and interpret the results.

II.19. A national magazine reported the following age–sex distribution of its readership (in percent):

	Age: 18–24	25–34	35–49	50–64	65–
Sex					
Male	6.1	6.6	13.1	10.8	7.2
Female	6.8	7.9	14.5	16.4	10.7

A merchant who advertised a mail-order product in the magazine obtained responses from 290 readers, classified by age and sex below:

	Age: 18–24	25–34	35–49	50–64	65–
Sex					
Male	1	6	19	38	30
Female	0	5	33	78	80

Could the advertiser's sample have been drawn at random from the readership? If not, is it the age distribution, the sex distribution, or both which are responsible for the difference?

II.20. The readers in Problem II.19 purchased the product advertised, but some of them returned it within 30 days. The data follow:

	*S*0				*S*1			
	*A*0	*A*1	*A*2	*A*3	*A*0	*A*1	*A*2	*A*3
*R*0	5	12	38	30	4	29	68	67
*R*1	2	7	5	4	1	4	10	13

The factors are:

$S =$ Sex
$S0 =$ Male
$S1 =$ Female

$A =$ Age
  $A0 = 18$–34 years
  $A1 = 35$–49
  $A2 = 50$–64
  $A3 = 65$–
$R =$ Returned
  $R0 =$ Kept Product
  $R1 =$ Returned Product

Is there any evidence that males behave differently than females, or that older people behave differently than younger, with regard to their tendency to return the product? (You may want to use Problems II.4 and II.16.)

II.21. Let $\{X_i : 1 \leqq i \leqq n\}$ be independent, X_i being $N(\mu_i, \sigma_i^2)$. Define the following quantities:

$$\sigma^{*2} = \frac{1}{\sum_{i=1}^{n} \frac{1}{\sigma_i^2}}, \qquad w_i^* = \frac{\sigma^{*2}}{\sigma_i^2}$$

$$\mu^* = \sum_i w_i^* \mu_i, \qquad X^* = \sum_i w_i^* X_i$$

Show the following:

1. X^* is $N(\mu^*, \sigma^{*2})$.
2. If $w_i \geq 0$ and $\Sigma w_i = 1$, then for $\tilde{X} = \Sigma_i w_i X_i$, we have

$$\sum w_i^*(X_i - \tilde{X})^2 = \sum w_i^*(X_i - X^*)^2 + (X^* - \tilde{X})^2$$

3. $$\frac{\Sigma w_i^*(X_i - \mu^*)^2}{\sigma^{*2}} = \frac{\Sigma w_i^*(X_i - X^*)^2}{\sigma^{*2}} + \frac{(X^* - \mu^*)^2}{\sigma^{*2}}$$

   Argue that the left side is $\chi^2(n, \Sigma(\mu_i - \mu^*)^2/\sigma_i^2)$ and that the extreme right term is $\chi^2(1)$, and thus conjecture that the remaining term is $\chi^2(n - 1, \Sigma(\mu_i - \mu^*)^2/\sigma_i^2)$.
4. $X_i - X^*$ is $N(\mu_i - \mu^*, \sigma_i^2 - \sigma^{*2})$.

II.22. Prove that

$$\mathbb{D}_2(\hat{p}, p) = \max\{M[u : \hat{p}]^2 : M[u : p] = 0, V[u : p] = 1\}$$

(Hint: for the functions u above, the Cauchy–Schwartz inequality is $C[u, w: p] \leqq \sqrt{V[w: p]}$.)

II.23. Consider two groups of individuals at risk for some event. Let $\pi_i(1)$ be the probability of the event occurring to a person in group i, and $\pi_i(0) = 1 - \pi_i(1)$. Under the assumption that $\pi_1(1) > \pi_0(1)$, Shep's measure of excess probability of the event in group 1 is

$$\frac{\pi_1(1) - \pi_0(1)}{\pi_0(0)}$$

which is intended to be the conditional probability of an individual in group 1 escaping the cause of the event acting in group 0, but experiencing the event nonetheless. When the event is death, this is a measure of the excess mortality in group 1. Determine the asymptotic behavior of the estimator

$$\frac{\hat{p}_1(1) - \hat{p}_0(1)}{\hat{p}_0(0)}$$

based on independent samples in the two groups.

REMARKS

The importance of function spaces in the analysis of contingency tables cannot be overemphasized. The linear properties of function spaces are used in proving theoretical results, the parameters associated with function-space descriptions of models determine the interpretations of those models, and the computational algorithm by which data are actually fitted is most easily described in function-space terms. Surprisingly, function spaces have not been emphasized as such in the contingency-table literature. In a particular case Bahadur (1961) made very effective use of a special class of functions for analyzing a 2^m table (see Problem III.18). Lancaster (1969) used systems of orthogonal functions as a means of defining interactions among factors, but he did not exhaust the potential of this approach [see O'Neill (1980) and Problem II.9].

The factorial indicators are particularly important for cross-classifications. We will follow the custom of using the factorial indicators of all but the lowest level of a factor. Thus if A has levels $A0, A1, \ldots, Am$, then by $\mathbf{u}_A$ we generally mean the vector with components $u_{A1}, u_{A2}, \ldots, u_{Am}$. The reason for doing this is so that $\mathbf{u}_A$ will have affinely linearly independent components, the importance of which will be seen in Chapter III, Section A.1. With this convention, the second-order factorial indicator $\mathbf{u}_{AB}$ can be defined as $\mathbf{u}_A \otimes \mathbf{u}_B$, in the notation of Problem II.9. In fact, one should check that Problem II.9 remains true if "maximal standard" is replaced throughout by "maximal with affinely linearly independent components." In this fashion the factorial structure of the table is built up by successively larger products of simpler terms. For example, $\mathbf{u}_{ABC}$ is $\mathbf{u}_A \otimes \mathbf{u}_B \otimes \mathbf{u}_C$. Spaces of factorial indicators will be used extensively in Chapter IV.

Standardized functions are important primarily for theoretical reasons, since they greatly simplify proofs. The important thing to remember here is that standardization leaves function spaces invariant; that is, if $\mathbf{u}$ is the standardization of $\mathbf{v}$, then $\mathbb{S}(1,\mathbf{u}) = \mathbb{S}(1,\mathbf{v})$. Thus, in principle, anything which depends only on the function space can be done with the standardized function instead of the original function.

One exceptional practical use of standardized functions is the decomposition of Pearson's Chi-square given after Proposition 3. From Problem II.10, a sum of independent Chi-squares is Chi-square, and the dfs add. Thus, given that $n\mathbb{D}_2(\hat{p}, q)$ is (asymptotically) Chi-square, it ought to follow that it can be decomposed into (asymptotically) independent component Chi-squares. This idea has been very important in the historical development of tests for contingency tables [Lancaster (1949) and (1950), Irwin (1949)]. The result is Proposition 9, but from Problem II.6 we can see that there are numerous possible decompositions of Pearson's Chi-square, and naturally some of these will be more interpretable than others.

The two results which have been "brought in from the outside" without proof are the MCLT and MWT. Versions of the MCLT have been published in many places, but the approach in Rao (1973) is particularly accessible. The MWT is essentially Theorem 5 of Mann and Wald (1943), and can be seen in a more modern context in Billingsley (1968).

As Section C shows, there are many ways of defining the distance between two probability distributions. For the most part $\mathbb{D}_0$ and $\mathbb{D}_1$ will be used here, because they have additive decompositions for the linear and exponential models to be introduced in Chapter III, Section A.1. Frequently we will use $\mathbb{D}_2$ in proofs, on account of the decomposition of Proposition 9, and then invoke the asymptotic equivalence of $\mathbb{D}_2$ with $\mathbb{D}_0$ and $\mathbb{D}_1$. It is conceivable that a practical choice between these discrepancies could be made on the basis that some have distributions approaching the asymptotic Chi-square more rapidly than others, but the evidence for this is not sufficiently strong as yet to make general statements [see Hutchinson (1979)].

The discrepancies $\mathbb{D}_\omega$ are related to various concepts of entropy [Vajda (1971), Kerridge (1961)]. Proposition 12 generalizes inequalities for these, the most famous of which is the "information inequality" pertaining to $\mathbb{D}_1$ [Kullback (1968)].

The approximation to the Normal distribution is due to Page (1977), and the Chi-square approximation is from Wilson and Hilferty (1931); see also Zar (1978).

Problem II.1 gives the important operation of projection, which amounts to weighted least squares. This shows that the LAMDA command FIT computes a weighted-least-squares fit.

CHAPTER III

Inference for a Single Model

If the only hypotheses we ever wanted to test were of the form $\pi = q$, then the method of the preceding chapter would be sufficient. However, it is much more common to want to test a hypothesis which only specifies that π satisfy certain conditions, for example, to test the independence of two factors in a two-way contingency table. The techniques of the preceding chapter are not adequate for testing such hypotheses, and so the present chapter develops the requisite procedure.

The idea of a distribution satisfying certain conditions is naturally expressed in terms of models. The central results of this chapter concern the construction of inferential procedures for handling three kinds of problems:

1. to test whether π lies in a model,
2. given that π lies in a model, to test $\pi = q$ for a particular q in the model,
3. to test whether π lies in a model, and then upon acceptance of the model to test a particular hypothesis $\pi = q$.

We will present the theoretical support for the simple ANALYSIS table and the most common procedures for estimating model parameters. The LAMDA command **LXM** will be described, and a number of data sets will be analyzed with its help.

One of the most important themes of this chapter is that the empirical distribution is often not the best basis for statistical inference. It can frequently be improved on by the use of a fitted distribution in a model, which explains why so much of discrete data analysis involves fitting models to data.

A. ASYMPTOTIC THEORY FOR MODELS

1. Linear and Exponential Models

This subsection introduces the two types of models that will be used for discrete data analysis throughout the rest of the book.

The *linear model* $\mathcal{L}[q:\mathbf{u}]$ is determined by a probability distribution $q \in \mathcal{P}$ and a function $\mathbf{u}: \mathcal{Y} \to \mathbb{R}^k$ according to the definition

$$\mathcal{L}[q:\mathbf{u}] = \{p \in \mathcal{P}_0 : p = (\alpha_0 + \boldsymbol{\alpha}'\mathbf{u})q \text{ for some } \alpha_0, \boldsymbol{\alpha}\}$$

The parameter in the model is $\boldsymbol{\alpha}$. From $1 = \Sigma p(y) = \alpha_0 + \boldsymbol{\alpha}'M[\mathbf{u}:q]$ we can see that α_0 is a function of $\boldsymbol{\alpha}$, and so α_0 is not free to vary independently of $\boldsymbol{\alpha}$. When it is necessary to emphasize the dependence of p on $\boldsymbol{\alpha}$ we will write $p^{\boldsymbol{\alpha}} = (\alpha_0 + \boldsymbol{\alpha}'\mathbf{u})q$. The distribution q is the *generating distribution*, and corresponds to the parameter value $\boldsymbol{\alpha} = \mathbf{0}$, that is, $q = p^{\mathbf{0}}$.

Clearly an alternate description of $\mathcal{L}[q:\mathbf{u}]$ is

$$\mathcal{L}[q:\mathbf{u}] = \{p \in \mathcal{P}_0 : p/q \in \mathcal{S}(1,\mathbf{u})\}$$

which indicates why the model is called linear. It also shows that the particular choice of the function $\mathbf{u}$ is not so important—what determines the model is the space $\mathcal{S}(1,\mathbf{u})$. Stated another way, if $\mathcal{S}(1,\mathbf{u}) = \mathcal{S}(1,\mathbf{v})$ then $\mathcal{L}[q:\mathbf{u}] = \mathcal{L}[q:\mathbf{v}]$.

In general, the parameter of a parametric model is *identifiable* if and only if different parameter values correspond to different distributions. If $\mathbf{u}$ has affinely linearly independent components, then $\boldsymbol{\alpha}$ is identifiable, which follows from Proposition 1. Unidentifiable parameters frustrate the attempt to interpret models, and so we will tacitly assume that the $\mathbf{u}$ in $\mathcal{L}[q:\mathbf{u}]$ has affinely linearly independent components. If we further assume that $\mathbf{u}$ is q-standard, then we can write $\mathcal{L}[q:\mathbf{u}] = \{p \in \mathcal{P}_0 : p = (1 + \boldsymbol{\alpha}'\mathbf{u})q\}$, and according to Proposition 3 the computation of $\boldsymbol{\alpha}$ is very easy: $\boldsymbol{\alpha} = M[\mathbf{u}:p]$. This provides us with an interpretation of the parameter $\boldsymbol{\alpha}$ directly in terms of p.

Frequently the most convenient way of describing a linear model is by specifying what the values of $M[\mathbf{u}:p]$ must be. For a fixed $\mathbf{u}$ we cannot specify $M[\mathbf{u}:p] = \boldsymbol{\mu}$ arbitrarily, since there might be no distribution satisfying this equation. Thus, for consistency we assume that there is at least one distribution q, and so our model condition is $M[\mathbf{u}:p] = M[\mathbf{u}:q]$, where $\mathbf{u}$ and q are fixed. In order to show that such a model is a linear model we need the following definition. We say that $\mathbf{v}$ is a *q-complement* of $\mathbf{u}$ provided (1) $\mathcal{S}(1,\mathbf{u},\mathbf{v})$ is the space of all functions on $\mathcal{Y}$, (2) $C[\mathbf{u},\mathbf{v}:q] = \mathbf{0}$.

Proposition 16. If $\mathbf{u}: \mathcal{Y} \to \mathbb{R}^k$, and $\mathbf{v}$ is a q-complement of $\mathbf{u}$, then

$$\mathcal{L}[q:\mathbf{v}] = \{p \in \mathcal{P}_0 : M[\mathbf{u}:p] = M[\mathbf{u}:q]\}$$

Proof. Any $p \in \mathcal{P}_0$ can be written $p = (\alpha_0 + \boldsymbol{\alpha}'\mathbf{v} + \boldsymbol{\beta}'\mathbf{u})q$, and clearly $\boldsymbol{\beta} = \mathbf{0}$ is equivalent to $p \in \mathcal{L}[q:\mathbf{v}]$. Now $1 = \alpha_0 + \boldsymbol{\alpha}'M[\mathbf{v}:q] + \boldsymbol{\beta}'M[\mathbf{u}:q]$, so that

$$p = \{1 + \boldsymbol{\alpha}'(\mathbf{v} - M[\mathbf{v}:q]) + \boldsymbol{\beta}'(\mathbf{u} - M[\mathbf{u}:q])\}q$$

It now follows from $C[\mathbf{u},\mathbf{v}:q] = \mathbf{0}$ that $M[\mathbf{u}:p] = M[\mathbf{u}:q] + C[\mathbf{u}:q]\boldsymbol{\beta}$, and so $\boldsymbol{\beta} = \mathbf{0}$ is equivalent to $M[\mathbf{u}:p] = M[\mathbf{u}:q]$. □

Obtaining q-complements is not difficult. If $\binom{\mathbf{u}}{\mathbf{v}}$ were maximal q-standard, then $\mathbf{u}$ and $\mathbf{v}$ would be q-complements. Note also from Problem II.1 that if $\mathcal{S}(1,\mathbf{u},\mathbf{v})$ were the space of all functions on $\mathcal{Y}$, and $\mathbf{v}^*$ were the q-projection of $\mathbf{v}$ on $\mathcal{S}(1,\mathbf{u})$, then $\mathbf{u}$ and $\mathbf{v} - \mathbf{v}^*$ would be q-complements.

Examples of Linear Models. The *symmetric distributions* on a one-way table form a linear model. Consider the case $\#\mathcal{Y} = 5$ shown in Table III.1. The condition $M[\mathbf{u}:p] = \mathbf{0}$ is equivalent to p being symmetric. Taking q to be the uniform distribution, since $M[\mathbf{u}\mathbf{v}':q] = \mathbf{0}$, we have that $\mathbf{u}$ and $\mathbf{v}$ are q-complements, and Proposition 16 is in effect. Any symmetric p can be written $p = (\alpha_0 + \boldsymbol{\alpha}'\mathbf{v})q$.

Another important model is that of *marginal homogeneity*. In Table III.2 we see two functions u_1 and u_2 on a 3^2 table which have the property that $M[\mathbf{u}:p] = \mathbf{0}$ if and only if the row and column marginal distributions of p coincide. One could determine an appropriate complement of $\mathbf{u}$, but this would require some work. Fortunately, this situation is illustrative of the usual case: the function $\mathbf{u}$ (for which $M[\mathbf{u}:p] = \mathbf{0}$ specifies the model

Table III.1

The Symmetric and Antisymmetric Functions on a Five-Cell Table

y	v_1	v_2	u_1	u_2
1	1	1	1	1
2	1	0	1	0
3	0	0	0	0
4	1	0	−1	0
5	1	1	−1	−1

Table III.2

Functions for Specifying Marginal Homogeneity on a 3^2 Table

u_1			u_2		
0	−1	−1	0	1	0
−1	0	0	−1	0	−1
−1	0	0	0	1	0

condition) is much easier to obtain than its complement, and as we will see later, in the required computations knowledge of **u** suffices.

For another example, imagine a two-way table with factors A and B. Let $\mathbf{u}_A$ be the factorial indicator of all but the lowest level of A, and similarly let $\mathbf{u}_B$ be the factorial indicator of B. The model $\mathcal{L}[q:\mathbf{u}_A, \mathbf{u}_B]$ contains all distributions of the form $p = (\alpha_0 + \boldsymbol{\alpha}'_A\mathbf{u}_A + \boldsymbol{\alpha}'_B\mathbf{u}_B)q$. We can think of $\boldsymbol{\alpha}_A$ and $\boldsymbol{\alpha}_B$ as representing the main effects of the factors on the ratios $p(y)/q(y)$. Another way of looking at this model is to observe that it specifies

$$\frac{p(ij)}{q(ij)} - \frac{p(ij')}{q(ij')} - \frac{p(i'j)}{q(i'j)} + \frac{p(i'j')}{q(i'j')} = 0$$

When q is the uniform distribution, this means that there is no linear interaction in p.

Exponential Models. The second kind of model that will be studied here is very widely used in contingency-table analysis. For a fixed $q \in \mathcal{P}$ and a vector-valued function u, we define the *exponential model*

$$\mathcal{E}[q:\mathbf{u}] = \{p \in \mathcal{P} : p = e^{\boldsymbol{\theta}'\mathbf{u} - \delta(\boldsymbol{\theta})}q \text{ for some } \boldsymbol{\theta}\}$$

In this expression for p, δ is the *normalizing factor* required to make $\Sigma p(y) = 1$, so that

$$\delta(\boldsymbol{\theta}) = \ln \sum_y e^{\boldsymbol{\theta}'\mathbf{u}(y)}q(y)$$

The parameter in $\mathcal{E}[q:\mathbf{u}]$ is $\boldsymbol{\theta}$, and any value of $\boldsymbol{\theta}$ is permissible. An alternate description of this type of model is

$$\mathcal{E}[q:\mathbf{u}] = \left\{ p \in \mathcal{P} : \ln \frac{p}{q} \in \mathcal{S}(1, \mathbf{u}) \right\}$$

Thus, we see that $\mathbb{S}(1,\mathbf{u})$ determines the model, and so $\mathbb{S}(1,\mathbf{u}) = \mathbb{S}(1,\mathbf{v})$ implies $\mathcal{E}[q:\mathbf{u}] = \mathcal{E}[q:\mathbf{v}]$. Again as a consequence of Proposition 1 we may say that $\boldsymbol{\theta}$ is identifiable whenever $\mathbf{u}$ has affinely linearly independent components.

When we want to emphasize the dependence of p on $\boldsymbol{\theta}$ we write $p^{\boldsymbol{\theta}} = e^{\boldsymbol{\theta}'\mathbf{u}-\delta(\boldsymbol{\theta})}q$. The *generating distribution* q corresponds to $\boldsymbol{\theta} = \mathbf{0}$, so $q = p^{\mathbf{0}}$. Any element of $\mathcal{E}[q:\mathbf{u}]$ can serve as the generating distribution. To see this, let $p^{\boldsymbol{\theta}_0} \in \mathcal{E}[q:\mathbf{u}]$; then

$$\ln \frac{p}{p^{\boldsymbol{\theta}_0}} = \ln \frac{p}{q} - \boldsymbol{\theta}_0'\mathbf{u} + \delta(\boldsymbol{\theta}_0)$$

so that $p \in \mathcal{E}[q:\mathbf{u}]$ if and only if $p \in \mathcal{E}[p^{\boldsymbol{\theta}_0}:\mathbf{u}]$.

Two very useful and easily proved facts are

$$\frac{\partial}{\partial \boldsymbol{\theta}}\delta(\boldsymbol{\theta}) = M\left[\mathbf{u}: p^{\boldsymbol{\theta}}\right] \quad \frac{\partial^2}{\partial \boldsymbol{\theta}\, \partial \boldsymbol{\theta}'}\delta(\boldsymbol{\theta}) = C\left[\mathbf{u}: p^{\boldsymbol{\theta}}\right]$$

We will use these two formulas repeatedly without special mention.

Binomial Models. Let $\mathcal{Y} = \{0, 1, \ldots, k\}$, and let q be the Binomial distribution

$$q(y) = \binom{k}{y} \cdot \left(\tfrac{1}{2}\right)^k,$$

so that $q(y)$ is the probability of y successes in k independent trials, with probability $\frac{1}{2}$ of success on each trial. Let $u(y) = y$, so that an element of $\mathcal{E}[q: u]$ is of the form

$$q^{\theta}(y) = e^{\theta y - \delta(\theta)}q(y)$$

If we define $\rho = 1/(1 + e^{-\theta})$, the *logistic* function, then $\ln[\rho/(1-\rho)] = \theta$, and so it may be computed that

$$q^{\theta}(y) = \binom{k}{y}\rho^{y}(1-\rho)^{k-y}$$

and so $q^{\theta}(y)$ is the probability of y successes in k independent trials with probability $\rho = 1/(1 + e^{-\theta})$ of success on each trial.

If for some reason the event $y = 0$ never comes to the attention of the observer, then we could let $\mathcal{Y} = \{1, 2, \ldots, k\}$,

$$q(y) = \frac{\binom{k}{y} \cdot \left(\frac{1}{2}\right)^k}{1 - \left(\frac{1}{2}\right)^k}$$

so that $q(y)$ is the conditional probability of y successes given that $y > 0$, and then $\mathfrak{S}[q : u]$ consists of all distributions of the form

$$q^{\theta}(y) = \frac{\binom{k}{y}\rho^{y}(1-\rho)^{k-y}}{1-(1-\rho)^{k}} \qquad \left(\rho = \frac{1}{1+e^{-\theta}}\right)$$

so that $q^{\theta}(y)$ is the conditional probability of y successes given that $y > 0$, where the probability of a success on one trial is ρ.

In fact, if A is any subset of $\{0, 1, \ldots, k\}$, we may define the *truncated model* by

$$q^{\theta}(y) = \frac{\binom{k}{y}\rho^{y}(1-\rho)^{k-y}}{\sum_{z \in A}\binom{k}{z}\rho^{z}(1-\rho)^{k-z}} \qquad (y \in A)$$

and observe that this is just $\mathfrak{S}[q : u]$ with

$$q(y) = \binom{k}{y} \Big/ \sum_{z \in A}\binom{k}{z}.$$

Truncated Inverse Binomial Models. If a number of independent success-or-failure trials are made, then the probability of exactly y failures before the sth success is

$$\binom{y+s-1}{s-1}\rho^{s}(1-\rho)^{y} \qquad (y = 0, 1, 2, \ldots)$$

where ρ is the probability of a success on each trial. Let q correspond to $\rho = \frac{1}{2}$, let $\theta = \ln(1-\rho)$, and note

$$\frac{\binom{y+s-1}{s-1}\rho^{s}(1-\rho)^{y}}{q(y)} = e^{\theta y}(2\rho)^{s}$$

so that

$$q^{\theta}(y) = \binom{y+s-1}{s-1}\rho^{s}(1-\rho)^{y} \qquad [\theta = \ln(1-\rho)]$$

specifies an exponential model. Formally y can take on any positive integer value, but in practice there is an integer k so large that $y > k$ may be deemed an event of negligible probability. Then

$$q^\theta(y) = \frac{\binom{y+s-1}{s-1}(1-\rho)^y}{\sum\limits_{0 \leqq z \leqq k} \binom{z+s-1}{s-1}(1-\rho)^z}$$

denotes the Truncated Inverse Binomial model.

Note that this latter model is identical in form to the Binomial model, that is, both are $\mathcal{E}[q:u]$ where $u(y) = y$. The difference lies entirely in the generating distribution.

Truncated Poisson Models. The Poisson distribution with parameter λ is

$$p(y) = \frac{\lambda^y e^{-\lambda}}{y!} \qquad (y = 0, 1, \ldots)$$

By letting $\theta = \ln \lambda$ and choosing q to correspond to $\lambda = 1$, we have $q^\theta(y)/q(y) = \lambda^y e^{1-\lambda} = e^{\theta y - \delta(\theta)}$, so that again we have an exponential model $\mathcal{E}[q:u]$ with $u(y) = y$. As before, truncations of the model produce

$$q^\theta(y) = \frac{\lambda^y q(y)}{\sum\limits_{z \in A} \lambda^z q(z)} \qquad (\theta = \ln \lambda)$$

Truncated Gamma Models. Letting $\mathcal{Y} = \{1, 2, \ldots, k\}$, we may specify the discrete analogue of the gamma distributions by

$$q^{\boldsymbol{\theta}}(y) = \frac{y^{\theta_1} e^{-\theta_2 y}}{\sum\limits_{1 \leqq z \leqq k} z^{\theta_1} e^{-\theta_2 z}}$$

Evidently the generating distribution $q^{\mathbf{0}}$ is the uniform distribution, and we can write the model as $\mathcal{E}[q:u_1, u_2]$ where q is uniform, $u_1(y) = \ln y$, and $u_2(y) = y$.

Two-Way Independence Models. Let $\mathcal{Y}$ be the two-way contingency table $\{ij : 0 \leq i \leq r, 0 \leq j \leq c\}$ with row factor A and column factor B. If q is any element of $\mathcal{P}$ under which A and B are independent, then setting

$\mathbf{u}'_A = (u_{A1}, u_{A2}, \ldots, u_{Ar})$ and $\mathbf{u}'_B = (u_{B1}, u_{B2}, \ldots, u_{Bc})$, the exponential model $\mathcal{E}[q : \mathbf{u}_A, \mathbf{u}_B]$ contains all elements of $\mathcal{P}$ under which A and B are independent. This follows from the computation of the normalizing factor

$$\delta(\boldsymbol{\theta}_A, \boldsymbol{\theta}_B) = \ln\left[\sum_y e^{\boldsymbol{\theta}'_A \mathbf{u}_A(y)} \cdot \sum_y e^{\boldsymbol{\theta}'_B \mathbf{u}_B(y)}\right]$$

when q is uniform, so that

$$q^{\boldsymbol{\theta}} = \frac{e^{\boldsymbol{\theta}'_A \mathbf{u}_A}}{\sum_y e^{\boldsymbol{\theta}'_A \mathbf{u}_A(y)}} \frac{e^{\boldsymbol{\theta}'_B \mathbf{u}_B}}{\sum_y e^{\boldsymbol{\theta}'_B \mathbf{u}_B(y)}}$$

The statement $p \in \mathcal{E}[q : \mathbf{u}_A, \mathbf{u}_B]$ is equivalent to

$$\frac{p(ij)p(i'j')}{p(ij')p(i'j)} = 1$$

for all choices of i, j, i', and j'.

Local-Independence Models. If we take $r = c$ in the two-way contingency table of the preceding example, and truncate the table so that it becomes

$$\mathcal{Y} = \{ij : 0 \leqq i \leqq r, 0 \leqq j \leqq r, |i - j| \leqq 1\}$$

then under any $p \in \mathcal{E}[q : \mathbf{u}_A, \mathbf{u}_B]$ we have local independence of A and B. This means that

$$\frac{p(ii)p(i+1, i+1)}{p(i, i+1)p(i+1, i)} = 1$$

for all $i < r$. Thus each of the adjacent 2^2 tables $\{AiBi, A(i+1)Bi, AiB(i+1), A(i+1)B(i+1)\}$ exhibits independence. This kind of model might be appropriate if $AiBj$ denoted a transition from state i at one time to state j at a later time, and transitions between adjacent states were independent, but transitions between nonadjacent states were dependent.

Note that the notation $\mathcal{E}[q : \mathbf{u}_A, \mathbf{u}_B]$ is ambiguous here, since it does not reflect the fact that we have truncated the set of outcomes. Thus the appropriate $\mathcal{Y}$ needs to be understood from the context when one uses the $\mathcal{E}$- or $\mathcal{L}$-notation.

2. Rapid Convergence

The empirical distribution $\hat{p}_n$ is such a natural estimator of π that it is hard to see why one might want to use a different estimator. Nevertheless, one of the main themes of discrete data analysis is that under certain frequently occurring circumstances there are estimators which do much better than $\hat{p}_n$. The reason is this. If it were known in advance of data collection that π belonged to a model $\mathfrak{M}$, then one ought to use this knowledge in estimating π. In particular, one ought to estimate π by a distribution which lies in $\mathfrak{M}$, and since $\hat{p}_n$ will nearly always be outside of $\mathfrak{M}$, it is an unreasonable estimator. Thus we should look for an alternative estimator $\hat{q}_n$ that satisfies two kinds of constraints: (1) $\hat{q}_n$ is heavily influenced by $\hat{p}_n$, and (2) $\hat{q}_n \in \mathfrak{M}$. In this subsection and in Section III.A.3 we will make precise how this may be done and what are the advantages of $\hat{q}_n$ over $\hat{p}_n$.

We begin by supposing that $\mathbf{u}$ and $\mathbf{v}$ are vector-valued functions for which $\mathcal{S}(1, \mathbf{u}, \mathbf{v})$ is the space of all functions defined on $\mathcal{Y}$. The sequence $\{\hat{q}_n\}$ of distributions on $\mathcal{Y}$ *converges rapidly* on $\mathcal{S}(\mathbf{v})$ provided

1. $\sqrt{n}(M[\mathbf{u} : \hat{p}_n] - M[\mathbf{u} : \hat{q}_n]) \to_p \mathbf{0}$,
2. $\sqrt{n}(M[\mathbf{v} : \hat{q}_n] - M[\mathbf{v} : \pi]) \to_p \mathbf{0}$.

Condition 1 expresses the fact that $\hat{q}_n$ is heavily influenced by $\hat{p}_n$, while condition 2 makes use of knowledge about π. It may not be obvious at this point why the definition is appropriate, but we will try to justify it in this and the following subsection.

In order to motivate the definition of rapid convergence in a particular case, suppose that $\pi \in \mathcal{L}[q : \mathbf{u}]$ were known. Suppose also that $\mathbf{u}$ and $\mathbf{v}$ are q-complements, so that

$$\mathcal{L}[q : \mathbf{u}] = \{p \in \mathcal{P}_0 : M[\mathbf{v} : p] = M[\mathbf{v} : q]\}$$

and so we are asserting that $M[\mathbf{v} : \pi]$ is known. Under these circumstances condition 2 is reasonable, since it makes little sense to permit $\sqrt{n}(M[\mathbf{v} : \hat{q}_n] - M[\mathbf{v} : \pi])$ to tend to a Normal distribution when $\hat{q}_n$ is supposed to estimate a distribution for which the value of $M[\mathbf{v} : \pi]$ is known. In fact, it would not seem unreasonable to require that $\hat{q}_n \in \mathcal{L}[q : \mathbf{u}]$ so that condition 2 would be trivially satisfied. As for condition 1, this is just a way of saying that $\hat{q}_n$ should be determined by $\hat{p}_n$ to the maximum extent possible, subject to condition 2.

We now prove two consequences of rapid convergence for the asymptotic distribution of $\mathbb{D}_2$.

Proposition 17. Assume that $\mathbf{u}, \mathbf{v}$ are maximal π-standard, and that $\hat{q}_n$ converges rapidly on $\mathbb{S}(\mathbf{v})$. Then $n\mathbb{D}_2(\hat{q}_n, \pi) \to_d \chi^2(\#\mathbf{u})$.

Proof. Write

$$\hat{q}_n = (1 + M[\mathbf{u} : \hat{q}_n]'\mathbf{u} + M[\mathbf{v} : \hat{q}_n]'\mathbf{v})\pi$$

so that

$$\mathbb{D}_2(\hat{q}_n, \pi) = \|M[\mathbf{u} : \hat{q}_n]\|^2 + \|M[\mathbf{v} : \hat{q}_n]\|^2$$

and so

$$n\mathbb{D}_2(\hat{q}_n, \pi) - n\|M[\mathbf{u} : \hat{q}_n]\|^2 \to_p 0$$

But then

$$n\left|\|M[\mathbf{u} : \hat{p}_n]\|^2 - \|M[\mathbf{u} : \hat{q}_n]\|^2\right| \leqq n\|M[\mathbf{u} : \hat{p}_n] - M[\mathbf{u} : \hat{q}_n]\|^2 \to_p 0$$

which implies that $n\|M[\mathbf{u} : \hat{q}_n]\|^2 \to_d \chi^2(\#\,\mathbf{u})$ and completes the proof. □

Proposition 18. Continuing the assumptions of the preceding proposition, if $\{p_n\}$ is a contiguous sequence satisfying $\sqrt{n}\, M[\mathbf{v} : p_n] \to 0$, then

$$n\mathbb{D}_2(\hat{q}_n, p_n) \to_d \chi^2(\#\mathbf{u}, \lim n\mathbb{D}_2(p_n, \pi)).$$

Proof. We have for

$$\mathbf{w} = \begin{pmatrix} \mathbf{u} \\ \mathbf{v} \end{pmatrix}$$

that

$$\frac{\hat{q}_n - p_n}{\pi} = \hat{\boldsymbol{\beta}}_n'\mathbf{w}$$

where $\hat{\boldsymbol{\beta}}_n = M[\mathbf{w} : \hat{q}_n] - M[\mathbf{w} : p_n]$. If v_i is a component of $\mathbf{v}$, then

$$\sqrt{n}\,\hat{\beta}_{ni} = \sqrt{n}\,(M[v_i : \hat{q}_n] - M[v_i : p_n]) \to_p 0$$

while if u_i is a component of $\mathbf{u}$, then

$$\sqrt{n}\,\hat{\beta}_{ni} = \sqrt{n}\,(M[u_i : \hat{q}_n] - M[u_i : \hat{p}_n]) + \sqrt{n}\,(M[u_i : \hat{p}_n] - M[u_i : p_n])$$

$$\to_d N(-\lim\sqrt{n}\, M[u_i : p_n], 1)$$

Moreover, terms of this latter type are asymptotically independent, so that

$$n\sum_y \frac{[\hat{q}_n(y) - p_n(y)]^2}{\pi(y)} = n\|\hat{\boldsymbol{\beta}}_n\|^2 \to_d \chi^2\left(\#\mathbf{u}, \lim n\|M[\mathbf{u}: p_n]\|^2\right)$$

Proceed now as in Proposition 14 to finish the proof. □

Asymptotic Power Comparisons. We will now see that the effect of basing a hypothesis test on a rapidly converging sequence of distributions is to raise the asymptotic power of the tests against a rapidly converging sequence of contiguous alternatives. Aside from the argument that rapid convergence is reasonable, this is the main formal justification of the notion.

Continue the assumptions and notation of the preceding two propositions. Thus we are considering a sequence of tests of hypotheses $\pi = p_n$ where $\{p_n\}$ is a contiguous sequence that satisfies $\sqrt{n}\,(M[\mathbf{v}: p_n] - M[\mathbf{v}: \pi]) \to_p \mathbf{0}$. There are two competing tests:

Test 1 (based on $\hat{p}_n$): reject $\pi = p_n$ if $n\mathbb{D}_2(\hat{p}_n, p_n) \geqq \xi$, where $F_m(\xi) = 1 - \alpha$, $m = \#\mathcal{Y} - 1$

Test 2 (based on $\hat{q}_n$): reject $\pi = p_n$ if $n\mathbb{D}_2(\hat{q}_n, p_n) \geqq \xi$, where $F_r(\xi) = 1 - \alpha$, $r = \#\mathbf{u}$.

We are assuming $\hat{q}_n$ converges rapidly on $\mathbb{S}(\mathbf{v})$, and it is fair to compare these tests, since they both have asymptotic significance level α. The issue is, which has greater power against π?

To answer this question we use the results of Propositions 17 and 18:

$$n\mathbb{D}_2(\hat{p}_n, p_n) \to_d \chi^2(m, \lambda), \qquad n\mathbb{D}_2(\hat{q}_n, p_n) \to_d \chi^2(r, \lambda)$$

where $\lambda = \lim n\mathbb{D}_2(p_n, \pi)$. Thus, the issue is: of two Chi-square tests that have the same noncentrality parameter but different degrees of freedom, which has the greater power? The most straightforward answer is obtained by considering the computations in a particular case. The results are presented in Table III.3, which shows that in the cases of significance levels .01 and .05 and for values of the noncentrality parameter running from 1 to 20 and df running from 1 to 10, the test with the smaller df is the more powerful.

In terms of the example cited earlier in this section, assuming that $\pi \in \mathcal{L}[q:\mathbf{u}]$ were known, the test of $\pi = q$ need only direct its power against alternatives that are also in $\mathcal{L}[q:\mathbf{u}]$, and so contiguous alternatives p_n satisfying $M[\mathbf{v}: p_n] = M[\mathbf{v}: \pi]$ are the only ones of interest. Proposition 18 and Table III.3 show that a test based on a rapidly converging estimator,

Test 2 above, will tend to be more powerful than the test based on $\hat{p}_n$, Test 1 above. This shows that a definite benefit will accrue to us if we can produce a sequence $\hat{q}_n \in \mathfrak{L}[q:\mathbf{u}]$ for which $\sqrt{n}(M[\mathbf{u}:\hat{p}_n] - M[\mathbf{u}:\hat{q}_n]) \rightarrow_p \mathbf{0}$.

The situation is similar, but more complex, in the case of models of the form $\mathfrak{S}[q:\mathbf{u}]$. However, in both cases the problem is not to find a rapidly converging estimator, it is that there are too many rapidly converging estimators, and so some additional principle is required to choose one. We provide such a principle in the next subsection.

Table III.3

Power of Chi-square tests (in percent); α = significance level, λ = noncentrality parameter, ν = degrees of freedom.

α	λ	ν = 1	2	3	4	5	6	7	8	9	10
	1	5	4	3	2	2	2	2	2	2	2
	2	12	8	6	5	5	4	4	3	3	3
	3	19	13	11	9	8	7	6	6	5	5
	4	28	20	16	13	12	10	9	8	8	7
	5	36	27	22	18	16	14	13	12	11	10
	6	44	34	28	24	21	19	17	16	14	13
	7	52	41	35	30	27	24	22	20	18	17
	8	59	48	41	36	32	29	27	25	23	21
	9	66	55	48	42	38	35	32	29	27	26
	10	72	61	54	48	44	40	37	34	32	30
.01	11	77	67	60	54	49	46	42	39	37	35
	12	81	72	65	59	55	51	47	44	42	40
	13	84	76	70	64	60	56	52	49	47	44
	14	87	80	74	69	65	61	57	54	51	49
	15	90	83	78	73	69	65	62	59	56	54
	16	92	86	81	77	73	69	66	63	60	58
	17	93	89	84	80	77	73	70	67	65	62
	18	95	91	87	83	80	77	74	71	68	66
	19	96	92	89	86	83	80	77	75	72	70
	20	97	94	91	88	85	83	80	78	75	73

Table III.3 (Continued)

α	λ	ν = 1	2	3	4	5	6	7	8	9	10
	1	17	13	11	10	9	9	8	8	8	8
	2	29	22	19	17	15	14	13	13	12	12
	3	40	32	27	24	22	20	19	18	17	16
	4	51	41	35	31	29	26	25	23	22	21
	5	60	50	44	39	36	33	31	29	28	26
	6	68	58	51	46	43	40	37	35	33	32
	7	75	65	58	53	50	46	44	41	39	37
	8	80	71	65	60	56	53	50	47	45	43
	9	85	77	71	66	62	58	55	53	51	48
	10	88	81	76	71	67	64	61	58	56	54
.05	11	91	85	80	76	72	69	66	63	61	59
	12	93	88	84	80	76	73	71	68	66	63
	13	95	90	87	83	80	77	75	72	70	68
	14	96	92	89	86	83	81	78	76	74	72
	15	97	94	91	89	86	84	82	79	77	75
	16	97	95	93	91	89	86	84	82	81	79
	17	98	96	94	92	91	89	87	85	83	82
	18	98	97	95	94	92	91	89	87	86	84
	19	99	98	96	95	94	92	91	89	88	87
	20	99	98	97	96	95	94	92	91	90	89

3. Minimum-Discrepancy Estimators

We have just seen that when some knowledge is available about π, it can turn out that $\hat{p}$ is a poor choice of an estimator of π. Specifically, if it were known that $\pi \in \mathfrak{L}[q:\mathbf{u}]$ (or $\pi \in \mathfrak{S}[q:\mathbf{u}]$), then it seems one should be able to improve on $\hat{p}$ by choosing an estimator $\hat{q}$ which is strongly influenced by $\hat{p}$ but satisfies $\hat{q} \in \mathfrak{L}[q:\mathbf{u}]$ (or $\hat{q} \in \mathfrak{S}[q:\mathbf{u}]$). This is in fact the case, as will be shown in this section.

If one can become convinced that a discrepancy $\mathbb{D}$ has the interpretation that $\mathbb{D}(p, q)$ represents how far the hypothesized distribution q is from the data-based distribution p, then the following principle seems reasonable:

MDE Principle. If $\mathfrak{M} \subset \mathcal{P}$ and it is known that $\pi \in \mathfrak{M}$, then choose as estimator of π that $\hat{q} \in \mathfrak{M}$ (if it exists) which minimizes $\mathbb{D}(\hat{p}, \hat{q})$. Such a $\hat{q}$ is called a *minimum-discrepancy estimator* (MDE).

Because of the wide variety of plausible discrepancy measures, there is a fertile field for establishing the existence and desirable properties of an MDE with respect to one's favorite discrepancy. There are, however, two methods of estimation which have provided very attractive inferential schemes, and we shall restrict attention to them:

1. MDE with respect to $\mathbb{D}_1$ for models $\mathcal{E}[q:\mathbf{u}]$;
2. MDE with respect to $\mathbb{D}_0$ for models $\mathcal{L}[q:\mathbf{u}]$.

We begin by characterizing $\mathbb{D}_1$-MDEs in exponential models.

Proposition 19. For $\hat{q} \in \mathcal{E}[q:\mathbf{u}]$ the following are equivalent:

1. $M[\mathbf{u}:\hat{q}] = M[\mathbf{u}:\hat{p}]$,
2. $\mathbb{D}_1(\hat{p}, q) = \mathbb{D}_1(\hat{p}, \hat{q}) + \mathbb{D}_1(\hat{q}, q)$,
3. $\hat{q}$ is a $\mathbb{D}_1$-MDE in $\mathcal{E}[q:\mathbf{u}]$.

Moreover, $\hat{q}$ is unique if it exists.

Proof. Assuming condition 1 and letting $\hat{q} = e^{\hat{\boldsymbol{\theta}}'\mathbf{u} - \delta(\hat{\boldsymbol{\theta}})}q$,

$$\begin{aligned}
\mathbb{D}_1(\hat{p}, q) &= 2\sum \hat{p} \ln \frac{\hat{p}}{q} = 2\sum \hat{p} \ln \frac{\hat{p}}{\hat{q}} + \hat{p} \ln \frac{\hat{q}}{q} \\
&= \mathbb{D}_1(\hat{p}, \hat{q}) + 2\sum \hat{p}\left[\hat{\boldsymbol{\theta}}'\mathbf{u} - \delta(\hat{\boldsymbol{\theta}})\right] \\
&= \mathbb{D}_1(\hat{p}, \hat{q}) + 2\sum \hat{q}\left[\hat{\boldsymbol{\theta}}'\mathbf{u} - \delta(\hat{\boldsymbol{\theta}})\right] \\
&= \mathbb{D}_1(\hat{p}, \hat{q}) + \mathbb{D}_1(\hat{q}, q)
\end{aligned}$$

proving condition 2. From condition 2, if $p \in \mathcal{E}[q:\mathbf{u}]$ is arbitrary, then since $\mathcal{E}[q:\mathbf{u}] = \mathcal{E}[p:\mathbf{u}]$ we have

$$\mathbb{D}_1(\hat{p}, p) = \mathbb{D}_1(\hat{p}, \hat{q}) + \mathbb{D}_1(\hat{q}, p) \geqq \mathbb{D}_1(\hat{p}, \hat{q})$$

which shows that $\hat{q}$ is MDE. Moreover, there is equality here if and only if $\hat{q} = p$, which proves the uniqueness assertion.

From (3) we have that the function $\mathbb{D}_1(\hat{p}, q^{\boldsymbol{\theta}})$ assumes its minimum at $\hat{\boldsymbol{\theta}}$ corresponding to $\hat{q}$. Since the set of possible $\boldsymbol{\theta}$'s is all of $\mathbb{R}^{\#\mathbf{u}}$, the derivative of $\mathbb{D}_1(\hat{p}, q^{\boldsymbol{\theta}})$ vanishes at $\hat{\boldsymbol{\theta}}$:

$$-\frac{\partial}{\partial \boldsymbol{\theta}}\mathbb{D}_1(\hat{p}, \hat{q}) = \frac{\partial}{\partial \boldsymbol{\theta}}\bigg|_{\boldsymbol{\theta}=\hat{\boldsymbol{\theta}}} 2\sum p[\boldsymbol{\theta}'\mathbf{u} - \delta(\boldsymbol{\theta})]$$

$$= 2(M[\mathbf{u} : \hat{p}] - M[\mathbf{u} : \hat{q}]) = \mathbf{0}$$

which is condition 1. □

According to Proposition 16, the set of $p \in \mathcal{P}$ satisfying $M[\mathbf{u} : p] = M[\mathbf{u} : \hat{p}]$ can be expressed in the form $\mathfrak{L}[\hat{p} : \mathbf{v}]$ for an appropriate $\mathbf{v}$. Thus, in Proposition 19, although we began with a model of the form $\mathfrak{E}[q : \mathbf{u}]$, we were led to consider the model $\mathfrak{L}[\hat{p} : \mathbf{v}]$ as well. A corollary of the above proposition in which $\mathfrak{L}$ and $\mathfrak{E}$ appear more symmetrically is

Corollary. If $\mathbf{u}, \mathbf{v}$ are p-complements, then for any q, $\mathfrak{E}[q : \mathbf{u}] \cap \mathfrak{L}[p : \mathbf{v}]$ contains at most one element.

The dual MDE proposition for $\mathbb{D}_0$ and $\mathfrak{L}$ is

Proposition 20. Let $\mathbf{u}, \mathbf{v}$ be q-complements, and suppose that $\hat{q} \in \mathfrak{L}[q : \mathbf{u}]$. Then the following are equivalent:

1. $\hat{q} \in \mathfrak{E}[\hat{p} : \mathbf{v}]$,
2. $\mathbb{D}_0(\hat{p}, q) = \mathbb{D}_0(\hat{p}, \hat{q}) + \mathbb{D}_0(\hat{q}, q)$,
3. $\hat{q}$ is a $\mathbb{D}_0$-MDE in $\mathfrak{L}[q : \mathbf{u}]$.

Moreover, $\hat{q}$ is unique if it exists.

Proof. From condition 1, $\hat{q} \in \mathfrak{E}[\hat{p} : \mathbf{v}] \cap \mathfrak{L}[q : \mathbf{u}]$ (proving uniqueness), and so $\mathbb{D}_1(q, \hat{p}) = \mathbb{D}_1(q, \hat{q}) + \mathbb{D}_1(\hat{q}, \hat{p})$, which is condition 2 because $\mathbb{D}_0(p, q) = \mathbb{D}_1(q, p)$. From point 2 use the fact that for any $p \in \mathfrak{L}[q : \mathbf{u}]$ we have $\mathfrak{L}[p : \mathbf{w}] = \mathfrak{L}[q : \mathbf{u}]$ for an appropriate $\mathbf{w}$ (see Problem III.9) and so we may conclude that $\mathbb{D}_0(\hat{p}, p) \geq \mathbb{D}_0(\hat{p}, \hat{q})$ with equality if and only if $\hat{q} = p$, establishing condition 3.

Assuming condition 3, there is no loss of generality in taking $\mathbf{u}$ to be q-standard, and since the set of parameters $\boldsymbol{\alpha}$ of $\mathfrak{L}[q : \mathbf{u}]$ form a convex set, define $p^{\boldsymbol{\alpha}} = (1 + \boldsymbol{\alpha}'\mathbf{u})q$ and observe that the derivative of $\mathbb{D}_0(p, p^{\boldsymbol{\alpha}})$ vanishes at the value $\hat{\boldsymbol{\alpha}}$ corresponding to $\hat{q}$, and so

$$\frac{\partial}{\partial \boldsymbol{\alpha}} \mathbb{D}_0(\hat{p}, p^{\boldsymbol{\alpha}}) = 2\sum \frac{\partial}{\partial \boldsymbol{\alpha}} p^{\boldsymbol{\alpha}} \ln \frac{p^{\boldsymbol{\alpha}}}{\hat{p}}$$

$$= 2\sum \mathbf{u} q \ln \frac{p^{\boldsymbol{\alpha}}}{\hat{p}} + p^{\boldsymbol{\alpha}} \frac{1}{1 + \boldsymbol{\alpha}'\mathbf{u}} \mathbf{u}$$

$$= 2\sum \mathbf{u} q \ln \frac{p^{\boldsymbol{\alpha}}}{\hat{p}}$$

equals $\mathbf{0}$ at $\hat{\boldsymbol{\alpha}}$, that is, $\ln(p^{\hat{\boldsymbol{\alpha}}}/\hat{p}) \in \mathbb{S}(1, \mathbf{v})$, which is the assertion of condition 1. □

The first important consequence of these propositions is that finding a $\mathbb{D}_1$-MDE in an exponential model is mathematically equivalent to finding a $\mathbb{D}_0$ – MDE in a linear model:

1. The $\mathbb{D}_1$-MDE in $\mathcal{E}[q : \mathbf{u}]$ is the unique element of $\mathcal{L}[\hat{p} : \mathbf{v}] \cap \mathcal{E}[q : \mathbf{u}]$.
2. The $\mathbb{D}_0$-MDE in $\mathcal{L}[q : \mathbf{u}]$ is the unique element of $\mathcal{L}[q : \mathbf{u}] \cap \mathcal{E}[\hat{p} : \mathbf{v}]$.

The pair $\mathbf{u}, \mathbf{v}$ is assumed to consist of $\hat{p}$-complements in statement 1, q-complements in statement 2.

Since LAMDA carries out the $\mathbb{D}_1$-inference for an exponential model, we will need to translate every problem of $\mathbb{D}_0$-inference for a linear model into the corresponding $\mathbb{D}_1$-problem. For this purpose the following formulation is convenient.

Proposition 21. For any $\mathbf{u} : \mathcal{Y} \to \mathbb{R}^r$, the $\mathbb{D}_0$-MDE in the model $\{p \in \mathcal{P}_0 : M[\mathbf{u} : p] = M[\mathbf{u} : q]\}$ (for q fixed) is the same as the $\mathbb{D}_1$-MDE for the model $\mathcal{E}[\hat{p} : \mathbf{u}]$ and "data" q.

Proof. Let $\mathbf{v}$ be the q-standardization of $\mathbf{u}$, and let $\binom{\mathbf{v}}{\mathbf{w}}$ be maximal q-standard. Then $M[\mathbf{u} : p] = M[\mathbf{u} : q]$ if and only if $M[\mathbf{v} : p] = \mathbf{0}$ (since $\mathbb{S}(1, \mathbf{v}) = \mathbb{S}(1, \mathbf{u})$), and this is in turn equivalent to $p \in \mathcal{L}[q : \mathbf{w}]$, and so the result follows from (2) above. □

Since the $\mathbb{D}_1$-MDE in $\mathcal{E}[q : \mathbf{u}]$ minimizes the likelihood discrepancy, it should be called a minimum-likelihood estimator, but in the traditional way of developing estimators one considers $\hat{q}$ as the maximizer of the log likelihood $\Sigma p \ln q$, and so we will conform to the usual terminology by calling $\hat{q}$ the *maximum-likelihood* (ML) distribution. Condition 1 of Proposition 19 is usually called the *likelihood equation*. The $\mathbb{D}_0$-MDE in $\mathfrak{L}[q:\mathbf{u}]$ is generally called the *minimum-discrimination information* (MDI) distribution.

Notice that in both the ML and MDI cases we have the decomposition

$$\mathbb{D}(\hat{p}, q) = \mathbb{D}(\hat{p}, \hat{q}) + \mathbb{D}(\hat{q}, q)$$

where $\mathbb{D} = \mathbb{D}_1$ in the ML case and $\mathbb{D} = \mathbb{D}_0$ in the MDI case. It seems clear from the way $\hat{q}$ is chosen that $\mathbb{D}(\hat{p}, \hat{q})$ is a measure of how far the data $\hat{p}$ are from the model, and so large values of $\mathbb{D}(\hat{p}, \hat{q})$ should cast doubt on the hypothesis that π lies in the model. However, if one believes that π is in the model, then $\hat{q}$ seems like a reasonable choice of an estimator of π because it is as close as possible to the data subject to the restriction that it be in the model. Thus a large value of $\mathbb{D}(\hat{q}, q)$ should cast doubt on $\pi = q$. In the remainder of this subsection we will establish the asymptotic behavior of $\mathbb{D}(\hat{p}, \hat{q})$ and $\mathbb{D}(\hat{q}, q)$ and thus provide the theoretical support for the use of these statistics in the manner just described. In order to facilitate the exposition, we will first deal with $\mathbb{D}_1$-inference for exponential models.

It turns out that the ML distribution based on a sample of size n, $\hat{q}_n$, eventually exists as $n \to \infty$. This result is established in Section A.6, since its presentation here would be a diversion from the main flow of ideas. Thus in what follows we assume n large enough that $\hat{q}_n$ exists.

Proposition 22. If $\pi \in \mathfrak{E}[q:\mathbf{u}]$ and $\hat{q}_n$ is the ML distribution in $\mathfrak{E}[q:\mathbf{u}]$, then $\hat{q}_n \to_p \pi$.

Proof. Since $\hat{p}_n \to_p \pi$, from the equation $\mathbb{D}_1(\hat{p}_n, \pi) = \mathbb{D}_1(\hat{p}_n, \hat{q}_n) + \mathbb{D}_1(\hat{q}_n, \pi)$ we conclude that $\mathbb{D}_1(\hat{q}_n, \pi) \to_p 0$, and we want to show that this implies consistency of $\hat{q}_n$.

Begin with a Taylor expansion of $x \ln x$ about 1:

$$x \ln x = (x - 1) + (x - 1)^2 \int_0^1 (1 - t) \frac{1}{1 + t(x - 1)} dt$$

Substitute p/q for x to get

$$2\frac{p}{q} \ln \frac{p}{q} = 2\left(\frac{p}{q} - 1\right) + \left(\frac{p}{q} - 1\right)^2 R(p, q)$$

where

$$R(p, q) = \int_0^1 \frac{2(1 - t)q}{(1 - t)q + tp} dt$$

Replace p/q by $p(y)/q(y)$ and average with respect to q to obtain

$$\mathbb{D}_1(p,q) = \sum_y \frac{[p(y)-q(y)]^2}{q(y)} R(p(y), q(y))$$

It is easy to see that $R(p,q) \geqq q$, and this completes the proof. □

The approximation of $\mathbb{D}_1$ by $\mathbb{D}_2$ developed in the above proof,

$$\mathbb{D}_1(p,q) = \mathbb{D}_2(p,q) + \sum_y \frac{[p(y)-q(y)]^2}{q(y)} [R(p(y), q(y)) - 1]$$

is quite useful for proving results about one of these discrepancies from the corresponding results about the other. We now establish the asymptotic behavior of the parameter estimate corresponding to the ML distribution.

Proposition 23. If $\pi = p^{\boldsymbol{\theta}} \in \mathfrak{S}[q:\mathbf{u}]$ and $\hat{q}_n = p^{\hat{\boldsymbol{\theta}}_n}$ is the ML distribution in $\mathfrak{S}[q:\mathbf{u}]$, then

$$\sqrt{n}\,(\hat{\boldsymbol{\theta}}_n - \boldsymbol{\theta}) \to_d N(\mathbf{0}, C[\mathbf{u}:\pi]^{-1})$$

Proof. We know that $\sqrt{n}\,(M[\mathbf{u}:\hat{p}_n] - M[\mathbf{u}:\pi]) \to_d N(\mathbf{0}, C[\mathbf{u}:\pi])$, and from the observations

$$\frac{\partial}{\partial \boldsymbol{\theta}} \delta(\boldsymbol{\theta}) = M[\mathbf{u}:p^{\boldsymbol{\theta}}], \qquad \frac{\partial^2}{\partial \boldsymbol{\theta}\, \partial \boldsymbol{\theta}'} \delta(\boldsymbol{\theta}) = C[\mathbf{u}:p^{\boldsymbol{\theta}}]$$

we may conclude

$$\sqrt{n}\left(\frac{\partial}{\partial \boldsymbol{\theta}} \delta(\hat{\boldsymbol{\theta}}_n) - \frac{\partial}{\partial \boldsymbol{\theta}} \delta(\boldsymbol{\theta})\right) \to_d N\left(\mathbf{0}, \frac{\partial^2}{\partial \boldsymbol{\theta}\, \partial \boldsymbol{\theta}'} \delta(\boldsymbol{\theta})\right)$$

From the fundamental theorem of calculus,

$$\frac{\partial}{\partial \boldsymbol{\theta}} \delta(\hat{\boldsymbol{\theta}}_n) - \frac{\partial}{\partial \boldsymbol{\theta}} \delta(\boldsymbol{\theta}) = \int_0^1 \frac{\partial^2}{\partial \boldsymbol{\theta}\, \partial \boldsymbol{\theta}'} \delta(\boldsymbol{\theta} + t(\hat{\boldsymbol{\theta}}_n - \boldsymbol{\theta}))\, dt\, (\hat{\boldsymbol{\theta}}_n - \boldsymbol{\theta})$$

The integral $\to_p (\partial^2/\partial \boldsymbol{\theta}\, \partial \boldsymbol{\theta}')\delta(\boldsymbol{\theta})$ because $\hat{q}_n \to_p \pi$ implies $\hat{\boldsymbol{\theta}}_n \to_p \boldsymbol{\theta}$, and then the inverse of this matrix exists because $\mathbf{u}$ has affinely linearly independent components. Thus, we can solve the above equation for $\hat{\boldsymbol{\theta}}_n - \boldsymbol{\theta}$, and by the MWT we have the result. □

We are now able to show that the ML distributions converge rapidly, and thus provide better estimators than the empirical distribution when the true distribution lies in the exponential model.

Proposition 24. If $\pi \in \mathcal{E}[q:\mathbf{u}]$ and $\binom{\mathbf{u}}{\mathbf{v}}$ is maximal π-standard, then the ML distributions converge rapidly on $\mathbb{S}(\mathbf{v})$.

Proof. Without loss of generality we may suppose $\pi = q$. Let u_i and v_j be components of $\mathbf{u}$ and $\mathbf{v}$. Compute

$$\frac{\partial}{\partial\theta_i} M\left[v_j : p^{\boldsymbol\theta}\right] = M\left[v_j\left(u_i - \frac{\partial}{\partial\theta_i}\delta(\boldsymbol\theta)\right) : p^{\boldsymbol\theta}\right] = C\left[u_i, v_j : p^{\boldsymbol\theta}\right]$$

which $= 0$ when $\boldsymbol\theta = \mathbf{0}$. Thus we have the Taylor expansion

$$M\left[v_j : p^{\boldsymbol\theta}\right] = \boldsymbol\theta' \int_0^1 (1-t)\frac{\partial^2}{\partial\boldsymbol\theta\,\partial\boldsymbol\theta'} M\left[v_j : p^{t\boldsymbol\theta}\right] dt\,\boldsymbol\theta$$

Multiply both sides by $\sqrt{n}$, replace $\boldsymbol\theta$ by $\hat{\boldsymbol\theta}_n$, and use Proposition 23 and the MWT to conclude $\sqrt{n}\,M[v_j : p^{\hat{\boldsymbol\theta}_n}] \to_p 0$. Together with Proposition 19, condition 1, this completes the proof. □

Together with the result of Proposition 17 and the approximation of $\mathbb{D}_1$ by $\mathbb{D}_2$, we have the following frequently used result:

Proposition 25. If $\hat{q}_n$ is the ML distribution in $\mathcal{E}[q:\mathbf{u}]$ and $\pi \in \mathcal{E}[q:\mathbf{u}]$, then

$$n\mathbb{D}_1(\hat{q}_n, \pi) \to_d \chi^2(\#\mathbf{u})$$

The behavior of the ML chi-square statistic under a contiguous alternative is given in the next proposition.

Proposition 26. If $\hat{q}_n$ is the ML distribution in $\mathcal{E}[q:\mathbf{u}]$, $\pi \in \mathcal{E}[q:\mathbf{u}]$ and p_n is a contiguous sequence in $\mathcal{E}[q:\mathbf{u}]$, then

$$n\mathbb{D}_1(\hat{q}_n, p_n) \to_d \chi^2(\#\mathbf{u}, \lim n\mathbb{D}_1(p_n, \pi))$$

Proof. On account of the approximation of $\mathbb{D}_1$ by $\mathbb{D}_2$, it is sufficient to establish the result for $\mathbb{D}_2$ in place of $\mathbb{D}_1$. From Proposition 18 it is enough to check $\sqrt{n}\,M[\mathbf{v} : p_n] \to \mathbf{0}$, where $(\mathbf{u}, \mathbf{v})$ is maximal π-standard. From the

Taylor expansion

$$\ln x = (x-1) - (x-1)^2 \int_0^1 (1-t) \frac{1}{(1-t+tx)^2}\, dt$$

we have

$$\ln \frac{p_n}{\pi} = \frac{p_n - \pi}{\pi} - \left(\frac{p_n - \pi}{\pi}\right)^2 \int_0^1 \frac{1-t}{[1-t+t(p_n/\pi)]^2}\, dt$$

so that, setting $p_n = p^{\boldsymbol{\theta}_n}$, we have

$$\sqrt{n}\,[\theta_n' u - \delta(\theta_n)] = \frac{\sqrt{n}\,(p_n - \pi)}{\pi}$$

$$- \sqrt{n}\,(p_n - \pi)\frac{p_n - \pi}{\pi^2} \int_0^1 \frac{1-t}{[1-t+t(p_n/\pi)]^2}\, dt$$

and so $\lim \sqrt{n}\,[\boldsymbol{\theta}_n'\mathbf{u} - \delta(\boldsymbol{\theta}_n)]$ exists. Now from the Taylor expansion

$$e^x = 1 + x + x^2 \int_0^1 (1-t) e^{tx}\, dt$$

we have

$$p_n = \left[1 + \boldsymbol{\theta}_n'\mathbf{u} - \delta(\boldsymbol{\theta}_n) + \{\boldsymbol{\theta}_n'\mathbf{u} - \delta(\boldsymbol{\theta}_n)\}^2 \int_0^1 (1-t) e^{t\{\boldsymbol{\theta}_n'\mathbf{u} - \delta(\boldsymbol{\theta}_n)\}}\, dt\right] \pi$$

so that

$$M[\mathbf{v} : p_n] = M\left[\mathbf{v}\{\boldsymbol{\theta}_n'\mathbf{u} - \delta(\boldsymbol{\theta}_n)\}^2 \int_0^1 (1-t) e^{t\{\boldsymbol{\theta}_n'\mathbf{u} - \delta(\boldsymbol{\theta}_n)\}}\, dt : \pi\right]$$

whence $\sqrt{n}\, M[\mathbf{v} : p_n] \to \mathbf{0}$. □

The following proposition is extremely important for testing the hypothesis that the true distribution lies in an exponential model.

Proposition 27. If $\pi \in \mathfrak{S}[q : \mathbf{u}]$ and $\hat{q}_n$ is ML in $\mathfrak{S}[q : \mathbf{u}]$, then

$$n\mathbb{D}_1(\hat{p}_n, \hat{q}_n) \to_d \chi^2(\#\mathcal{Y} - 1 - \#\mathbf{u})$$

Proof. Again because of the approximation of $\mathbb{D}_1$ by $\mathbb{D}_2$, it is sufficient to establish the result for $\mathbb{D}_2$. To begin, there is no loss of generality in supposing that $(\mathbf{u}, \mathbf{v})$ is maximal π-standard. Then $n\mathbb{D}_2(\hat{p}_n, \hat{q}_n) = n\|M[\mathbf{v}_n : \hat{p}_n]\|^2$, where $(\mathbf{u}_n, \mathbf{v}_n)$ is the $\hat{q}_n$-standardization of $(\mathbf{u}, \mathbf{v})$. Consequently it will be enough to show that $\sqrt{n}\, M[\mathbf{v}_n : \hat{p}_n] \to_d N(\mathbf{0}, \mathbf{I})$, and by the MWT, for this it is sufficient to show $\sqrt{n}\,(M[\mathbf{v}_n : \hat{p}_n] - M[\mathbf{v} : \hat{p}_n]) \to_p \mathbf{0}$. Recalling the Gram–Schmidt steps as they pertain to $\mathbf{v}_n$, we have

$$w_{n1} = v_1 - M[v_1 : \hat{q}_n] - \sum_i M[v_1 u_{ni} : \hat{q}_n] u_{ni}$$

$$v_{n1} = \frac{w_{n1}}{\sqrt{V[w_{n1} : \hat{q}_n]}}$$

since $\hat{q}_n \to_p \pi$, we see that $w_{n1} \to_p v_1$ and so $V[w_{n1} : \hat{q}_n] \to_p 1$. Further, it is not hard to see that

$$\sqrt{n} \sum M[v_1 u_{ni} : \hat{q}_n] M[u_{ni} : \hat{p}_n] \to_p 0$$

Thus

$$\sqrt{n}\,(M[v_{n1} : \hat{p}_n] - M[v_1 : \hat{p}_n] + M[v_1 : \hat{q}_n]) \to_p 0$$

and since $\hat{q}_n$ converges rapidly on $\mathbb{S}(\mathbf{v})$,

$$\sqrt{n}\,(M[v_{n1} : \hat{p}_n] - M[v_1 : \hat{p}_n]) \to_p 0$$

The general result now follows by induction, v_{ni} replacing v_{n1} above. □

A great deal of the importance of this result rests on the fact that the limiting distribution of $n\mathbb{D}_1(\hat{p}_n, \hat{q}_n)$ does not depend on the particular value of π; that is, the limiting distribution is the same regardless of which distribution in $\mathfrak{E}[q : \mathbf{u}]$ is true. Thus the test based on this statistic is much preferred to the procedure used in Chapter II, Section B.5.

The next result is useful for computing the power of the serial test which will be introduced in the following section.

Proposition 28. If $\pi \in \mathfrak{E}[q : \mathbf{u}]$ and $\hat{q}_n$ is ML in $\mathfrak{E}[q : \mathbf{u}]$, and if p_n is any contiguous sequence in $\mathfrak{E}[q : \mathbf{u}]$, then $n\mathbb{D}_1(\hat{p}_n, \hat{q}_n)$ and $n\mathbb{D}_1(\hat{q}_n, p_n)$ are asymptotically independent.

Proof. We need only gather the elements of preceding proofs, and again on account of the asymptotic equivalence of $\mathbb{D}_1$ and $\mathbb{D}_2$ we can prove the result for $\mathbb{D}_2$. In the notation of the preceding propositions, assuming $(\mathbf{u}, \mathbf{v})$

to be maximal π-standard,

$$n\mathbb{D}_2(\hat{p}_n, \hat{q}_n) = n\|M[\mathbf{v} : \hat{p}_n]\|^2 + n(\|M[\mathbf{v}_n : \hat{p}_n]\|^2 - \|M[\mathbf{v} : \hat{p}_n]\|^2)$$

and the second term on the right $\to_p 0$. Then from the proofs of Propositions 18 and 26

$$n\mathbb{D}_2(\hat{q}_n, p_n) - n\|M[\mathbf{u} : \hat{q}_n] - M[\mathbf{u} : p_n]\|^2$$
$$- n\|M[\mathbf{v} : \hat{q}_n] - M[\mathbf{v} : p_n]\|^2 \to_p 0$$

and the last of these three terms $\to_p 0$. Thus $n\mathbb{D}_2(\hat{q}_n, p_n)$ depends asymptotically on $M[\mathbf{u} : \hat{p}_n]$, while $n\mathbb{D}_2(\hat{p}_n, \hat{q}_n)$ depends asymptotically on $M[\mathbf{v} : \hat{p}_n]$, and by Proposition 5 they are asymptotically independent. □

To each of the above results for ML inference there is a corresponding result for MDI inference:

Proposition 29. Let $\hat{q}_n$ be MDI in $\mathfrak{L}[q : \mathbf{u}]$.

1. If $\pi \in \mathfrak{L}[q : \mathbf{u}]$, then $\hat{q}_n \to_p \pi$,
2. and if $\pi = q$, $\hat{q}_n = (\hat{\alpha}_0 + \hat{\alpha}_n'\mathbf{u})$, then $\hat{\boldsymbol{\alpha}}_n \to_p \mathbf{0}$,
3. and if $\mathbf{u}$ is π-standard, then $\sqrt{n}\,\hat{\boldsymbol{\alpha}}_n \to_d N(\mathbf{0}, \mathbf{I})$,
4. and $n\mathbb{D}_0(\hat{q}_n, \pi) \to_d \chi^2(\#\mathbf{u})$.
5. If p_n is a contiguous sequence in $\mathfrak{L}[q : \mathbf{u}]$, then $n\mathbb{D}_0(\hat{q}_n, p_n) \to_d \chi^2(\#\mathbf{u}, \lim n\mathbb{D}_0(p_n, \pi))$,
6. and $n\mathbb{D}_0(\hat{p}_n, \hat{q}_n) \to_d \chi^2(\#\mathcal{Y} - 1 - \#\mathbf{u})$, and this statistic is asymptotically independent of $n\mathbb{D}_0(\hat{q}_n, p_n)$.

Proof. From Proposition 20(2) we have

$$\mathbb{D}_0(\hat{p}_n, \pi) = \mathbb{D}_0(\hat{p}_n, \hat{q}_n) + \mathbb{D}_0(\hat{q}_n, \pi)$$

and since $\hat{p}_n \to_p \pi$, we have $\mathbb{D}_0(\hat{p}_n, \pi) \to_p 0$, so that in particular $\mathbb{D}_0(\hat{q}_n, \pi) \to_p 0$, proving condition 1.

There is no harm in supposing $(\mathbf{u}, \mathbf{v})$ to be maximal π-standard. Since there is a $\hat{\boldsymbol{\theta}}_n$ for which $\hat{q}_n = \exp[\hat{\boldsymbol{\theta}}_n'\mathbf{v} - \delta(\hat{\boldsymbol{\theta}}_n)]\hat{p}_n$ according to Proposition 20, we have

$$\hat{\boldsymbol{\alpha}}_n = M[\mathbf{u} : \hat{q}_n] = \sum_y \mathbf{u}(y)\exp[\hat{\boldsymbol{\theta}}_n'\mathbf{v}(y) - \delta(\hat{\boldsymbol{\theta}}_n)]\,\hat{p}_n(y)$$

From $\hat{p}_n \to_p \pi$, $\hat{q}_n \to_p \pi$, it follows that $\exp[\hat{\boldsymbol{\theta}}_n'\mathbf{v} - \delta(\hat{\boldsymbol{\theta}}_n)] \to_p 1$, and so by the

MWT $\sqrt{n}\,\hat{\boldsymbol{\alpha}}_n - \sqrt{n}\,M[\mathbf{u}:\hat{p}_n] \to_p \mathbf{0}$, which simultaneously proves conditions 2 and 3.

Since $\mathbb{D}_2(\hat{q}_n, \pi) = \|\hat{\boldsymbol{\alpha}}_n\|^2$, the equivalence of $\mathbb{D}_0$ and $\mathbb{D}_2$ establishes condition 4.

Again using this equivalence, we can repeat the arguments of Propositions 24, 25, and 26 once we have checked that $\hat{q}_n$ converges rapidly on $\mathbb{S}(\mathbf{v})$ and that $\sqrt{n}\,M[\mathbf{v}:p_n] \to \mathbf{0}$. The latter is trivial because $p_n \in \mathfrak{L}[q:\mathbf{u}]$ is equivalent to $M[\mathbf{v}:p_n] = \mathbf{0}$. Likewise, $M[\mathbf{v}:\hat{q}_n] = \mathbf{0}$, and we showed in proving condition 3 above that $\sqrt{n}\,(M[\mathbf{u}:\hat{q}_n] - M[\mathbf{u}:\hat{p}_n]) \to_p \mathbf{0}$, and so condition 5 is proved.

From $\mathbb{D}_0(\hat{p}_n, \hat{q}_n) = 2[\hat{\boldsymbol{\theta}}_n' M[\mathbf{v}:\hat{q}_n] - \delta(\hat{\boldsymbol{\theta}}_n)] = -2\delta(\hat{\boldsymbol{\theta}}_n)$ and the observation made at the beginning of the proof, we have $\delta(\hat{\boldsymbol{\theta}}_n) \to_p 0$, and as a consequence $\hat{\boldsymbol{\theta}}_n \to_p \mathbf{0}$. Write

$$(1 + \hat{\boldsymbol{\alpha}}_n'\mathbf{u})\pi = \exp\left[\hat{\boldsymbol{\theta}}_n'\mathbf{v} - \delta(\hat{\boldsymbol{\theta}}_n)\right]\hat{p}_n$$

and using the Taylor expansion

$$e^x = 1 + x + x^2 \int_0^1 (1-t)e^{tx}\,dt$$

write

$$(1 + \hat{\boldsymbol{\alpha}}_n'\mathbf{u})\pi = \left\{1 + \left[\hat{\boldsymbol{\theta}}_n'\mathbf{v} - \delta(\hat{\boldsymbol{\theta}}_n)\right]R_n\right\}\hat{p}_n$$

where

$$R_n = 1 + \left[\hat{\boldsymbol{\theta}}_n'\mathbf{v} - \delta(\hat{\boldsymbol{\theta}}_n)\right]\int_0^1 (1-t)\exp\{t\left[\hat{\boldsymbol{\theta}}_n'v - \delta(\hat{\boldsymbol{\theta}}_n)\right]\}\,dt$$

Multiplying by $\mathbf{v}$ and summing over y, we obtain

$$\mathbf{0} = M[\mathbf{v}:\hat{p}_n] + \hat{\boldsymbol{\theta}}_n' M[\mathbf{v}\mathbf{v}'R_n:\hat{p}_n] - \delta(\hat{\boldsymbol{\theta}}_n)M[\mathbf{v}R_n:\hat{p}_n]$$

and since $R_n \to_p 1$, we may eventually write

$$\hat{\boldsymbol{\theta}}_n = M[\mathbf{v}\mathbf{v}'R_n:\hat{p}_n]^{-1}\{\delta(\hat{\boldsymbol{\theta}}_n)M[\mathbf{v}R_n:\hat{p}_n] - M[\mathbf{v}:\hat{p}_n]\}$$

Multiply by $\sqrt{n}$ and see that $\sqrt{n}\,\hat{\boldsymbol{\theta}}_n + \sqrt{n}\,M[\mathbf{v}:\hat{p}_n] \to_p \mathbf{0}$, so that $\sqrt{n}\,\hat{\boldsymbol{\theta}}_n \to_d N(\mathbf{0}, \mathbf{I})$, and from Proposition 5 and the MWT we conclude

$$-n\hat{\boldsymbol{\theta}}_n' M[\mathbf{v}:\hat{p}_n] \to_d \chi^2(\#\mathbf{v})$$

$$n\|\boldsymbol{\theta}_n\|^2 + n\hat{\boldsymbol{\theta}}_n' M[\mathbf{v}:\hat{p}_n] \to_p \mathbf{0}$$

From the Taylor expansion

$$\delta(\hat{\boldsymbol{\theta}}_n) = \hat{\boldsymbol{\theta}}_n' M[\mathbf{v} : \hat{p}_n] + \hat{\boldsymbol{\theta}}_n' \int_0^1 (1-t) C\left[\mathbf{v} : q^{t\hat{\boldsymbol{\theta}}_n}\right] dt\, \hat{\boldsymbol{\theta}}_n$$

where $q^{t\boldsymbol{\theta}} = \exp[t\boldsymbol{\theta}' v - \delta(\boldsymbol{\theta})]\, \hat{p}_n$, we conclude by the MWT that

$$n\delta(\hat{\boldsymbol{\theta}}_n) + n\|\hat{\boldsymbol{\theta}}_n\|^2 - \frac{n}{2}\|\hat{\boldsymbol{\theta}}_n\|^2 \to_p 0$$

and so

$$-2n\delta(\hat{\boldsymbol{\theta}}_n) - n\|\hat{\boldsymbol{\theta}}_n\|^2 \to_p 0$$

completing the first part of condition 6.

Finally, since $\sqrt{n}\,\hat{\boldsymbol{\theta}}_n + \sqrt{n}\, M[\mathbf{v} : \hat{p}_n] \to_p \mathbf{0}$ and $\sqrt{n}\,\hat{\boldsymbol{\alpha}}_n - \sqrt{n}\, M[\mathbf{u} : \hat{p}_n] \to_p \mathbf{0}$, from Proposition 5 we may deduce that $\hat{\boldsymbol{\theta}}_n$ and $\hat{\boldsymbol{\alpha}}_n$ are asymptotically independent. Since we have shown that $\mathbb{D}_0(\hat{p}_n, \hat{q}_n)$ depends only on $\hat{\boldsymbol{\theta}}_n$ while $\mathbb{D}_0(\hat{q}_n, \pi)$ depends asymptotically only on $\hat{\boldsymbol{\alpha}}_n$, these two discrepancies are asymptotically independent. The proof is now completed as in Proposition 26, since $p_n \in \mathfrak{L}[q : \mathbf{u}]$ means $M[\mathbf{v} : p_n] = \mathbf{0}$. □

Summary. This subsection contains a number of results which should be clearly understood in order to recognize that the statistical procedures of the next subsection are justified. The first collection of facts pertains to both linear and exponential models. Thus, we assume that $\mathfrak{M}[q : \mathbf{u}]$ stands for either $\mathfrak{E}[q : \mathbf{u}]$ or $\mathfrak{L}[q : \mathbf{u}]$, and that $\hat{q}_n$ is the $\mathbb{D}$-MDE in $\mathfrak{M}[q : \mathbf{u}]$, where $\mathbb{D}$ is $\mathbb{D}_1$ if $\mathfrak{M} = \mathfrak{E}$, and $\mathbb{D}$ is $\mathbb{D}_0$ if $\mathfrak{M} = \mathfrak{L}$.

1. $\hat{q}_n$ is consistent, that is, $\hat{q}_n \to_p \pi$.
2. $\hat{q}_n$ converges rapidly on $\mathfrak{S}(\mathbf{v})$, where $\mathbf{v}$ is any π-complement of $\mathbf{u}$.
3. $n\mathbb{D}(\hat{q}_n, \pi) \to_p \chi^2(\#\mathbf{u})$.
4. If $\{p_n\}$ is a contiguous sequence in $\mathfrak{M}[q : \mathbf{u}]$, then $n\mathbb{D}(\hat{q}_n, p_n) \to_d \chi^2(\#\mathbf{u}, \lim n\mathbb{D}(p_n, \pi))$.
5. If $\pi \in \mathfrak{M}[q : \mathbf{u}]$, then $n\mathbb{D}(\hat{p}_n, \hat{q}_n) \to_d \chi^2(\#\mathfrak{Y} - 1 - \#\mathbf{u})$, and this statistic is asymptotically independent of $n\mathbb{D}(\hat{q}_n, p_n)$ for any contiguous sequence $\{p_n\}$ in $\mathfrak{M}[q : \mathbf{u}]$.

There are two additional facts that must be stated differently, depending on whether $\mathfrak{M} = \mathfrak{E}$ or $\mathfrak{M} = \mathfrak{L}$.

6a. For $\mathfrak{M} = \mathfrak{E}$: if it exists, $\hat{q}_n$ is the unique distribution satisfying

$$\hat{q}_n \in \mathfrak{E}[q : \mathbf{u}], \qquad M[\mathbf{u} : \hat{q}_n] = M[\mathbf{u} : \hat{p}_n]$$

6b. For $\mathfrak{M} = \mathfrak{L}$: if $\hat{p}_n \in \mathfrak{P}$, then $\hat{q}_n$ is the unique distribution satisfying

$$\hat{q}_n \in \mathfrak{E}[\hat{p}_n : \mathbf{v}], \qquad M[\mathbf{v} : \hat{q}_n] = M[\mathbf{v} : q]$$

where $\mathbf{v}$ is any q-complement of $\mathbf{u}$.

7a. For $\mathfrak{M} = \mathfrak{E}$: if $\hat{q}_n$ corresponds to $\hat{\boldsymbol{\theta}}_n$ and π corresponds to $\boldsymbol{\theta}$, then

$$\sqrt{n}\,(\hat{\boldsymbol{\theta}}_n - \boldsymbol{\theta}) \to_d N(\mathbf{0}, C[\mathbf{u} : \pi]^{-1})$$

7b. For $\mathfrak{M} = \mathfrak{L}$: if $\hat{q}_n$ corresponds to $\hat{\boldsymbol{\alpha}}_n$ and π corresponds to $\boldsymbol{\alpha}$, then

$$\sqrt{n}\,(\hat{\boldsymbol{\alpha}}_n - \boldsymbol{\alpha}) \to_d N(\mathbf{0}, C[\mathbf{u} : q]^{-1} C[\mathbf{u} : \pi] C[\mathbf{u} : q]^{-1})$$

(the covariance matrix is obtained in Problem III.1)

4. The ANALYSIS Table

It is convenient to summarize the analysis of actual data in an array which we will call the ANALYSIS table. The purpose of the table is to display in compact form the inferential conclusions of the data analysis, and we will see in Chapter IV that this idea may be extended to quite complicated situations.

The basic form of the simple ANALYSIS table is given in Table III.4. Here $\mathfrak{M}$ stands for either $\mathfrak{E}$ or $\mathfrak{L}$, and the discrepancy statistics are the ones appropriate for these types of models. Also, $\hat{q}$ is either the ML or the MDI estimator, whichever is appropriate. According to Propositions 19 and 20 the χ^2 entries in the first two rows do add to the χ^2 in the "total" row. This gives the table a nice interpretation as a partition of the overall chi-square into pieces reflecting the departure from the model and the departure from the generating distribution. We now turn to the use of the table, which proceeds differently according to the hypothesis of interest.

Table III.4

The Simple ANALYSIS *Table*

Model	df	χ^2	$\langle P \rangle$
$\mathfrak{M}[q : \mathbf{u}]$	$\#\mathfrak{Y} - 1 - \#\mathbf{u}$	$n\mathbb{D}(\hat{p}, \hat{q})$	$\langle\ \rangle$
q	$\#\mathbf{u}$	$n\mathbb{D}(\hat{q}, q)$	$\langle\ \rangle$
Total	$\#\mathfrak{Y} - 1$	$n\mathbb{D}(\hat{p}, q)$	$\langle\ \rangle$

$\pi = q$. If our only interest were in testing $\pi = q$, then we look only at the "total" row of the table, and according to the results of Chapter II, Section C.3 and Proposition 29, condition 4, an appropriate test of the hypothesis is made by treating $n\mathbb{D}(\hat{p}, q)$ as $\chi^2\ (\#\mathcal{Y} - 1)$, and thus rejecting whenever the associated P-value is too small. The procedure for computing the P-value is given in Chapter II, Section B.4. The power against a particular alternative p would be based on Propositions 26 and 29 (condition 5), by treating $n\mathbb{D}(\hat{p}, q)$ as $\chi^2\ (\#\mathcal{Y} - 1, n\mathbb{D}(q, p))$ and using the approximation given in Chapter II, Section C.4.

$\pi \in \mathfrak{M}[q:\mathbf{u}]$. If we are interested in testing $\pi \in \mathfrak{M}[q:\mathbf{u}]$, then we use only the first line of the table, since by Propositions 27 and 29 (condition 6) we may treat $n\mathbb{D}(\hat{p}, \hat{q})$ as an observation from $\chi^2\ (\#\mathcal{Y} - 1 - \#\mathbf{u})$ for the determination of a P-value for this hypothesis. Notice the important point that the asymptotic distribution of $n\mathbb{D}(\hat{p}, \hat{q})$ does not depend on which element of $\mathfrak{M}[q:\mathbf{u}]$ is the true distribution.

$\pi = q$, ***Given*** $\pi \in \mathfrak{M}[q:\mathbf{u}]$. If we were again interested in testing $\pi = q$, but we have very strong reason to believe that $\pi \in \mathfrak{M}[q:\mathbf{u}]$ is true, then we use the second line of the table, since by Propositions 25 and 29 (condition 4) we may refer $n\mathbb{D}(\hat{q}, q)$ to the $\chi^2\ (\#\mathbf{u})$ distribution for P-values. The power against a particular alternative p would be computed by considering $n\mathbb{D}(\hat{q}, q)$ to be an observation from the $\chi^2\ (\#\mathbf{u}, n\mathbb{D}(q, p))$ distribution, according to Propositions 26 and 29 (condition 5). The justification for using this procedure rather than using the overall chi-square is that the test with the smaller df is more powerful, according to the discussion in Section A.2.

$\pi = q$, ***Using the Serial Test.*** We have seen a test of $\pi = q$ under no additional assumptions and a test of $\pi = q$ under the additional assumption that $\pi \in \mathfrak{M}[q:\mathbf{u}]$. The first procedure is less powerful against alternatives in $\mathfrak{M}[q:\mathbf{u}]$, but also does not depend on an additional, perhaps unwarranted, assumption. The *serial test* is designed to take advantage of the larger power of a test with smaller df, while at the same time providing some protection against making an unwarranted assumption that $\pi \in \mathfrak{M}[q:\mathbf{u}]$.

The procedure is to use $n\mathbb{D}(\hat{p}, \hat{q})$ to test $\pi \in \mathfrak{M}[q:\mathbf{u}]$ at level α_1, then upon confirmation of $\pi \in \mathfrak{M}[q:\mathbf{u}]$ proceed to test $\pi = q$ using $n\mathbb{D}(\hat{q}, q)$ at level α_2. If $\pi \in \mathfrak{M}[q:\mathbf{u}]$ is rejected, then implicitly $\pi = q$ is rejected also.

To compute the actual level, $\bar{\alpha}$, of this procedure, let R_1 and R_2 stand for the events that rejection will occur at stages 1 and 2 of the test, respectively, and let $\bar{R}_1$ denote the event that confirmation occurs at the first stage. Then by Propositions 28 and 29 (condition 6) we see that the overall probability of rejecting $\pi = q$ when in fact $\pi \in \mathfrak{M}[q:\mathbf{u}]$ is

$$P[R_1] + P[R_2|\bar{R}_1]P[\bar{R}_1] = P[R_1] + P[R_2]P[\bar{R}_1]$$
$$= \alpha_1 + P[R_2](1 - \alpha_1)$$

In particular, if $\pi = q$ we have $P[R_2] = \alpha_2$ and so $\bar{\alpha} = \alpha_1 + \alpha_2(1 - \alpha_1)$. It will nearly always be true in practice that α_1 is small enough that effectively we have $\bar{\alpha} = \alpha_1 + \alpha_2$.

Because the first part of the above computation was made under the sole assumption that $\pi \in \mathfrak{M}[q:\mathbf{u}]$, we also have the expression $\alpha_1 + P[R_2](1 - \alpha_1)$ for the power of the serial test, where $P[R_2]$ is the power of the second stage of the test against an alternative in $\mathfrak{M}[q:\mathbf{u}]$. In Table III.5 we see the power of the overall test of $\pi = q$ based on $n\mathbb{D}(\hat{p}, q)$ and two versions of the serial test, in the particular case $\#\mathcal{Y} - 1 = 8$, $\#\mathbf{u} = 2$, $\alpha = .05$. The serial test with $\alpha_1 = .04$ and $\alpha_2 = .01$ would be appropriate if there were doubt about the correctness of $\pi \in \mathfrak{M}[q:\mathbf{u}]$. The power in this case would be rather close to that of the overall test, but we must keep in mind that the results of the analysis would give us information about $\pi \in \mathfrak{M}[q:\mathbf{u}]$ as well as whether $\pi = q$. The serial test with $\alpha_1 = .01$ and $\alpha_2 = .04$ would be used if there were less doubt about $\pi \in \mathfrak{M}[q:\mathbf{u}]$, but some amount of protection were desired against making the assumption when it is unwarranted. In this case the power is a substantial amount greater than that of the overall test.

It should be clear that as $\alpha_1 \to 0$ and $\alpha_2 \to .05$, the serial test tends to the test based on $n\mathbb{D}(\hat{q}, q)$, while as $\alpha_1 \to .05$ and $\alpha_2 \to 0$, the procedure tends to the test based on $n\mathbb{D}(\hat{p}, \hat{q})$. Thus in choosing the values of α_1 and α_2 the statistician must deal with the tradeoff between the desire for a powerful test and the desire to avoid making an unwarranted assumption.

It would appear that even if one had quite substantial prior belief that $\pi \in \mathfrak{M}[q:\mathbf{u}]$, a very small loss of power would result if instead of using $n\mathbb{D}(\hat{q}, q)$ one used the serial test with values like $\alpha_1 = .001$, $\alpha_2 = .049$, in order to insure some minimal protection against being mistaken in that prior belief.

Serial P-values. Since a hypothesis test can, in general, be expressed in terms of either a test statistic or a P-value, a natural question to ask is whether one can compute a P-value for the serial test. If we let P_1 and P_2 be the P-values for the first and second stages of the test, then the serial test is equivalent to rejecting whenever $P_1 \leq \alpha_1$ or $P_2 \leq \alpha_2$. In order to find a P-value for the serial test, we need a quantity $\bar{P}$ which is uniformly distributed on the interval $[0, 1]$ when $\pi = q$, monotonically related to P_1 and P_2, and such that $\bar{P} \leq \bar{\alpha}$ corresponds to the occurrence of either $P_1 \leq \alpha_1$

Table III.5

	Overall Test	Serial Tests	
		$\alpha_1 = .04$	$\alpha_1 = .01$
λ	$\alpha = .05$	$\alpha_2 = .01$	$\alpha_2 = .04$
1	8	7	12
2	13	12	20
3	18	17	29
4	23	23	38
5	29	30	47
6	35	37	55
7	41	43	62
8	47	50	68
9	53	57	74
10	58	62	79
11	63	68	83
12	68	73	86
13	72	77	89
14	76	81	91
15	79	84	93
16	82	87	94
17	85	89	95
18	87	91	96
19	89	93	97
20	91	94	98

or $P_2 \leqq \alpha_2$. There are very many such quantities, but one simple and useful class can be obtained by setting

$$\bar{P} = 1 - (1 - \min^{1/w_1})(1 - \min^{1/w_2})$$

where $\min = \min\{P_1^{w_1}, P_2^{w_2}\}$ and $w_i = -1/\ln \alpha_i$. Then the rejection procedure $\bar{P} \leqq \bar{\alpha}$ corresponds to that of the serial test with $\bar{\alpha} = \alpha_1 + \alpha_2(1 - \alpha_1)$. We will call $\bar{P}$ the *serial P-value*. Note that $\alpha_1 = \alpha_2$ implies $w_1 = w_2$, so that we have the simple

$$\bar{P} = 1 - (1 - \min)^2, \qquad \min = \min\{P_1, P_2\}$$

which might be used when one has no particular prior ideas concerning $\pi \in \mathcal{M}[q : \mathbf{u}]$. Our convention will be that if no α_i are mentioned, it is assumed that they are equal.

The use of the ANALYSIS table is exemplified throughout Section B, while the computation based on the LAMDA instruction **LXM** is the subject of the following subsection.

5. LAMDA Notes

In order to use the LAMDA command **LXM** to compute the ML or MDI estimates, it is necessary to have the following information in the C-file:

1. the observed frequencies,
2. the u-functions.

Each row in the C-file stands for a $y \in \mathcal{Y}$. For an ML analysis, the weight variable should contain the observed frequencies N, and the other variables should contain the values of the u-functions. Thus in a typical case the C-file will have the appearance of Table III.6.

For an ML analysis the generating distribution may be specified by

m**GEN**

$$q(y_1), q(y_1), \ldots, q(y_m)$$

Table III.6

The Form of the C-file for an ML Analysis

Var 1	Var 2	Var 3	⋯	Var $k + 1$
$N(y_1)$	$u_1(y_1)$	$u_2(y_1)$	⋯	$u_k(y_1)$
$N(y_2)$	$u_1(y_2)$	$u_2(y_2)$	⋯	$u_k(y_2)$
⋮	⋮	⋮		⋮
$N(y_m)$	$u_1(y_m)$	$u_2(y_m)$	⋯	$u_k(y_m)$

where m is the number of cells in the table and $q(y_i)$ are proportional to the generating probabilities. Another way is to have the generating distribution stored in variable xx and then to specify

$$xx\texttt{G = V}$$

In either case, the first operation performed on the generating frequencies is to norm them to sum to one. If no generating distribution is entered, then LAMDA assumes the uniform distribution.

In order to perform the analysis, do

$$k\texttt{SEL 2 3 4} \cdots k + 1$$

$$k\texttt{LXM}$$

Here **SEL** selects the k column numbers of the variables which contain the u-functions. Note that the number k preceding **SEL** must be the same as that preceding **LXM** (this will not be true for the covariate models of Chapter V).

In accordance with the discussion in Section A.3, fitting an $\mathfrak{L}$-model by MDI is mathematically equivalent to fitting an $\mathfrak{E}$-model by ML. Since **LXM** fits the $\mathfrak{E}$-model assuming that the data appear as the weight variable and the generating distribution is declared by **GEN** or **G = V**, one can fit the $\mathfrak{L}$-model by reversing these two distributions, that is, by placing the generating distribution in the weight variable and declaring the data to be the generating distribution (by **GEN** or **G = V**). The specified u-functions then play the role described in Proposition 21. This procedure is summarized in Table III.7. Notice especially that for an MDI fit it is necessary to have the sum of the **WGT** frequencies be n.

The typical sequence of operations for setting up an analysis would be

1. dimension the C-file large enough to hold all the variables,
2. put the required u-functions in the C-file,
3. read the data and/or generating distribution into the C-file, and
4. perform one or more analyses using **SEL** and **LXM**.

Table III.7

LAMDA *Specifications for* **ML** *and* **MDI** *Fits*

Type of Fit	Model	WGT	GEN
ML	$\mathfrak{E}[q:\mathbf{u}]$	N	q
MDI	$M[\mathbf{u}:p] = M[\mathbf{u}:q]$	nq	N

The **CRT** command is especially useful for creating the *u*-functions for cross-classifications, and can often replace steps 1 and 2 above.

Our first example output from **LXM** is in Figure III.3 below. The reader should see the description associated with that figure, and should use Appendix A when necessary for a review of **LXM** output.

6. Existence of ML and MDI Estimates

In this subsection we will see a condition which is equivalent to the existence of the ML distribution. Since fitting the MDI distribution is mathematically equivalent to fitting the ML distribution, the result presented here will have implications for MDI as well. The arguments to be used here are more sophisticated than those used earlier, and so one may wish to skip over this subsection while keeping in mind the following fact: *if the empirical distribution is strictly positive, then the ML distribution exists.* Thus the use of the pseudo-empirical distributions described in Chapter II, Section B.3 removes the need to worry about existence. Note also that to use **LXM** for an MDI fit the empirical distribution must be strictly positive as a matter of definition.

For any $\boldsymbol{\theta}$, in the model

$$q^{\boldsymbol{\theta}}(y) = e^{\boldsymbol{\theta}'\mathbf{u}(y) - \delta(\boldsymbol{\theta})} q(y)$$

if $\boldsymbol{\theta} \neq \mathbf{0}$ define the *plateau*

$$\mathcal{Y}(\boldsymbol{\theta}) = \left\{ y \in \mathcal{Y} : \boldsymbol{\theta}'\mathbf{u}(y) = \max_{z \in \mathcal{Y}} \boldsymbol{\theta}'\mathbf{u}(z) \right\}$$

The main result of this subsection is that there is a ML distribution in the model if and only if $\hat{p}$ does not give probability 1 to any plateau.

First, suppose that $\hat{p}$ satisfies this condition. Then write for $\boldsymbol{\theta} \neq \mathbf{0}$

$$\delta(t\boldsymbol{\theta}) - t\boldsymbol{\theta}'\mathbf{u}(y) = \ln\left[\sum_{z \in \mathcal{Y}(\boldsymbol{\theta})} e^{t\boldsymbol{\theta}'[\mathbf{u}(z) - \mathbf{u}(y)]} q(z) + \sum_{z \notin \mathcal{Y}(\boldsymbol{\theta})} e^{t\boldsymbol{\theta}'[\mathbf{u}(z) - \mathbf{u}(y)]} q(z) \right]$$

If $y \in \mathcal{Y}(\boldsymbol{\theta})$ then the second sum in the brackets $\to 0$ as $t \to \infty$, while the first is identically $\Sigma\, q(z)$ for $z \in \mathcal{Y}(\boldsymbol{\theta})$. Therefore,

$$\delta(t\boldsymbol{\theta}) - t\boldsymbol{\theta}'\mathbf{u}(y) \to \ln \sum_{z \in \mathcal{Y}(\boldsymbol{\theta})} q(z)$$

On the other hand, if y is not in $\mathcal{Y}(\boldsymbol{\theta})$, then the first sum in the brackets $\to \infty$ as $t \to \infty$, and so

$$\delta(t\boldsymbol{\theta}) - t\boldsymbol{\theta}'\mathbf{u}(y) \to \infty$$

Since $\mathbb{D}_1(\hat{p}, q^{t\boldsymbol{\theta}}) - \mathbb{D}_1(\hat{p}, q) = 2\Sigma\, \hat{p}(y)[\delta(t\boldsymbol{\theta}) - t\boldsymbol{\theta}'\mathbf{u}(y)]$ and $\hat{p}$ does not give probability 1 to $\mathcal{Y}(\boldsymbol{\theta})$, we have $\mathbb{D}_1(\hat{p}, q^{t\boldsymbol{\theta}}) \to \infty$ as $t \to \infty$. It will be convenient now to think of the general parameter as being of the form $t\boldsymbol{\theta}$ with $\|\boldsymbol{\theta}\| = 1$ and $0 \leqq t < \infty$. Since $\mathbb{D}_1(\hat{p}, q^{t\boldsymbol{\theta}})$ is continuous in t and tends to ∞ as $t \to \infty$, it follows that it actually attains its smallest value for each fixed value of $\boldsymbol{\theta}$. If it assumes its smallest value at a $t > 0$, then the derivative must vanish at that point, which gives $\boldsymbol{\theta}'M[\mathbf{u} : \hat{p}] = \boldsymbol{\theta}'M[\mathbf{u} : q^{t\boldsymbol{\theta}}]$ at the minimizing point.

Let D be the largest of the values which are $\leqq \mathbb{D}_1(\hat{p}, q^{\boldsymbol{\theta}})$ for all $\boldsymbol{\theta}$. To avoid trivialities, assume that $\mathbb{D}_1(\hat{p}, q) > D$. We can then choose a sequence $\boldsymbol{\theta}_i$ with $\|\boldsymbol{\theta}_i\| = 1$ and a sequence t_i such that $\mathbb{D}_1(\hat{p}, q^{t_i\boldsymbol{\theta}_i}) \to D$, and clearly there is no loss in generality in supposing that t_i is the minimizing point corresponding to $\boldsymbol{\theta}_i$. The strategy is to show that $t_i\boldsymbol{\theta}_i$ (or some subsequence of it) converges, say to $t_0\boldsymbol{\theta}_0$, since then by the continuity of $\mathbb{D}_1$ it will have the value D at $t_0\boldsymbol{\theta}_0$.

Now since all of the coordinates of each $\boldsymbol{\theta}_i$ are bounded between -1 and 1, there is a subsequence of $\boldsymbol{\theta}_i$ which converges, say to $\boldsymbol{\theta}_0$. To avoid complicating the notation, we retain $\boldsymbol{\theta}_i$ for this subsequence. Since there are only finitely many plateaus, one of the plateaus $\mathcal{Y}(\boldsymbol{\theta}_i)$ must appear infinitely often; call it $\mathcal{Y}_0$. Again without disturbing the notation, pass to a subsequence for which $\mathcal{Y}(\boldsymbol{\theta}_i) = \mathcal{Y}_0$.

If t_i (or some subsequence of it) drifts out to ∞, then as at the beginning of the proof examine the behavior of $\delta(t_i\boldsymbol{\theta}_i) - t_i\boldsymbol{\theta}_i'\mathbf{u}(y)$ as $t_i \to \infty$ and $\boldsymbol{\theta}_i \to \boldsymbol{\theta}_0$. By passing to another subsequence if necessary, we can assume that each $q^{t_i\boldsymbol{\theta}_i}(y)$ converges, and it is then clear that this limiting distribution can give positive probability only to points in $\mathcal{Y}(\boldsymbol{\theta}_0) \supset \mathcal{Y}_0$. As a consequence

$$\boldsymbol{\theta}_i'M\left[\mathbf{u} : q^{t_i\boldsymbol{\theta}_i}\right] \to \max_{z\in\mathcal{Y}} \boldsymbol{\theta}_0'\mathbf{u}(z)$$

and since t_i is a sequence of minimizing points,

$$\boldsymbol{\theta}_0'M[\mathbf{u} : \hat{p}] \leftarrow \boldsymbol{\theta}_i'M[\mathbf{u} : \hat{p}] \to \max_{z\in\mathcal{Y}} \boldsymbol{\theta}_0'\mathbf{u}(z)$$

which implies that $\hat{p}$ gives probability 1 to the plateau $\mathcal{Y}(\boldsymbol{\theta}_0)$. This is contrary to hypothesis, and so we conclude that t_i has no subsequence drifting out to ∞. But then $t_i\boldsymbol{\theta}_i$ has all of its coordinates bounded, and so has a convergent subsequence, as was to be shown. This completes the proof that the condition on $\hat{p}$ implies the existence of the ML distribution.

To prove the converse, suppose that $\hat{\boldsymbol{\theta}}$ is the ML estimate. Assume that there is a $\boldsymbol{\theta} \neq \mathbf{0}$ for which $\hat{p}$ gives probability 1 to $\mathcal{Y}(\boldsymbol{\theta})$. This gives us

$$\boldsymbol{\theta}'M[\mathbf{u} : \hat{p}] = \max_{z\in\mathcal{Y}} \boldsymbol{\theta}'\mathbf{u}(z)$$

and from Proposition 19, condition 1, we infer

$$\boldsymbol{\theta}' M\left[\mathbf{u} : q^{\hat{\boldsymbol{\theta}}}\right] = \max_{z \in \mathcal{Y}} \boldsymbol{\theta}' \mathbf{u}(z)$$

and this implies that the ML distribution $q^{\hat{\boldsymbol{\theta}}}$ gives probability 1 to $\mathcal{Y}(\boldsymbol{\theta})$. But since the ML distribution is strictly positive, this is a contradiction. Thus the assumption about $\hat{p}$ must have been false, and this completes the proof.

Unfortunately it is true that even for models of only moderate complexity the condition on $\hat{p}$ is difficult to check because it is difficult to determine the collection of plateaus. However, it does suggest a way of proceeding when the ML distribution does not exist. Specifically, one might truncate $\mathcal{Y}$ to the smallest plateau to which $\hat{p}$ gives probability 1, and try to refit the model. It will be necessary to discard some of the components of **u** in order to have an identifiable parameter. The computational aspects of this approach are discussed in Chapter IV, Section A.2.

B. DATA EXAMPLES

The best way to understand how the theory of discrete linear statistical analysis is applied in actual situations is to study examples of its application. In this section we will analyze a number of data sets which occurred in real experiments or surveys, with the aim of illustrating how techniques discussed up to this point can be used. The reader is invited to repeat these analyses and to take them further when appropriate. To facilitate this, all LAMDA runs are listed in Appendix C. Many of the data sets will be reanalyzed in Chapter IV and in later sections.

Appropriate inferential techniques were outlined in Section A.5 for cases in which hypotheses about a model or about a particular distribution exist before the data are collected. It frequently happens that there are no prior hypotheses, and so the statistical analysis then consists of summarizing for the purpose of compact description of the data and generation of hypotheses for future testing. The ANALYSIS table is also used for this purpose, although then the P-values do not have their interpretation with regard to significance levels, but rather as statistics which were used to guide the process of choosing a model which fits the data reasonably well. We will try to maintain the distinction between inference and description, while using whichever technique seems most useful in a given situation.

1. One-Way Tables

In this section we will study several contingency tables involving a single factor.

Table III.8

Parameter Estimates for the SIDS Data

u-function	$\hat{\theta}$	$\hat{\theta}/\sqrt{\hat{V}[\hat{\theta}]}$
u_1	$-.3734$	-4.46
u_4	.1203	1.50
v_5	.0573	0.72
v_2	.0797	0.99

Using Periodic Functions. Reconsider the SIDS data presented in Table II.5. Again letting $q(y)$ be the population probability of being born in month y, we consider the model $\mathcal{E}[q: u_1, u_4, v_5, v_2]$. As we remarked before, the hypothesis-test assumptions have not been followed precisely, because the model was suggested by the data, and so we will proceed with the purposes in mind of description and generation of new hypotheses.

The ML parameter estimates appear in Table III.8. Dividing each parameter estimate by its estimated standard deviation yields a "t-value" which may in large samples be treated as a standard Normal observation, under the hypothesis that the true value of the associated parameter is zero. These estimated standard deviations are obtained by evaluating the expression for the covariance matrix of $\hat{\boldsymbol{\theta}}$ (given in Proposition 23) with π replaced by the ML distribution. From the analysis in Table III.9, it is clear that one should reject $\pi = q$, but not $\pi \in \mathcal{E}[q: u_1, u_4, v_5, v_2]$. The distributions q, $\hat{p}$ and $\hat{q}$ are displayed in Table III.10. In order to display the results of the fitted model, the ratio $\hat{q}/q$ could be taken as a measure of the incidence of SIDS, one which captures more of its systematic variation and less of the purely random variation than does the similar measure $\hat{p}/q$. The former values are plotted in Figure III.1, showing a low incidence in spring and summer with a high and variable incidence in fall and winter. In order to assess the model fit, the ratio $\hat{p}/\hat{q}$ could be used to see whether there are some trends which have been ignored in the model-selection process. There

Table III.9

Analysis of the SIDS Data

Model	df	χ^2	$\langle P \rangle$
$\mathcal{E}[q: u_1, u_4, v_5, v_2]$	7	2.685	$\langle .912 \rangle$
q	4	25.008	$\langle .00005 \rangle$

Table III.10

Month of Birth (**Y**),
Population Probability of Being Born in That Month (**Q**), *and Empirical Probability* (**P − HAT**) *and Fitted Probability* (**Q − HAT**) *of Birth Month for* SIDS *Infants*

L A M D A --- VERSION 4.0

*** RUN III.B.2 ***

Y	Q	P-HAT	Q-HAT
1.0000	0.0770	0.0898	0.0849
2.0000	0.0763	0.0681	0.0669
3.0000	0.0863	0.0433	0.0535
4.0000	0.0816	0.0588	0.0499
5.0000	0.0854	0.0557	0.0616
6.0000	0.0837	0.0712	0.0658
7.0000	0.0860	0.0867	0.0846
8.0000	0.0858	0.1115	0.1206
9.0000	0.0850	0.0836	0.0950
10.0000	0.0846	0.1207	0.1091
11.0000	0.0807	0.1176	0.1176
12.0000	0.0875	0.0929	0.0905

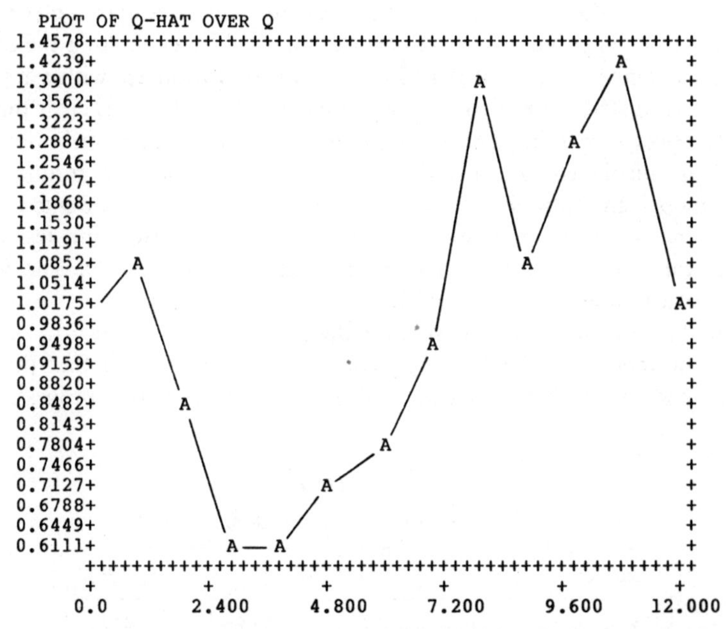

Fig. III.1 Plot of the ratio of the fitted probability of being born in a particular month for an SIDS *infant to the population probability of being born in that month, against birth month (Jan. = 1, Dec. = 12).*

does not appear to be any clear trend in Figure III.2 where these ratios are displayed, but the extreme March–April observations might raise some questions.

The LAMDA program which fitted the model appears in Run III.B.1. The N-file read at line 3 contains the sines and cosines, in the order u_1–u_6, alternating with lines of v_1–v_6. Thus each y-value corresponds to two rows in the N-file. At line 4 a check is made that the correct number of lines was read, and at line 7 the C-file is re-dimensioned. Before this line was executed the C-file had 24 rows and 6 columns, and afterwards it had 12 rows and 12 columns. As always with re-dimensioning, the data is read row-wise into the re-dimensioned C-file, so that the first two 6-entry rows are combined into one 12-entry row. At line 8 blank variables are supplied in columns 1 and 14, and then in lines 10–12 the population frequencies are read into variable 1 and in lines 13–15 the observed frequencies are read into variable 14. After some displaying of the functions, at line 24 the population frequencies are declared to be the generating distribution, and at line 27 the observed frequencies are specified as the weights. The u-functions to be included in the model are given at line 26, and the instruction to fit the model is at line 28. The program which produced the plots is given in Run III.B.2.

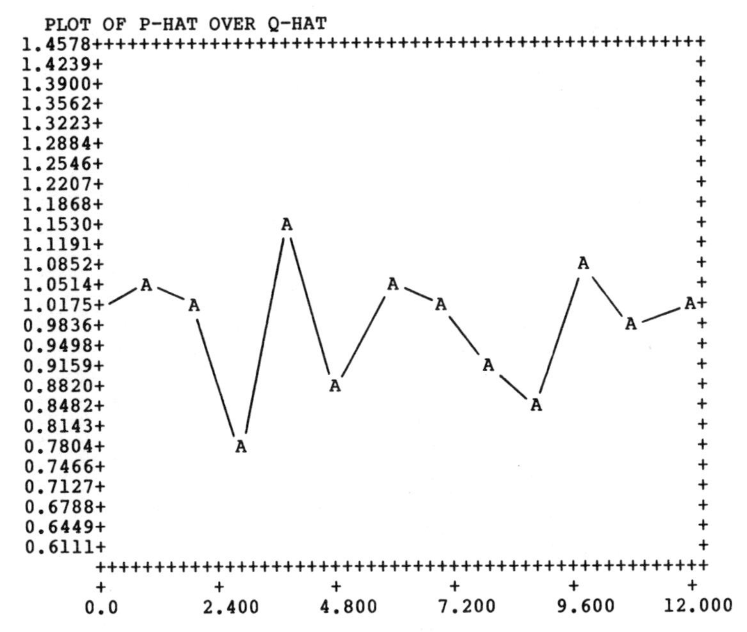

Fig. III.2 Plot of the ratio of the empirical probability of being born in a particular month for an SIDS *infant to the fitted probability, against birth month (Jan. = 1, Dec. = 12).*

The output from the **LXM** command appears in Figure III.3, and this is an appropriate place to indicate what various parts of the output mean. The ANALYSIS table appears on lines 9–12, where **D** stands for $\mathbb{D}_1$, **P** stands for the empirical distribution, **Q** stands for the generating distribution, and **QP** stands for the fitted distribution. Under **COEFFICIENTS** we see $\hat{\theta}$ written as a row vector, and under **T-VALUES** the parameter estimates divided by their estimated standard deviations. In lines 22–35 we find one line for each cell in the table, with the generating frequencies nq (**GEN**), the observed frequencies $n\hat{p}$ (**OBS**), and the fitted frequencies $n\hat{q}$ (**FIT**). Under **RES** we see the residuals, defined as observed minus fitted frequency, and then some standardized residuals which are explained in Chapter IV. In lines 37–42 we see the covariance matrix (estimated) of $\hat{\theta}$, in lower-triangular form.

In lines 3–7 appear some other pieces of information. In line 3 is the "unlikelihood" of the model fit, which is explained in Chapter V. The two **DELTA**s refer to measures of how well the iterative computational algorithm

```
 1:  FIT MODEL WITH SIN1,SIN4,COS5,COS2
 2:  === LXM ===
 3:  U(P,QP)    =      1580.795
 4:  DELTA(C) =  0.00000040
 5:  DELTA(U) =  0.00024414
 6:   NO OF CASES =    323.000
 7:   4 ITERATIONS
 8:
 9:  - - - A N A L Y S I S - - -
10:  D(P,QP)    =          2.685
11:  D(QP,Q)    =         25.008
12:  D(P,Q)     =         27.693
13:
14:  - - - COEFFICIENTS - - -
15:
16:       -0.3734       0.1203       0.0573       0.0797
17:
18:  - - - T-VALUES - - -
19:
20:       -4.4590       1.5001       0.7186       0.9926
21:
22:  POS     GEN        OBS        FIT        RES        F-RES  G-RES  RES-F  RES-G
23:
24:   1    24.885     29.000     27.437      1.563      0.304  0.509  0.312  0.532
25:   2    24.647     22.000     21.607      0.393      0.087 -0.677  0.088 -0.637
26:   3    27.868     14.000     17.275     -3.275     -0.895 -2.620 -0.810 -2.099
27:   4    26.368     19.000     16.113      2.887      0.683 -2.621  0.738 -2.084
28:   5    27.596     18.000     19.897     -1.897     -0.460 -1.782 -0.439 -1.532
29:   6    27.033     23.000     21.247      1.753      0.379 -1.299  0.394 -1.163
30:   7    27.783     28.000     27.317      0.683      0.135 -0.093  0.137 -0.093
31:   8    27.698     36.000     38.952     -2.952     -0.522  1.923 -0.504  2.236
32:   9    27.459     27.000     30.691     -3.691     -0.742  0.613 -0.700  0.645
33:  10    27.323     39.000     35.234      3.766      0.643  1.412  0.672  1.582
34:  11    26.062     38.000     37.993      0.007      0.001  2.061  0.001  2.438
35:  12    28.277     30.000     29.237      0.763      0.146  0.186  0.148  0.189
36:
37:  - - - COVARIANCE MATRIX - - -
38:
39:      0.00701
40:      0.00013      0.00643
41:      0.00028     -0.00125      0.00635
42:     -0.00121      0.00001     -0.00016      0.00645
```

Fig. III.3 The LAMDA *output for the* **LXM** *command for analyzing the* SIDS *birth-month data.*

converged. These are explained in Chapter VI, Section A, and so long as they are both close to zero the output can be trusted. The **NO OF CASES** is the number of observations specified in the weight variable, and **ITERATIONS** gives the number of iterations required to attain convergence of the approximations to the fitted distribution.

Using Indicators. In Table III.11 the $nq(y)$ row gives the age distribution of a small Eskimo population, and the $n\hat{p}(y)$ row gives the age distribution of polio cases during an epidemic in 1949. If we imagine that the cases consist of those who reported to a doctor, so that it is not known whether those who failed to report did or did not contract polio, then it would be appropriate to ask whether $\pi = q$, where π refers to the true age distribution of reported polio cases and q refers to the population age distribution.

In the absence of any particular theory for the relationship between age and susceptibility to polio in this population, we begin by fitting the exponential model with generating distribution q and u-functions $u_i(y) = 1$ when $y = i$ ($= 0$ otherwise) for $1 \leqq i \leqq 6$. This model is saturated, which means that every element of $\mathcal{P}$ is in the model. This is not for the purpose of model fitting, of course (since $\hat{p}$ is the ML distribution), but to view the parameter estimates in search of a reasonable model. The estimates and the t-values appear in Table III.12. The fact that each of the t-values is statistically significant does not mean that all the u-functions are required for a well-fitting model. Obviously,

$$\hat{\theta}_i = \ln\frac{\hat{p}(i)}{q(i)} - \ln\frac{\hat{p}(0)}{q(0)}$$

so that the evidence points to $\pi(y)/q(y) \neq \pi(0)/q(0)$ for all $y \neq 0$. Examination of the values of $\hat{\theta}_i$ suggests that

$$\frac{\pi(1)}{q(1)} = \frac{\pi(2)}{q(2)} = \frac{\pi(4)}{q(4)} = \frac{\pi(5)}{q(5)} = \frac{\pi(6)}{q(6)}$$

Table III.11

Age Distribution of Polio Cases in an Eskimo Population

Age	0–4	5–9	10–14	15–19	20–29	30–49	50–
y	0	1	2	3	4	5	6
$nq(y)$	53	56	33	26	30	52	25
$n\hat{p}(y)$	2	13	8	11	6	12	5

Table III.12

Parameter Estimates for the Eskimo Polio Data

i	1	2	3	4	5	6
$\hat{\theta}_i$	1.82	1.86	2.42	1.67	1.81	1.67
t_i	2.39	2.35	3.14	2.04	2.37	1.99

and so we might fit $\mathfrak{S}[q: v_1, v_2]$, where v_1 is the indicator of {3} and v_2 is the indicator of {1, 2, 4, 5, 6}, since every element of this model has the property of the equation given above for π. The parameter estimates are $\hat{\theta}_1 = 2.4169$ $\langle .001 \rangle$ and $\hat{\theta}_2 = 1.7832$ $\langle .007 \rangle$ and we have the ANALYSIS of Table III.13. Since the model was suggested by the data, we regard this as a description of the data rather than an inferential analysis. The LAMDA programs are Run III.B.3 and Run III.B.4.

Using Polynomials. The data in Table III.14 are from a diphtheria outbreak in 1970. We might try to fit the Finite Gamma model (Chapter III, Section A.1) to these data, but in order to allow departures from the model

Table III.13

Analysis of the Eskimo Polio Data

Model	df	χ^2	$\langle P \rangle$
$\mathfrak{S}[q: v_1, v_2]$	4	0.222	$\langle .992 \rangle$
q	2	15.906	$\langle .00035 \rangle$

Table III.14

Age Distribution of Diphtheria Cases

Age	y	$n\hat{p}$	$\hat{p}$	$\hat{q}$
1	1	2	.0102	.0100
1–4	2	32	.1633	.1614
5–9	3	53	.2704	.2868
10–14	4	55	.2806	.2432
15–19	5	23	.1173	.1510
20–29	6	18	.0918	.0809
30–39	7	7	.0357	.0400
40–49	8	5	.0255	.0186
50	9	1	.0051	.0081

we will fit $\mathfrak{S}[q: u_0, u_1, u_2, u_3]$, in which q is the finite gamma

$$q(y) = \frac{ye^{-y}}{\sum_z ze^{-z}}$$

$u_0(y) = \ln y$, and $u_i(y) = (-y)^i$ for $i = 1, 2, 3$. The LAMDA program is Run III.B.5, and the resulting analysis from the first **LXM** command is in Table III.15. Since the values of t_2 and t_3 are so modest, we might well try to fit $\mathfrak{S}[q: u_0, u_1]$. This is done by the second **LXM** command, with the results in Table III.16. We would confirm $\pi \in \mathfrak{S}[q: u_0, u_1]$ and estimate π with the Finite Gamma

$$\hat{q}(y) = \frac{y^{5.5612}e^{-1.6717y}}{\sum_z z^{5.5612}e^{-1.6717z}}$$

which is also given in Table III.14.

The advantage in having fitted this model to the data is that now we have a two-parameter description of the age distribution of diphtheria cases, rather than the eight-parameter description provided by the raw data. In addition to the simplification, there is another benefit to be obtained if this distribution were to be compared to a similar distribution from the same population. For example, if the age distribution of another epidemic in this population could also be fitted by the two-parameter Finite Gamma model, then the test of equality of age distributions would be based on a 2-df test rather than on an 8-df test. As we have seen in Section A.2, the test based on the smaller df is the more powerful.

Table III.15

Analysis and Parameter Estimates for the Diphtheria Data

Model	df	χ^2	$\langle P \rangle$
$\mathfrak{S}[q: u_0, u_1, u_2, u_3]$	4	3.932	$\langle .416 \rangle$
q	4	299.909	$\langle .000 \rangle$

i	0	1	2	3
$\hat{\theta}_i$	7.9239	3.0251	0.2261	0.0076
t_i	1.83	0.84	0.49	0.33

Table III.16

Analysis and Parameter Estimates for the Diphtheria Data

Model	df	χ^2	$\langle P \rangle$
$\mathcal{G}[q: u_0, u_1]$	6	5.660	$\langle .463 \rangle$
q	2	298.181	$\langle .000 \rangle$
i	0	1	
$\hat{\theta}_i$	4.5612	0.6717	
t_i	6.85	3.74	

Another issue in fitting a model to these data is whether the coding of the Finite Gamma is sensible. One might argue that the age-class boundaries are to a certain extent arbitrary, and it would be better to use actual age (rather than age category) in model fitting. Thus in place of u_0 and u_1 one might take $u_1(y) =$ the midpoint of the yth age category [with $u_1(9) = 55$, say] and $u_0(y) = \ln u_1(y)$. This would be especially useful if a comparison epidemic involved age categories with boundaries different from those used in the diphtheria data.

Finally, we note that this analysis was undertaken without regard to the population age distribution, and so it would be inappropriate to use the fitted Finite Gamma distribution in making comparisons with data from another population having a different age structure.

Symmetry. There are a number of models which specify certain kinds of symmetry in a contingency table. The simplest involves a one-way table, an example of which appears in Table III.17. Here each subject was asked to report how many events of a certain type occurred to him or her in the past year, and then one year later was asked to recall again how many events had occurred during the period in question. The responses were coded as $y =$ (second response) − (first response), and since the possible responses were 0, 1, 2, and 3, y assumes the values shown in Table III.17. The model consisting of all symmetric distributions is $\{p: M[u_i: p] = M[u_i: q],\ i = 1, 2, 3\}$, a linear model which could be written $\mathcal{L}[q: \mathbf{w}]$ for an appropriate $\mathbf{w}$ having three components. In the analysis (Table III.18) we have augmented the basic ANALYSIS table by including a column for the serial P-value.

We might confirm the symmetry model while certainly rejecting $\pi = q$. Since $y > 0$ indicates that the person "recalled" an event not reported earlier, while $y < 0$ stands for "forgetting" an earlier-reported event, by

Table III.17

The Symmetric Functions on a One-Way Table

y	u_1	u_2	u_3	q	$n\hat{p}$
−3	0	0	−1	.05	4
−2	0	−1	0	.10	8
−1	−1	0	0	.20	18
0	0	0	0	.30	40
1	1	0	0	.20	10
2	0	1	0	.10	2
3	0	0	1	.05	3

confirming the symmetry model we conclude that there is no net tendency to make one of these types of errors more than the other.

The LAMDA run for the analysis of these data is Run III.B.6. The generating distribution (variable 5) is to play the role of "data" here, so that at line 15 it is multiplied by the sample size. At line 17 the observed frequencies are identified as the generating distribution, while the following line identifies variable 5 as the "data." Note that since we are fitting an $\mathfrak{L}$ model, the **COEFFICIENTS**, **T-VALUES** and **COVARIANCE MATRIX** on the output should be ignored.

2. Multi-way Tables

Statistical analysis based on a fitted distribution is much more widely used for investigating the association of several factors than for modeling a distribution along one factor, as was done in the preceding section. Here we will consider a number of multi-way tables with the aim of illustrating some of the possibilities inherent in the exponential and linear models for several factors.

We will use the notation of Chapter I with regard to factors and levels. Thus A, B and C stand for factors, while Ai, Bj and Ck stand for levels of

Table III.18

Analysis of Symmetry on the One-Way Table

Model	df	χ^2	$\langle P \rangle$	$\langle \bar{P} \rangle$
$\mathfrak{L}[q:\mathbf{w}]$	3	6.605	$\langle .084 \rangle$	$\langle .084 \rangle$
q	3	15.637	$\langle .001 \rangle$	$\langle .002 \rangle$

these factors. This simplifies the writing of the factorial indicators $\mathbf{u}_A$, $\mathbf{u}_B$ and $\mathbf{u}_C$, and permits us in many cases to reduce certain model descriptions to a shorthand involving only the letters of the factors whose u-functions appear in the model. Thus, for example, the exponential model generated by q and containing all the indicators of the form u_{Ai} and u_{Bj} would be called "$A\ B$." Recall that to maintain affine linear independence of the components of the u-functions, it is necessary to delete the indicator of one of the levels of each factor (which we have usually designated $A0$ and $B0$). The more complex model containing in addition the products $u_{AiBj} = u_{Ai}u_{Bj}$ would be written as either "$A\ B\ AB$," or the more succinct $[AB]$. The brackets mean that all products involving subsets of the factors appear in the model. For example, $[ABC]$ means the same thing as the saturated model "$A\ B\ C\ AB\ AC\ BC\ ABC$," and $[AB][BC]$ means the same as "$A\ B\ C\ AB\ BC$." We will use "0" for the model which contains the generating distribution as its only element.

2^2 ***Independence.*** In Chapter II, Section B.5 we saw a 2^2 contingency table whose factors were Tenure (present or absent) and Doctorate (present or absent). We recast the setup for analyzing this data in Table III.19. Notice that the factors T (Tenure) and D (Doctorate) are independent under q, and that as a consequence $\mathcal{E}[q: u_D, u_T]$ contains all the distributions under which T and D are independent (see Section A.1). Thus the test of $\pi \in \mathcal{E}[q: u_D, u_T]$ is a test of the independence of T and D.

The independence model on a 2^2 table is so simple that it is not necessary to use a computer program to carry out the analysis. We will show how the computations go, while advocating the use of Run III.B.7 because it is no harder than doing the computations by hand and is somewhat less error-prone.

One easily verifies that the distribution

$$\hat{q}(ij) = \hat{p}(i+)\hat{p}(+j)$$

Table III.19

Indicators for the Independence Model on a Two-Way Table

y	u_D	u_T	q	$n\hat{p}$
00	0	0	.046	18
01	0	1	.119	50
10	1	0	.232	78
11	1	1	.604	373

renders D and T independent and has the same marginal distributions as $\hat{p}$. Since $M[u_D: \hat{p}]$ and $M[u_T: \hat{p}]$ determine those marginal distributions, it follows from Proposition 19 that $\hat{q}$ is the ML distribution. In order to compute and interpret the parameters, begin with

$$q^{\boldsymbol{\theta}} = e^{\theta_T u_T + \theta_D u_D - \delta(\boldsymbol{\theta})} q$$

and compute

$$\ln \frac{q^{\boldsymbol{\theta}}(10)}{q^{\boldsymbol{\theta}}(00)} = \theta_D + \ln \frac{q(10)}{q(00)}$$

$$\ln \frac{q^{\boldsymbol{\theta}}(01)}{q^{\boldsymbol{\theta}}(00)} = \theta_T + \ln \frac{q(01)}{q(00)}$$

so that

$$\hat{\theta}_D = \ln \frac{\hat{q}(10)\, q(00)}{\hat{q}(00)\, q(10)} = .2687$$

$$\hat{\theta}_T = \ln \frac{\hat{q}(01)\, q(00)}{\hat{q}(00)\, q(01)} = .5270$$

From

$$C\left[\begin{pmatrix} u_D \\ u_T \end{pmatrix} : q^{\boldsymbol{\theta}}\right] = \begin{bmatrix} q^{\boldsymbol{\theta}}(1+)\,q^{\boldsymbol{\theta}}(0+) & 0 \\ 0 & q^{\boldsymbol{\theta}}(+1)\,q^{\boldsymbol{\theta}}(+0) \end{bmatrix}$$

using Proposition 23 we may estimate the covariance matrix of $\hat{\boldsymbol{\theta}}$ by

$$\frac{1}{n} C\left[\begin{pmatrix} u_D \\ u_T \end{pmatrix} : \hat{q}\right]^{-1} = \begin{bmatrix} \dfrac{1}{n\hat{q}(1+)\,\hat{q}(0+)} & 0 \\ 0 & \dfrac{1}{n\hat{q}(+1)\,\hat{q}(+0)} \end{bmatrix}$$

This leads to the t-values $t_D = 2.065\ \langle .039 \rangle$ and $t_T = 4.662\ \langle .000 \rangle$. The analysis is in Table III.20, and we might confirm the independence of D and T but reject $\pi = q$. Since

$$q^{\boldsymbol{\theta}}(1+) = \frac{e^{\theta_D} q(1+)}{q(0+) + e^{\theta_D} q(1+)}$$

Table III.20

Analysis of the 2^2 Table

Model	df	χ^2	$\langle P \rangle$
DT	1	2.979	$\langle .085 \rangle$
0	2	28.731	$\langle .000 \rangle$

and the expression on the right equals $q(1+)$ only when θ_D is zero, we see that θ_D reflects a departure from the q-marginal D-distribution, and by similar calculations θ_T reflects a departure from the q-marginal T-distribution. Consequently, we may say that the evidence is in favor of a difference between the π- and q-marginal T-distributions.

In Run III.B.7 we first put u_D and u_T in two columns, then enlarge the C-file and put the observed frequencies in column 1 and the generating distribution in column 2. At line 12 the **G = V** command establishes q as the generating distribution, and then u_D and u_T (variables 3 and 4) are **SEL**ected as the u-functions for the **LXM** command.

3×4 and 3×2^2 Independence. The data in Table III.21 are from a sample of women during the Great Depression, and concern the factors

E = Amount of full-time employment
- $E0$ = None
- $E1$ = 1–4 years
- $E2$ = > 4 years

C = Class
- $C0$ = Middle Class
- $C1$ = Working Class

D = Deprivation
- $D0$ = Not Deprived
- $D1$ = Deprived

To begin with, we may consider (C, D) to be a single factor having four levels, so that this is a 3×4 table. Letting q be the uniform distribution, the model $\mathcal{E}[q: u_{E1}, u_{E2}, u_{C1}, u_{D1}, u_{C1D1}]$ (that is, $E[CD]$) consists of all distributions under which E and (C, D) are independent factors. It is easy to verify that the ML distribution is $\hat{q}(ijk) = \hat{p}(i++)\hat{p}(+jk)$, and from Table III.22 we confirm that the independence model fits the data.

Table III.21

A Three-Way Table Relating Employment, Class, and Deprivation

	$C0$		$C1$	
	$D0$	$D1$	$D0$	$D1$
$E0$	4	5	2	3
$E1$	7	8	5	13
$E2$	2	7	4	9

Now considering the C and D factors to be separate, the model $\mathfrak{S}[q: u_{E1}, u_{E2}, u_{C1}, u_{D1}]$ (that is, $E\,C\,D$) is the model containing all distributions under which E, C, and D are independent. This model is more restrictive than the preceding one, reflecting the fact that independence of E, C, and D implies independence of E and (C, D). The ML distribution under this model is

$$\hat{q}(ijk) = \hat{p}(i++)\hat{p}(+j+)\hat{p}(++k)$$

and from Table III.23 we would confirm independence. The computation was done by Run III.B.8.

2^3 Conditional Independence. Surveys carried out in 1975 and 1976 asked individuals whether they favored registration of guns. The question was asked either in a neutral form ($F0$) or in a form ($F1$) slanted against gun registration. The data appear in Table III.24, where $G0$ = Opposes Gun Registration and $G1$ = Favors Gun Registration.

We do not expect these three factors to be independent, but it is of interest in examining time trends of opinion to see whether G and Y are independent within levels of F. That is, if we let ijk denote the cell $GiYjFk$,

Table III.22

Analysis of the Three-Way Table as a Two-Way Table

Model	df	χ^2	$\langle P \rangle$
$E[CD]$	6	3.924	$\langle .689 \rangle$
0	5	15.188	$\langle .010 \rangle$

Table III.23

Analysis and Parameter Estimates for the Three-Way Table

	Model	df	χ^2	$\langle P \rangle$
	ECD	7	4.517	$\langle .721 \rangle$
	0	4	14.595	$\langle .006 \rangle$
i	$E1$	$E2$	$C1$	$D1$
$\hat{\theta}_i$	0.8575	0.4520	0.0871	0.6286
t_i	2.69	1.32	0.36	2.49

we want to ask whether G and Y are conditionally independent given F:

$$\frac{\pi(ijk)}{\pi(++k)} = \frac{\pi(i+k)}{\pi(++k)} \frac{\pi(+jk)}{\pi(++k)}$$

The exponential model which embodies this condition is $[GF][YF]$, and Run III.B.9 produced the analysis of Table III.25. We conclude that π lies in the model, and thus that attitude toward gun registration did not change over the one-year period within subsamples who were administered the same form of the question.

2^4 *Constant Odds Ratio.* In any 2^2 table we know that the odds ratio $p(00)p(11)/[p(01)p(10)]$ may be considered a measure of the interdependence of the row and column factors, since it is 1 only when they are independent. In Table III.26 we have four 2^2 tables relating marijuana use ($M0$ = Used, $M1$ = Not Used) and religion ($R1$ = Protestant or Catholic,

Table III.24

Attitude Towards Gun Registration Related to Year and Whether the Question was Slanted

	$Y0 = 1975$		$Y1 = 1976$	
	$F0$	$F1$	$F0$	$F1$
$G0$	126	141	152	182
$G1$	319	290	463	403

Table III.25

Analysis of the Gun-Registration Data

Model	df	χ^2	$\langle P \rangle$
$[GF][YF]$	2	2.015	$\langle .366 \rangle$
0	5	439.239	$\langle .000 \rangle$

$R0$ = Other). These tables are themselves indexed by the factors of geographical region ($G0$ = San Francisco, $G1$ = Contra Costa) and family status ($F1$ = Married with Children, $F0$ = Unmarried or Without Children). The model which stipulates a constant odds ratio in each of the four RM-subtables is $[FGM][FGR][RM]$. It takes a moment to check this assertion, but basically it comes down to seeing that θ_{RM} is the ln odds ratio in each RM-subtable under the model, and the inclusion of u_{GRM}, u_{FRM}, or u_{GRFM} in the model would permit the odds ratio to vary among these subtables (see Problem III.21).

Run III.B.10 produced the analysis in Table III.27. If we had expected to see the same relationship between religion and marijuana use, then we might have set the level of the test at .01, and thus we would confirm the hypothesis that the model fits. Of course, if we were model-fitting for the purpose of describing the data, we would probably conclude that the model was not adequate. In any case, the estimate of the supposed common value of the ln odds ratio in the four subtables is $\hat{\theta}_{RM} = 1.0757$, with $t_{RM} = 5.84$ $\langle .000 \rangle$. These data will be reanalyzed in Chapter V, Section B.5.

4^2 Marginal Homogeneity. In Table III.28 we see the responses of a sample of lung patients to a question about the severity of their dyspnea

Table III.26

The Marijuana Data

		$G0$		$G1$	
		$M0$	$M1$	$M0$	$M1$
$F0$	$R0$	52	35	23	23
	$R1$	37	130	69	109
$F1$	$R0$	3	15	12	35
	$R1$	9	67	17	136

Table III.27

Analysis of the Marijuana Data

Model	df	χ^2	$\langle P \rangle$
$[FGM][FGR[RM]$	3	8.528	$\langle .036 \rangle$
0	12	526.269	$\langle .000 \rangle$

(shortness of breath.) This question was asked once upon the patient's entry into the study ($Z0$ = None, $Z1$ = Mild, $Z2$ = Moderate, and $Z3$ = Severe), and again one year later (Y, with the same levels). One hypothesis of interest here is whether the Z-marginal distribution and Y-marginal distribution coincide. Letting ij stand for $ZiYj$, this amounts to $\pi(i+) = \pi(+i)$ for $i = 0, 1, 2, 3$. Defining $u_{Hi} = u_{Yi} - u_{Zi}$ ($i = 0, 1, 2, 3$), we see that the model of marginal homogeneity can be expressed as $\{p : M[\mathbf{u}_H : p] = M[\mathbf{u}_H : q]\}$, where q is the uniform distribution (or in fact any distribution having the same Z- and Y-marginals). It is not hard to check that $M[u_{Hi} : p] = 0$ for $i = 0, 1, 2$ implies that $M[u_{H3} : p] = 0$, and so we should redefine

$$\mathbf{u}_H = \begin{pmatrix} u_{H0} \\ u_{H1} \\ u_{H2} \end{pmatrix}$$

We have seen how linear models are analyzed in Chapter III, Section A.5, and the LAMDA program for the current analysis appears in Run III.B.11. Note that since $\hat{p}$ is to be the generating distribution, it must be in $\mathcal{P}$, and so we modify $\hat{p}$ by adding the arbitrary 0.5 to each of the empty cells in the original table of frequencies. The analysis is in Table III.29, and so we confirm the H-model and conclude that there is not evidence of a change in the distribution of dyspnea over the one year period. The reader may wish

Table III.28

Severity of Dyspnea Initially (Z) and Upon Follow-up (Y)

	$Y0$	$Y1$	$Y2$	$Y3$
$Z0$	5	5	4	1
$Z1$	8	11	4	3
$Z2$	0	3	8	5
$Z3$	0	1	5	9

Table III.29

Analysis of Dyspnea Data

Model	df	χ^2	$\langle P \rangle$
H	3	3.119	$\langle .374 \rangle$
0	12	46.477	$\langle .000 \rangle$

to examine the **LXM** output to see that the fitted Z- and Y-marginals are the averages of the observed Z- and Y-marginals.

5^2 Paired Comparisons. The data in Table III.30 were obtained by choosing 200 entries at random from a university course evaluation book. In each entry the instructor and course were rated in five areas:

$A0$ = Motivation of Students,
$A1$ = Clarity,
$A2$ = Worth of Exams,
$A3$ = Worth of Text,
$A4$ = Emphasis of Important Material.

The 200 entries were broken into groups of 20, each group was assigned to a pair of the A-factors, and then the number of times one of these factors exceeded the other was recorded. For example, of the 20 times $A0$ and $A1$ were compared, $A1$ was rated higher 19 times and $A0$ was rated higher one time, as can be seen from the 01 and 10 cells of the table. Of course, the diagonal cells are not possible.

Table III.30

The Paired-Comparison Data

		Rated Higher				
		$A0$	$A1$	$A2$	$A3$	$A4$
Rated Lower	$A0$	—	19	18	16	20
	$A1$	1	—	9	9	10
	$A2$	2	11	—	15	13
	$A3$	4	11	5	—	8
	$A4$	0	10	7	12	—

This is called a *paired-comparison* experiment, because the underlying data consist of pairs with a designated "winner" and "loser" in each pair. A measure of the tendency for Aj to "win" over Ai is $\ln[p(ij)/p(ji)]$, which is the ln odds favoring Aj given that the comparison is between Ai and Aj. The model we have in mind for these data says that it is possible to order the factors $A0, \ldots, A4$ according to their "strengths," as measured by the above ln odds.

This notion is embodied in the assumption that there are parameters γ_i such that

$$\ln\frac{p(ij)}{p(ji)} = \gamma_j - \gamma_i$$

In order to have the γ's identifiable, we must impose a restriction, and the one we will choose is $\gamma_0 = 0$, so that $\ln[p(0i)/p(i0)] = \gamma_i$ determines the γ's. Thus the γ's can be thought of as scale measures of the strengths of the factors with the strength of $A0$ arbitrarily taken as the reference point 0. In order to fit this model we must find a maximal set of u-functions which does not destroy the relations

$$\ln\frac{p(ij)}{p(ji)} = \ln\frac{p(ik)}{p(ki)} + \ln\frac{p(kj)}{p(jk)}$$

which are equivalent to the existence of the γ's. If we denote the above relation by $R(ijk)$, then the complete set of restrictions is equivalent to $R(210)$, $R(310)$, $R(410)$, $R(320)$, $R(420)$, and $R(430)$, since the others may be obtained by addition and subtraction. Thus the model imposes six restrictions. An appropriate set of functions is the symmetries

$$u_{Sij}(i'j') = \begin{cases} 1 & \text{if} \quad \{i, j\} = \{i', j'\} \\ 0 & \text{otherwise} \end{cases}$$

and the functions

$$u_{Bi} = u_{Ci} - u_{Ri}, \qquad i = 1, 2, 3, 4$$

where u_{Ci} is the indicator of the ith column and u_{Ri} is the indicator of the ith row of the table. Since $\sum u_{Sij} = 1$, we must discard one of the symmetries in determining $\mathbf{u}_S$, and then $\#\mathbf{u}_S = 9$, $\#\mathbf{u}_B = 4$, and $\#\mathcal{Y} = 20$, so that the df for testing the associated model are six, and this checks with our earlier observation about the number of restrictions imposed by the model. It is true in general that the symmetries and differences of corresponding row and column indicators determine the paired-comparison model for tables of any size.

The LAMDA program is in Run III.B.12. At lines 14–17 the $\mathbf{u}_B$-function is created, and then the symmetries are entered directly in lines 21–74. Note that when $n\hat{p}$ is entered (lines 76–80) the diagonal cells are entered as -1. The **BDY** at line 81 omits all rows with negative values in column 1.

The output appears in Table III.31 and shows that the model fits ⟨.150⟩. The first four parameters $\hat{\theta}_{Bi}$ ($i = 1, 2, 3, 4$) and their estimated covariance matrix are also shown in this table. We can compute

$$\gamma_i = \ln \frac{p(0i)}{p(i0)} = 2\theta_{Bi}$$

so that the relationships between the θ's and γ's and all differences among γ's are given by

$$\begin{pmatrix} \gamma_1 \\ \gamma_2 \\ \gamma_3 \\ \gamma_4 \\ \gamma_2 - \gamma_1 \\ \gamma_3 - \gamma_2 \\ \gamma_3 - \gamma_1 \\ \gamma_4 - \gamma_3 \\ \gamma_4 - \gamma_2 \\ \gamma_4 - \gamma_1 \end{pmatrix} = 2 \begin{bmatrix} 1 & 0 & 0 & 0 \\ 0 & 1 & 0 & 0 \\ 0 & 0 & 1 & 0 \\ 0 & 0 & 0 & 1 \\ -1 & 1 & 0 & 0 \\ 0 & -1 & 1 & 0 \\ -1 & 0 & 1 & 0 \\ 0 & 0 & -1 & 1 \\ 0 & -1 & 0 & 1 \\ -1 & 0 & 0 & 1 \end{bmatrix} \boldsymbol{\theta}_B$$

Table III.31

Analysis of the Paired-Comparison Data

Model	df	χ^2	⟨P⟩
SB	6	9.429	⟨.150⟩
0	13	68.781	⟨.000⟩

i	$B1$	$B2$	$B3$	$B4$
$\hat{\theta}_i$	1.2508	0.9682	1.2747	1.2508
t_i	5.64	4.50	5.73	5.64
	.04921			
	.03565	.04638		
	.03742	.03571	.04949	
	.03732	.03565	.03742	.04921

Table III.32

Differences and P-Values for the Differences Between γ_j (j = Column) and γ_i (i = Row)

	1	2	3	4
0	2.506 ⟨.000⟩	1.936 ⟨.000⟩	2.550 ⟨.000⟩	2.502 ⟨.000⟩
1		−0.566 ⟨.070⟩	0.048 ⟨.876⟩	0.000 ⟨1.00⟩
2			0.614 ⟨.025⟩	0.566 ⟨.070⟩
3				−0.048 ⟨.876⟩

From this equation and the covariance matrix of the $\hat{\theta}$'s we can obtain the covariance matrix of the $\hat{\gamma}$'s and all differences of $\hat{\gamma}$'s, and thus compute t-values and P-values for all these latter statistics. The results appear in Table III.32. If we adopt the arbitrary .05 cutoff point for P-values, we assert differences in strengths as indicated by the solid arrows in Figure III.4, with the dashed arrows indicating the direction of nonsignificant differences. Thus we would conclude that the worth of the text ($A3$) was rated higher than the worth of the exams ($A2$), which was in turn rated higher than motivation ($A0$). All the other factors were rated higher than motivation.

The above model is very useful in any situation in which the data consist of contests between pairs, such as a comparison of two therapeutic agents within pairs of patients who are matched with regard to relevant characteristics, or athletic contests involving two teams or players.

3 × 6 Linear Model with an Ordinal Factor. The data in Table III.33 consist of the subsample of smoking males in a study of arteriosclerosis. The

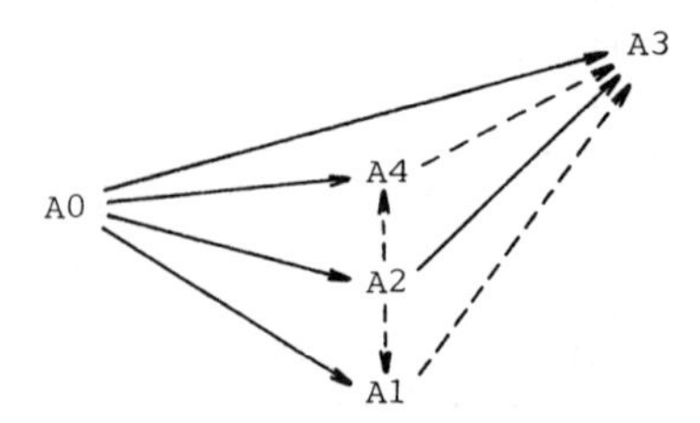

Fig. III.4 Ordering of aspects of course rating, according to which tended to be rated higher than others (see text for definition of the symbols).

Table III.33

Data Relating Smoking, Arteriosclerosis, and Age

	A0		A1		A2	
	D0	D1	D0	D1	D0	D1
S0	4	9	3	2	14	4
S1	9	20	14	4	10	2
S2	20	24	34	10	15	3

factors are

A = Age
- $A0$ = 35–54
- $A1$ = 55–65
- $A2$ = > 65

D = Disease Status
- $D0$ = Case
- $D1$ = Control

S = Smoking
- $S0$ = < 1 Pack
- $S1$ = 1 Pack
- $S2$ = > 1 Pack

For the present we will view this table by considering (A, D) to be a single factor with indicator function $\mathbf{u}_{(A,D)}$. Since S implies an ordering, it is not unreasonable to be interested in the function $v(ij) = i - 1$, where ij stands for $Si(AD)j$. The model $\mathfrak{L}[q: v, \mathbf{u}_{(A,D)}]$, with q uniform, contains all distributions of the form

$$p = \left(\alpha_0 + \alpha_S v + \boldsymbol{\alpha}'_{(AD)}\mathbf{u}_{(AD)}\right)q$$

The components of $\boldsymbol{\alpha}_{(AD)}$ reflect the different A, D-marginal probabilities, since

$$p(+j) = M\left[u_{(AD)j}: p\right] = \frac{\alpha_0 + \alpha_{(AD)j}}{6}$$

The parameter α_S stands for a smoking gradient common to all A, D-classes, since

$$p(2j) - p(1j) = p(1j) - p(0j) = \tfrac{1}{18}\alpha_S$$

We could take $M[vu_{(AD)j} : p]$ to be an index of the smoking rate in the $(AD)j$ column, and then

$$M[vu_{(AD)j} : p] = \tfrac{1}{9}\alpha_S$$

shows another sense in which the model embodies the condition that smoking behavior is not dependent on the A, D-class into which the individual falls.

In order to fit the model, let us observe that the vector $(v, \mathbf{u}'_{(AD)}, v\mathbf{u}'_{(AD)}, v^2, v^2\mathbf{u}'_{(AD)})$ has affinely linearly independent components, and since it has 17 components it is maximal. We may then q-standardize, obtaining

$$\begin{pmatrix} v \\ \mathbf{u}_{(AD)} \end{pmatrix} \to \mathbf{w}_1, \qquad \begin{pmatrix} v\mathbf{u}_{(AD)} \\ v^2 \\ v^2\mathbf{u}_{(AD)} \end{pmatrix} \to \mathbf{w}_2$$

and then use the fact that MDI fitting of $\mathcal{L}[q : \mathbf{w}_1]$ is the same as ML fitting of $\mathcal{E}[\hat{p} : \mathbf{w}_2]$. In Run III.B.13 the **C*C** at line 22 accomplishes the following correspondence:

Function	Variable
v	3
v^2	4
$\mathbf{u}_{(AD)}$	5–9
$v\mathbf{u}_{(AD)}$	10–14
$v^2\mathbf{u}_{(AD)}$	15–19

At line 23 the q-standardization of these functions is done in the order specified before. Note that the **WGT** variable is variable 1, which is the constant 1 (see line 5). At line 32 the components of $\mathbf{w}_2$ are **SEL**ected for **LXM**, with $\hat{p}$ as the generating distribution and the uniform distribution as the "data." From the output in Table III.34 we reject the model. We will review these data in search of an adequate model in Chapter IV, Section B.3.

Table III.34

Analysis of the Smoking and Arteriosclerosis Data

Model	df	χ^2	$\langle P \rangle$
$\mathfrak{L}[q::v, \mathbf{u}_{(AD)}]$	11	42.661	⟨.000⟩
q	6	91.534	⟨.000⟩

Truncated Geometric with Two Latent Classes. The data in Table III.35 show how many syphilis patients remained on a clinical roster at time periods $y = 1$ to 11, and how many stayed beyond $y \geqq 12$. Note that each individual appeared in as many cells as there are time periods for which he remained on the roster. Our first step is to recalculate the counts so that each person is only in the cell at which he last appeared at the clinic, listed under N.

The nature of the treatment suggests that the proportion of new cures in a time period should be a constant fraction ρ of the subjects who appeared in the preceding time period. Thus the period y at which a person last appeared should be Geometric, $p(y) = \rho(1 - \rho)^{y-1}$. The first problem is that we have data only for $1 \leqq y \leqq 11$, but this can be handled by fitting the Truncated Geometric. There is, however, an additional difficulty. Among those who appear in the first time period are some "transients" who will not

Table III.35

Clinically Treated Syphilis Patients

y	Number of Patients	N
1	232	61
2	171	23
3	148	14
4	134	13
5	121	17
6	104	14
7	90	12
8	78	11
9	67	6
10	61	5
11	56	4
$\geqq 12$	52	

return regardless of whether they are cured. Moreover, there are some who are "resistant" to the treatment, and will essentially never be cured.

We assume that the transients appear only in the $y = 1$ cell, and clearly the $y \geqq 12$ cell contains all the resistants along with some patients who will eventually be cured. We approach this by fitting the model $\mathcal{E}[q: u_1, u_2]$, where q is the Truncated Geometric with $\rho = .25$, $\mathcal{Y} = \{1, 2, \ldots, 11\}$, and

$$u_1(y) = y$$

$$u_2(y) = \begin{cases} 1 & \text{if } \ y = 1 \\ 0 & \text{otherwise} \end{cases}$$

The program in Run III.B.14 and output in Table III.36 lead us to confirm the model. For $y \neq 1$ we have

$$p(y) = \exp\left[y \ln \frac{1-\rho}{.75} - \delta\right] q(y)$$

so that

$$\theta_1 = \ln \frac{1-\rho}{.75}, \qquad \rho = 1 - .75e^{\theta_1}$$

and so we estimate

$$1 - .75e^{.1348} = .1418 = \hat{\rho}$$

with a .95 confidence interval

$$1 - .75e^{.1348 \pm 2\sqrt{.00114}} = [.0818, .1978]$$

Table III.36

Analysis of the Syphilis Data

Model	df	χ^2	$\langle P \rangle$
$\mathcal{E}[q: u_1, u_2]$	8	5.022	$\langle .757 \rangle$
q	2	20.119	$\langle .000 \rangle$

i	1	2
$\hat{\theta}_i$	0.1348	0.8880
t_i	3.99	4.15

Table III.37

Truncated Geometric Probabilities of Being Cured, and Probabilities of Being Transient or Resistant

y	$\hat{q}(y)$	Transient or Resistant
1	.1082	.1547
2	.0928	
3	.0797	
4	.0684	
5	.0587	
6	.0504	
7	.0432	
8	.0371	
9	.0318	
10	.0273	
11	.0235	
$\geqq 12$	.1155	.1086

For the Truncated Geometric we have the relation $p(2) = (1 - \rho)p(1)$, and consequently $\hat{q}(2)/(1 - \hat{\rho})$ is the part of $\hat{q}(1)$ which is due to the Truncated Geometric model, and so $n[\hat{q}(1) - \hat{q}(2)/(1 - \hat{\rho})] = 35.9$ is the estimated number of transients. The number of cures for $y \geqq 12$ is estimated by $(n - 35.9)(1 - \hat{\rho})^{11} = 26.8$, and so we estimate 25.2 resistant patients. Putting the transients and resistants to the right, we have the fitted distribution of Table III.37.

PROBLEMS

III.1. 1. Show that if $p \in \mathfrak{L}[q : \mathbf{w}]$, $p = (\alpha_0 + \boldsymbol{\alpha}'\mathbf{w})q$, then for any $\mathbf{u}$

$$M[\mathbf{u} : p] = M[\mathbf{u} : q] + C[\mathbf{u}, \mathbf{w}; q]\boldsymbol{\alpha}$$

2. Let $\hat{q}_n$ be MDI in $\mathfrak{L}[q : \mathbf{w}]$, and assume that $\pi \in \mathfrak{L}[q : \mathbf{w}]$. If $\hat{q}_n = (\hat{\alpha}_{0n} + \hat{\boldsymbol{\alpha}}_n'\mathbf{w})q$ and $\pi = (\alpha_0 + \boldsymbol{\alpha}'\mathbf{w})q$, then show that

$$\sqrt{n}\,(\hat{\boldsymbol{\alpha}}_n - \boldsymbol{\alpha}) \to_d N\left(\mathbf{0}, C[\mathbf{w} : q]^{-1}C[\mathbf{w} : \pi]C[\mathbf{w} : q]^{-1}\right)$$

(Hint: in the proof of Proposition 29 it was shown that $\hat{q}_n$ converges rapidly on any π-complement of $\mathbf{w}$.)

III.2. 1. Suppose that $p = (\alpha_0 + \boldsymbol{\alpha}'\mathbf{w})q$, where $\mathbf{w}' = (w_1, w_2, \ldots, w_k)$. Let w_1^* be the q-projection of w_1 on $\mathbb{S}(1, w_2, \ldots, w_k)$. Show that

$$\alpha_1 = \frac{M[w_1 - w_1^* : p]}{V[w_1 - w_1^* : q]}$$

2. Let $\hat{q}_n$ be MDI in $\mathcal{L}[q : \mathbf{w}]$, $\hat{q}_n = (\hat{\alpha}_{0n} + \hat{\boldsymbol{\alpha}}_n'\mathbf{w})q$, and assume $\pi \in \mathcal{L}[q : \mathbf{w}]$, $\pi = (\alpha_0 + \boldsymbol{\alpha}'\mathbf{w})q$. Show that

$$\sqrt{n}\,(\hat{\alpha}_{1n} - \alpha_1) \to_d N\left(0, \frac{V[w_1 - w_1^* : \pi]}{V[w_1 - w_1^* : q]^2}\right)$$

(Hint: same as in Problem III.1.)

III.3. Show that if $\mathbf{u}$ is p-standard and $\mathbf{w}^*$ is the p-projection of $\mathbf{w}$ on $\mathbb{S}(1, \mathbf{u})$, then

$$C[\mathbf{w}^* : p] = M[\mathbf{w}\mathbf{u}' : p]M[\mathbf{u}\mathbf{w}' : p]$$

III.4. Show that the following interpretations are correct:

1. $A\,B\,C$ means that the three factors are independent.
2. $[AB][BC]$ means that A and C are conditionally independent given B.
3. $[AB]C$ means that A and B are both independent of C.
4. $[AB][BC][AC]$ means that there is the same conditional association between any two factors, given the third (that is, the conditional relationship does not depend on the level of the third factor).

III.5. Extend the marginal homogeneity model to cover 3 factors; to cover k factors.

III.6. Using the sines and cosines of period 6 and 12 (thus four functions), reanalyze the data of Table II.5 by changing the beginning month from January to February to March and so on. How many distinct models are being fitted? What would you conclude about the data?

III.7. 1. For a square table, the model of *symmetry* requires that $p(ij) = p(ji)$. Formulate this as an exponential model. Formulate it as a linear model. Determine the ML distribution in the exponential model and the MDI distribution in the linear model.

2. Generalize the symmetry model to higher dimensions.
3. Define the model of conditional symmetry of two factors given a third. Does symmetry imply conditional symmetry, or vice versa?

III.8. Let f be a strictly increasing function. Show that the parameter in the model $\{p : f(p) - f(q) \in \mathbb{S}(1, \mathbf{u})\}$ is identifiable if and only if $\mathbf{u}$ has affinely linearly independent components.

III.9. Suppose $q \in \mathcal{L}[p : \mathbf{v}]$. Show how to find a $\mathbf{w}$ for which $\mathcal{L}[p : \mathbf{v}] = \mathcal{L}[q : \mathbf{w}]$.

III.10. Show that the set of $\boldsymbol{\alpha}$ corresponding to $p = (\alpha_0 + \boldsymbol{\alpha}'\mathbf{u})q \in \mathcal{L}[q : \mathbf{u}]$ is convex.

III.11. Let $\hat{p}_{1n_1}$ and $\hat{p}_{2n_2}$ be the empirical distributions from independent samples of sizes n_1 and n_2 on the same contingency table. Find the asymptotic distribution of

$$n \sum_y \frac{\left[\hat{p}_{1n_1}(y) - \hat{p}_{2n_2}(y)\right]^2}{\hat{p}_n(y)}$$

where $n = n_1 + n_2$ and $\hat{p}_n$ is the empirical distribution of the pooled sample.

III.12. A set of 66 faculty members were classified according to their teaching, once by peers and once by students, with the results

		Peers	
		Bad	Good
Students	Bad	19	2
	Good	8	37

Consider the u-functions as follows;

y	$N(y)$	u_1	u_2	u_3
0	19	1	−1	0
1	2	0	0	−1
2	8	0	0	1
3	37	1	1	0

1. Interpret the parameters θ_1, θ_2 and θ_3 in the model $\mathcal{E}[q : u_1, u_2, u_3]$, where q is uniform.
2. Interpret and fit the model $\mathcal{E}[q : u_1, u_2]$, and give the ANALYSIS table.

III.13. In Table I.5, consider the factors

S = Grandson's Status
  $S0$ to $S3$
F = Grandfather's Status
  $F0$ to $F3$

Let v_i be the indicator of $SiFi$ for $i = 1, 2, 3$, and $\mathbf{v}' = (v_1, v_2, v_3)$. Fit the model $\mathcal{E}[q : \mathbf{u}_F, \mathbf{u}_S, \mathbf{v}]$, and give the ANALYSIS table.

III.14. In Table I.6 consider the factors

B = Breathlessness
  $B0$ = Absent
  $B1$ = Present
W = Wheeze
  $W0$ = Absent
  $W1$ = Present
A = Age
  $A0$ to $A8$

Interpret the model $\mathcal{E}[q : \mathbf{u}_A, u_B, u_W, \mathbf{u}_{AB}, \mathbf{u}_{AW}, u_{BW}, u_L]$, that is, $[AB][AW][BW]L$, where q is the uniform distribution and $u_L(BiWjAk) = ijk$. Fit the model and give the ANALYSIS table.

III.15. Let $\hat{\boldsymbol{\theta}}$ be the ML estimator in $\mathcal{E}[\pi : \mathbf{u}]$. Show that the statistics

$$n\mathbb{D}_1\left(q^{\hat{\boldsymbol{\theta}}}, \pi\right) \quad \text{and} \quad n\hat{\boldsymbol{\theta}}'C\left[\mathbf{u} : q^{\hat{\boldsymbol{\theta}}}\right]\hat{\boldsymbol{\theta}}$$

are asymptotically equivalent.

III.16. Prove the following. The model $\mathcal{E}[q : \mathbf{u}]$ subject to the consistent nonredundant constraints $\mathbf{H}\boldsymbol{\theta} = \mathbf{h}$ is equivalent to the model $\mathcal{E}[q_0 : \mathbf{G}\mathbf{u}]$, where (1) $\mathbf{H}\boldsymbol{\theta}_0 = \mathbf{h}$ and q_0 has parameter $\boldsymbol{\theta}_0$ in $\mathcal{E}[q : \mathbf{u}]$, (2) $\mathbf{H}\mathbf{G}' = \mathbf{0}$, and (3) the matrix $\begin{pmatrix} \mathbf{H} \\ \mathbf{G} \end{pmatrix}$ is nonsingular.

III.17. Let q be the uniform distribution on an (A, B)-table. In addition to the usual vectorial indicators $\mathbf{u}_A$ and $\mathbf{u}_B$, define $v_A(ij) = i$ and $v_B(ij) = j$.

1. Show that $\mathcal{E}[q : \mathbf{u}_A, \mathbf{u}_B, v_A, v_A\mathbf{u}_B]$ is equivalent to $\mathcal{E}[q : \mathbf{u}_A, \mathbf{u}_B, v_A\mathbf{u}_B]$. Which model is preferred?
2. $\mathcal{E}[q : \mathbf{u}_A, \mathbf{u}_B, v_A\mathbf{u}_B]$ is a model for a *singly ordered* table (that is, A is an ordinal factor). Interpret the parameters in the model.

3. Construct the corresponding model for a *doubly ordered* table (that is, B is an ordinal factor too).
4. Write out the likelihood equations explicitly for the singly and doubly ordered models. Also, show that

$$\sum_i i\hat{p}(AiB0) = \sum_i i\hat{q}(AiB0)$$

is a consequence of the equations for the singly ordered model.
5. Compute the degrees of freedom for the independence and the singly and doubly ordered models in the context of an ANALYSIS table. Discuss the special cases of 2^2, 3^2, and 4^2.

III.18. Let $\mathcal{Y}$ be a 2^m table, so that $y = (y_1, \ldots, y_m)$ with each factor having levels 0 and 1. Let q be a complete independence distribution on $\mathcal{Y}$, and let q_i denote the marginal distribution of the ith factor, derived from q. For each subset A of $\{1, 2, \ldots, m\}$ define

$$u_A(y) = \prod_{i \in A} \frac{y_i - q_i(1)}{\sqrt{q_i(1) q_i(0)}}$$

1. Show that $\{u_A : A \text{ nonvoid}\}$ is maximal q-standard.
2. Define $\alpha_A = M[u_A : p]$, and show that if

$$p_m = \left(1 + \sum_{\#A \leqq m} \alpha_A u_A\right) q$$

is a probability distribution, then p_m and p give the same marginal distribution to any selection of m or fewer factors.
3. A distribution p is permutation-invariant if and only if for each permutation σ of $\{1, 2, \ldots, m\}$, letting y_σ have ith component $y_{\sigma(i)}$, we have $p(y_\sigma) = p(y)$. Show that this happens if and only if each α_A depends only on $\#A$.
4. Let $z = \sum_{i=1}^m y_i$, and assume q permutation-invariant, with $q_i(1) = \rho$. Define

$$v_i = \sum_{\#A = i} u_A \binom{m}{\#A}^{-1/2}$$

Show that $\mathbf{v} = (v_1, \ldots, v_m)'$ is q-standard and a function of z, so that $\mathbf{v}$ is maximal standard with respect to the distribution of z. Show that under q, z is Binomial with parameters m and ρ, and derive its distribution in terms of $\mathbf{v}$ otherwise.

5. Show that if p is permutation-invariant, then $p(y|z)$ is a uniform distribution.

III.19. If some of the cells of a cross-classification are impossible, then the table is called *incomplete*. The *quasi-independence* model on an incomplete table has the uniform generating distribution and all the u-functions of the independence model. (Note that some components may need to be discarded in order to have an identifiable parameter.)

1. Consider a paired-comparison experiment. Consider also the derived table with factors C at levels $\{ij : i \neq j\}$ indicating which two treatments are involved in the comparison, and W with levels indicating the winner. Show that the paired-comparison model given in Section B.2 is equivalent to the quasi-independence model on the (C, W)-table.
2. Consider a triplet-comparison experiment. Let C have the levels $\{ijk : i \neq j, j \neq k, i \neq k\}$ indicating which three treatments are being compared, and let W have levels $\{ij : i \neq j\}$ indicating the winner and runner-up. Consider the quasi-independence model on the (C, W)-table. Determine the corresponding model in the original three-way table having cells $FiSjTk$ indicating that treatment i was first, j was second, and k was third.

III.20. In Proposition 22 Taylor's theorem was used to derive an approximation of $\mathbb{D}_1$ by $\mathbb{D}_2$. Carry out the corresponding approximation of an arbitrary $\mathbb{D}_\omega$.

III.21. Consider the constant-odds-ratio example in Section B.2. Show that under the saturated model the RM ln odds ratio can be written as

θ_{RM}	in the $F0G0$-subtable
$\theta_{RM} + \theta_{FRM}$	in the $F1G0$-subtable
$\theta_{RM} + \theta_{GRM}$	in the $F0G1$-subtable
$\theta_{RM} + \theta_{FRM} + \theta_{GRM} + \theta_{FGRM}$	in the $F1G1$-subtable

Repeat the analysis of the data, using the results of Problems II.8 and II.21.

III.22. Reconsider the data on paired comparisons in Section B.2. Fit a linear version of the model, in which there are strength parameters γ_i satisfying $p(ij) - p(ji) = \gamma_j - \gamma_i$.

III.23. Reanalyze the SIDS birth-month data of Table II.5, using the u-functions of Problem III.6, but now fitting a linear rather than exponential model. Plot the functions $\hat{q}/q$ and $\hat{p}/\hat{q}$.

III.24. Let $\hat{q}_n$ be ML in $\mathfrak{E}[\pi : \mathbf{u}]$. Let w be any function on the table, and w^* be its π-projection onto $\mathfrak{S}(1, \mathbf{u})$. Show that

$$\sqrt{n}\,(M[w : \hat{q}_n] - M[w : \pi]) \to_d N(0, V[w^* : \pi])$$

Generalize to vector-valued $\mathbf{w}$.

III.25. Let $(\mathbf{u}', \mathbf{v}')$ be maximal π-standard. Let $\hat{q}_n$ be the ML distribution in $\mathfrak{E}[\pi : \mathbf{u}]$, and let $(\hat{\mathbf{u}}'_n, \hat{\mathbf{v}}'_n)$ be the $\hat{q}_n$-standardization of $(\mathbf{u}', \mathbf{v}')$. Show that

$$\sqrt{n}\, M[\hat{\mathbf{v}}_n : \hat{p}_n] \to_d N(\mathbf{0}, \mathbf{I})$$

REMARKS

The choice to restrict consideration to exponential and linear models is based on a number of considerations. Perhaps the most important is that they have simple structure, but are very flexible and cover a wide variety of useful applications. The theory which underlies their use can be developed in a relatively clean way; any more general models would likely involve theorems stated in terms of conditions that would need to be verified in any particular case, which could become difficult as well as tedious. In Chapter VI, Section E we will see how exponential models can be used to approximate more general models.

The simple concept of rapid convergence does not appear to have been identified in the literature, although some version of it probably underlies various proofs of asymptotic results. The fact that both ML and MDI (and other) estimators converge rapidly permits results to be stated simultaneously for several kinds of estimators.

Proposition 20 is the discrete version of the general minimum-discrimination-information theorem, proved most succinctly in Kullback and Khairat (1966). The asymptotic Normality of MLEs (Proposition 23) is proved more generally in too many places to cite here. Most of the other asymptotic results given here also appear in many places in the literature, although our proofs can be relatively simple because there is no necessity to obtain the most general results. However, I do not believe that Proposition 29 has ever appeared in this form.

It is "well known" that ML estimates have important asymptotic optimality properties over other methods of estimation. However, as Neyman (1949) showed, there is no lack of estimators, and thus estimation principles, which are asymptotically equivalent to ML [see also Rao (1973) in this regard]. We have not used optimality here as an argument in favor of ML or MDI estimation.

The idea of the ANALYSIS table goes back to the analysis of variance (ANOVA) table, familiar to any student of experimental statistics. The idea of using such a table in discrete data analysis is due to Kullback (1968). The use of the serial test and serial *P*-value is new. The fact that smaller df is better was proved by Fix, Hodges, and Lehmann (1959) in an article which has dropped into undeserved obscurity. So far as I know, their results have not been used before to justify the prevalent practice of fitting models to discrete data.

The material in Section A.6 is from Aickin (1979), part of which appears in a less explicit form in Haberman (1974). A most ingenious result for generalized exponential families is proved in Lauritzen (1975).

Problems III.1 and III.2 contain material which is new. Problem III.16 shows how linear constraints on the parameter of an exponential model can be taken into account. Note that help on several of the problems in this and later chapters can be found in the Data and Problem References.

CHAPTER IV

Inference for Model Chains

The basic ideas of the preceding two chapters are developed much further in this one. Here we consider an array of models, either as a network for inference based on hypothesis testing, or as a series of probes to try to understand the patterns of variation in the data. In either case we will see that proper analysis generally requires fitting a number of models to the data and looking in some detail at the results of each fitting.

The simplest kind of analysis involves a chain of models, each of which is less complex than the next. In Section A.1 some of the necessary asymptotic theory is presented, and in Section A.2 the extension of the ANALYSIS table is made to cover chains of models. An important statistic for both testing and fitting is the serial P-value, which is also extended to cover model chains. Also in Section A.2 we present the general ANALYSIS table for an array of models, which is very useful in complex situations, especially when it is not clear before the experiment which of the possible simple ANALYSIS tables ought to be used. In this subsection we will develop the serial P-values in such a way that the overall probability of rejecting any true model is no more than α for all the rejection–confirmation decisions made in the ANALYSIS table.

The classical approach to the analysis of cross-classifications by hierarchical models is briefly surveyed in Section A.3.

A statistical analysis is generally incomplete unless some attention has been paid to the differences between the fitted and observed distributions. When appropriately scaled these are called residuals. The asymptotic theory is given in Section A.4 for the four types of residuals which LAMDA computes and displays. Plotting the P-values of the residuals provides a useful, if not very rigorous, display of the adequacy of the fitted distribution over individual cells.

In Section A.5 we provide two solutions to the problem of testing multiple moments of a distribution. Finally, in Section B a number of data sets are analyzed with these various techniques. As before, the computation is by LAMDA and the runs are in Appendix C.

A. NESTED MODELS

In Chapter II we considered inference concerning a single distribution, that is, a model of the form $\{q\}$, and in Chapter III we investigated inference about the model pairs $\{q\}$ and $\mathfrak{S}[q:\mathbf{u}]$ or $\{q\}$ and $\mathfrak{L}[q:\mathbf{u}]$. In this section we extend this same kind of inference to a *chain of nested models*

$$\{q\} \subset \mathfrak{M}_k \subset \mathfrak{M}_{k-1} \subset \cdots \subset \mathfrak{M}_2 \subset \mathfrak{M}_1$$

in which all of the $\mathfrak{M}_i$ are models of the form $\mathfrak{S}$ or $\mathfrak{L}$.

We let $\mathbf{u}_0$ be maximal with affinely linearly independent components, and let $\mathbf{u}_i$ $(1 \leqq i \leqq k)$ be such that all the components of $\mathbf{u}_i$ appear among the components of $\mathbf{u}_{i-1}$. Thus

$$\mathbb{S}(1,\mathbf{u}_k) \subset \mathbb{S}(1,\mathbf{u}_{k-1}) \subset \cdots \subset \mathbb{S}(1,\mathbf{u}_2) \subset \mathbb{S}(1,\mathbf{u}_1)$$

and consequently we have the chain of models

$$\mathfrak{S}[q:\mathbf{u}_k] \subset \mathfrak{S}[q:\mathbf{u}_{k-1}] \subset \cdots \subset \mathfrak{S}[q:\mathbf{u}_2] \subset \mathfrak{S}[q:\mathbf{u}_1]$$

and

$$\mathfrak{L}[q:\mathbf{u}_k] \subset \mathfrak{L}[q:\mathbf{u}_{k-1}] \subset \cdots \subset \mathfrak{L}[q:\mathbf{u}_2] \subset \mathfrak{L}[q:\mathbf{u}_1]$$

The main objective of this section is to show how to test the hypothesis $\pi \in \mathfrak{M}_i$ given that $\pi \in \mathfrak{M}_{i-1}$, and to carry out the serial testing procedure along a chain.

1. Asymptotic Behavior of Discrepancies

Consider a chain of models

$$\{q\} \subset \mathfrak{S}[q:\mathbf{u}_k] \subset \mathfrak{S}[q:\mathbf{u}_{k-1}] \subset \cdots \subset \mathfrak{S}[q:\mathbf{u}_2] \subset \mathfrak{S}[q:\mathbf{u}_1]$$

where, as above, $\mathbf{u}_0$ is maximal with affinely linearly independent components, and all the components of $\mathbf{u}_i$ are also components of $\mathbf{u}_{i-1}$. We may thus assume that

$$\mathbf{u}_0 = \begin{pmatrix} \mathbf{u}_i \\ \mathbf{v}_i \end{pmatrix}, \qquad \mathbf{u}_{i-1} = \begin{pmatrix} \mathbf{u}_i \\ \mathbf{w}_i \end{pmatrix}$$

for each i. Note that $\#\mathbf{w}_i = \#\mathbf{u}_{i-1} - \#\mathbf{u}_i$. We have established in earlier sections all of the technical facts to assert the following.

Proposition 30. Let $\hat{q}_i$ be ML in $\mathfrak{S}[q:\mathbf{u}_i]$. Then

$$\mathbb{D}_1(\hat{p}, q) = \mathbb{D}_1(\hat{p}, \hat{q}_1) + \mathbb{D}_1(\hat{q}_1, \hat{q}_2) + \cdots + \mathbb{D}_1(\hat{q}_k, q)$$

If $\pi \in \mathfrak{S}[q:\mathbf{u}_i]$, then $n\mathbb{D}_1(\hat{q}_{j-1}, \hat{q}_j) \to_d \chi^2(\#\mathbf{u}_{j-1} - \#\mathbf{u}_j)$ for all $1 \leqq j \leqq i$ (where $\hat{q}_0 = \hat{p}$), and these statistics are asymptotically independent. If $\pi = q$, then also $n\mathbb{D}_1(\hat{q}_k, q) \to_d \chi^2(\#\mathbf{u}_k)$, and this statistic is asymptotically independent of all the others.

Proof. From Proposition 19 we have $M[\mathbf{u}_i : \hat{p}] = M[\mathbf{u}_i : \hat{q}_i] = M[\mathbf{u}_i : \hat{q}_{i-1}]$, so that $\hat{q}_i$ is the ML distribution in $\mathfrak{S}[q:\mathbf{u}_i]$ for the "data" $\hat{q}_{i-1}$, and so by the same proposition we have $\mathbb{D}_1(\hat{q}_{i-1}, q) = \mathbb{D}_1(\hat{q}_{i-1}, \hat{q}_i) + \mathbb{D}_1(\hat{q}_i, q)$. An inductive argument now completes the proof of the first part.

For the second part, write

$$\hat{q}_i = (1 + M[\mathbf{u}_0 : \hat{q}_i]'\mathbf{u}_0)\pi$$

where we assume that $\mathbf{u}_0$ is maximal π-standard. Then

$$\mathbb{D}_2(\hat{q}_{i-1}, \hat{q}_i) = \sum_y \frac{\{(1 + M[\mathbf{u}_0 : \hat{q}_i]'\mathbf{u}_0)\pi - (1 + M[\mathbf{u}_0 : \hat{q}_{i-1}]'\mathbf{u}_0)\pi\}^2}{\hat{q}_i(y)}$$

$$= \sum_y \{(M[\mathbf{u}_0 : \hat{q}_i] - M[\mathbf{u}_0 : \hat{q}_{i-1}])'\mathbf{u}_0\}^2 \pi(y) \frac{\pi(y)}{\hat{q}_i(y)}$$

From Propositions 6 and 22 conclude that $n\mathbb{D}_2(\hat{q}_{i-1}, \hat{q}_i)$ has the same asymptotic distribution as $n\|M[\mathbf{u}_0 : \hat{q}_i] - M[\mathbf{u}_0 : \hat{q}_{i-1}]\|^2$. Now

$$\begin{aligned} n\|M[\mathbf{u}_0 : \hat{q}_{i-1}] - M[\mathbf{u}_0 : \hat{q}_i]\|^2 &= n\|M[\mathbf{u}_i : \hat{q}_i] - M[\mathbf{u}_i : \hat{q}_{i-1}]\|^2 \\ &\quad + n\|M[\mathbf{w}_i : \hat{q}_i] - M[\mathbf{w}_i : \hat{q}_{i-1}]\|^2 \\ &\quad + n\|M[\mathbf{v}_{i-1} : \hat{q}_i] - M[\mathbf{v}_{i-1} : \hat{q}_{i-1}]\|^2 \end{aligned}$$

The last of these terms $\to_p 0$ because of Proposition 24, while the first term vanishes according to the first line of the proof. Again from Proposition 24, $\sqrt{n}\, M[\mathbf{w}_i : \hat{q}_i] \to_p 0$, and since $M[\mathbf{w}_i : \hat{q}_{i-1}] = M[\mathbf{w}_i : \hat{q}_i]$ we have

$$n\mathbb{D}_2(\hat{q}_{i-1}, \hat{q}_i) - n\|M[\mathbf{w}_i : \hat{p}]\|^2 \to_p 0$$

and so the proof is finished by application of the asymptotic equivalence of $\mathbb{D}_2$ and $\mathbb{D}_1$. □

For a chain of $\mathfrak{L}$-models we have the corresponding result.

Proposition 31. Let $\hat{q}_i$ be MDI in $\mathfrak{L}[q:\mathbf{u}_i]$. Then

$$\mathbb{D}_0(\hat{p}, q) = \mathbb{D}_0(\hat{p}, \hat{q}_1) + \mathbb{D}_0(\hat{q}_1, \hat{q}_2) + \cdots + \mathbb{D}_0(\hat{q}_k, q)$$

If $\pi \in \mathfrak{L}[q:\mathbf{u}_i]$, then $n\mathbb{D}_0(\hat{q}_{j-1}, \hat{q}_j) \rightarrow_d \chi^2(\#u_{j-1} - \mathbf{u}_j)$ for all $1 \leqq j \leqq i$ (where $\hat{q}_0 = \hat{p}$), and these statistics are asymptotically independent. If $\pi = q$, then also $n\mathbb{D}_0(\hat{q}_k, q) \rightarrow_d \chi^2(\#\mathbf{u}_k)$, and this statistic is asymptotically independent of all the others.

Proof. Repeat the notation and argument of the preceding proposition. Assume that $\mathbf{u}_i, \mathbf{v}_i$ is q-standard, so that $\hat{q}_{i-1} \in \mathfrak{S}[\hat{p}: v_{i-1}]$ and so $\mathfrak{S}[\hat{p}:\mathbf{v}_i] = \mathfrak{S}[\hat{q}_{i-1}:\mathbf{v}_i]$. Thus $\hat{q}_i \in \mathfrak{S}[\hat{q}_{i-1}:\mathbf{v}_i]$, which shows by Proposition 20 that $\hat{q}_i$ is MDI in $\mathfrak{L}[q:\mathbf{u}_i]$ with respect to the "data" $\hat{q}_{i-1}$. Consequently $\mathbb{D}_0(\hat{q}_{i-1}, q) = \mathbb{D}_0(\hat{q}_{i-1}, \hat{q}_i) + \mathbb{D}_0(\hat{q}_i, q)$, and induction now finishes the first part of the proof.

For the second part, we have the decomposition of $\mathbb{D}_2(\hat{q}_{i-1}, \hat{q}_i)$ of the previous proposition, and there we have $M[\mathbf{v}_{i-1}:\hat{q}_{i-1}] = \mathbf{0} = M[\mathbf{v}_{i-1}:\hat{q}_i]$ because $\hat{q}_{i-1}$ and $\hat{q}_i$ are in $\mathfrak{L}[q:\mathbf{u}_{i-1}]$. From Proposition 29 we have

$$\sqrt{n}\,(M[\mathbf{u}_i:\hat{q}_i] - M[\mathbf{u}_i:\hat{p}]) \rightarrow_p \mathbf{0}$$

$$\sqrt{n}\,(M[\mathbf{u}_i:\hat{q}_{i-1}] - M[\mathbf{u}_i:\hat{p}]) \rightarrow_p \mathbf{0}$$

so that

$$\sqrt{n}\,(M[\mathbf{u}_i:\hat{q}_{i-1}] - M[\mathbf{u}_i:\hat{q}_i]) \rightarrow_p \mathbf{0}$$

From the same proposition, $\sqrt{n}\,(M[\mathbf{w}_i:\hat{q}_{i-1}] - M[\mathbf{w}_i:\hat{p}]) \rightarrow_p \mathbf{0}$, and since $M[\mathbf{w}_i:\hat{q}_i] = \mathbf{0}$, this proves that

$$n\mathbb{D}_2(\hat{q}_{i-1}, \hat{q}_i) - n\|M[\mathbf{w}_i:\hat{p}]\|^2 \rightarrow_p 0$$

and so the equivalence of $\mathbb{D}_0$ and $\mathbb{D}_2$ finishes the proof. □

2. The General ANALYSIS Table

As a consequence of the propositions in the preceding section, we may extend the simple ANALYSIS table of Chapter III, Section A.4 to a display of

Table IV.1

The General ANALYSIS *Table for a Chain of Models*

Model	df	χ^2	$\langle P \rangle$	$\bar{\alpha}$	$\langle \bar{P} \rangle$
$\mathfrak{M}_1$	$\#\mathcal{Y} - 1 - \#\mathbf{u}_1$	$n\mathbb{D}(\hat{p}, \hat{q}_1)$			
$\mathfrak{M}_2$	$\#\mathbf{u}_1 - \#\mathbf{u}_2$	$n\mathbb{D}(\hat{q}_1, \hat{q}_2)$			
$\vdots$	$\vdots$	$\vdots$			
$\mathfrak{M}_{k-1}$	$\#\mathbf{u}_{k-2} - \#\mathbf{u}_{k-1}$	$n\mathbb{D}(\hat{q}_{k-2}, \hat{q}_{k-1})$			
$\mathfrak{M}_k$	$\#\mathbf{u}_{k-1} - \#\mathbf{u}_k$	$n\mathbb{D}(\hat{q}_{k-1}, \hat{q}_k)$			
$\{q\}$	$\#\mathbf{u}_k$	$n\mathbb{D}(\hat{q}_k, q)$			

the results of the analysis of a chain of models. This more general form appears in Table IV.1, where all the $\mathfrak{M}_i$ are either $\mathcal{E}[q:\mathbf{u}_i]$ or $\mathcal{L}[q:\mathbf{u}_i]$, the q_i are correspondingly either ML or MDI distributions, and $\mathbb{D}$ is either $\mathbb{D}_1$ or $\mathbb{D}_0$.

We know by Propositions 30 and 31 that if $\pi \in \mathfrak{M}_i$, then all $n\mathbb{D}(\hat{q}_{j-1}, \hat{q}_j)$ for $j \leqq i$ tend asymptotically to $\chi^2(\#\mathbf{u}_{j-1} - \#\mathbf{u}_j)$ distributions. Thus P-values may be obtained by referring these discrepancy statistics to their distributions as described in Chapter II, Section B.4. As in Chapter III, Section A.4, the appropriate tests for various hypotheses depend on what one is willing to assume. A test of $\pi = q$ would be based on $n\mathbb{D}(\hat{p}, q)$, whereas if it were known that $\pi \in \mathfrak{M}_i$, then $n\mathbb{D}(\hat{q}_i, q)$ would provide an appropriate test, and as discussed in Chapter III, Section A.2, the latter test would be more powerful. A test of $\pi \in \mathfrak{M}_i$ would be based on $n\mathbb{D}(\hat{p}, \hat{q}_i)$, but if $\pi \in \mathfrak{M}_j$ were known for some $j < i$, then $n\mathbb{D}(\hat{q}_j, \hat{q}_i)$ would be the appropriate statistic, referred to the $\chi^2(\#\mathbf{u}_j - \#\mathbf{u}_i)$ distribution.

Serial P-values. The serial test of Chapter III, Section A.4 generalizes to a chain of models $\mathfrak{M}_1, \mathfrak{M}_2, \ldots, \mathfrak{M}_k$ as follows. We begin at the top of the ANALYSIS table and successively test each hypothesis $\pi \in \mathfrak{M}_i$ at level α_i. We terminate the procedure as soon as the first rejection occurs, and if $\pi \in \mathfrak{M}_i$ is the first hypothesis to be rejected, then implicitly all $\pi \in \mathfrak{M}_j$ for $j > i$ are also rejected. The overall level of this procedure is $1 - \prod(1 - \alpha_i)$, the product being taken over $1 \leqq i \leqq k$, and we will always use the approximation $\bar{\alpha} = \sum \alpha_i$. For any j, $1 \leqq j \leqq k$, the level of the serial test along the chain which terminates at $\mathfrak{M}_j$ is $1 - \prod(1 - \alpha_i)$, the product now being taken for $1 \leqq i \leqq j$. We approximate this by $\bar{\alpha}_j = \sum \alpha_i$ $(1 \leqq i \leqq j)$, which is called the *serial level* of $\mathfrak{M}_j$. Note that $\bar{\alpha}_k$ is the overall level of the serial test.

The serial levels carry the following interpretation: if $\pi \in \mathfrak{M}_j$, then the probability of rejecting $\mathfrak{M}_j$ by the serial test is no more than $\bar{\alpha}_j$. No matter where the true distribution is, the probability of rejecting any model anywhere in the chain which contains the true distribution is no more than $\bar{\alpha}$. These facts are easily deduced the same way as in Chapter III, Section A.4.

An appropriate set of P-values for the various hypotheses is defined by

$$\bar{P}_j = 1 - \prod_{i=1}^{j} (1 - \min{}^{1/w_i}), \qquad \min = \min_{1 \leq i \leq j} P_i^{w_i}$$

where P_i is the P-value of the test of $\pi \in \mathfrak{M}_i$ based on $n\mathbb{D}(\hat{q}_{i-1}, \hat{q}_i)$, $w_i = -1/\ln \alpha_i$, and $\bar{P}_j$ is the serial P-value of $\mathfrak{M}_j$. It can be shown that if $\pi \in \mathfrak{M}_j$, then $\bar{P}_j$ is uniformly distributed on $[0, 1]$, and that the serial test along the chain is identical to rejecting all models $\mathfrak{M}_j$ for which $\bar{P}_j \leq \bar{\alpha}_j$ and confirming the others. Thus one can quickly determine the inferential conclusions from an ANALYSIS table by looking under the serial level and serial P-value columns.

As discussed before, the individual levels α_i would be chosen quite small for those hypotheses in which we had substantial prior belief, with the larger α_i's being reserved for the interesting hypotheses about which there was doubt. In the absence of any prior ideas about the hypotheses we might use equal α_i's:

$$\bar{P}_j = 1 - (1 - \min)^j, \qquad \min = \min_{1 \leq i \leq j} P_j$$

Model Fitting. The structure of the ANALYSIS table for a chain of models can be used for the purpose of model fitting as well as for hypothesis testing. The P-values are then used as diagnostic guides, and do not have their significance-testing interpretations. The point of view in model fitting is somewhat different than in hypothesis testing, and deserves some comment.

When $\pi \in \mathfrak{M}_i$ represents a formal hypothesis, the burden of proof is on some alternate hypothesis, that is, the test statistic must be truly extreme before we abandon the original hypothesis $\pi \in \mathfrak{M}_i$. However, in model fitting the burden of proof is on the statement $\pi \in \mathfrak{M}_i$, and thus a model including $\mathfrak{M}_i$ is preferred to $\mathfrak{M}_i$ except when the discrepancy statistic is in the middle or lower range of its distribution. Stated in terms of P-values, one would reject the hypothesis $\pi \in \mathfrak{M}_i$ only if the P-value were below (say) .05, whereas if one wanted a well-fitting model one might reject $\mathfrak{M}_i$ only if the P-value were below (say) .5.

One reason for model fitting in this fashion is the following. If $n\mathbb{D}(\hat{p}, \hat{q}_A)$ were to be used for assessing the fit of $\mathfrak{M}_A$, with df $= \#\mathcal{Y} - 1 - \#\mathbf{u}_A$, in the cases we are considering there are models $\mathfrak{M}_i$ such that

$$\mathfrak{M}_A \subset \mathfrak{M}_i \subset \mathfrak{M}_{i-1} \subset \cdots \subset \mathfrak{M}_1$$

and the accompanying discrepancy statistics $n\mathbb{D}(\hat{q}_{i-1}, \hat{q}_i)$ each have one df. Thus, although the overall "test" of $\mathfrak{M}_A$ based on $n\mathbb{D}(\hat{p}, \hat{q}_A)$ may not lead to rejection, there may be a $n\mathbb{D}(\hat{q}_{i-1}, \hat{q}_i)$ which is quite large. In other words, an overall Chi-square statistic which is insignificant may contain a specific Chi-square which is enormously significant. Since in model fitting the aim is to avoid omitting any model features which contribute to a good fit, one wants $n\mathbb{D}(\hat{p}, \hat{q}_A)$ to be rather small before $\mathfrak{M}_A$ is adopted as a fitted model.

The General* ANALYSIS *Table. The concepts of a chain of models and a serial test along a chain are very useful and flexible, but they are not quite broad enough. In the general situation, we have a collection of models $\mathbb{M} = \{\mathfrak{M}_\gamma\}$, where γ is an appropriate index for identifying the model $\mathfrak{M}_\gamma$. We will write $\mathfrak{M}_\gamma \to \mathfrak{M}_\eta$ to mean that $\mathfrak{M}_\eta$ is a submodel of $\mathfrak{M}_\gamma$ (that is, $\mathfrak{M}_\eta \subset \mathfrak{M}_\gamma$) and that there are no models in $\mathbb{M}$ which are between $\mathfrak{M}_\gamma$ and $\mathfrak{M}_\eta$. The arrow is intended to remind us that it is sensible to test $\pi \in \mathfrak{M}_\eta$ given that $\pi \in \mathfrak{M}_\gamma$, and we will sometimes use "$\mathfrak{M}_\gamma \to \mathfrak{M}_\eta$" to stand for this test. A model $\mathfrak{M}_\gamma$ is *maximal* if there is no $\mathfrak{M}_\eta$ for which $\mathfrak{M}_\eta \to \mathfrak{M}_\gamma$. A chain looks like

$$\mathfrak{M}_1 \to \mathfrak{M}_2 \to \mathfrak{M}_3 \to \cdots \to \mathfrak{M}_k$$

and it is a *maximal chain* if and only if $\mathfrak{M}_1$ is a maximal model.

We will assign a level to each test $\mathfrak{M}_\gamma \to \mathfrak{M}_\eta$, and this will be done so that for any model $\mathfrak{M}_\gamma$, the serial level of each maximal chain pointing to $\mathfrak{M}_\gamma$ is the same, $\bar{\alpha}_\gamma$. We call $\bar{\alpha}_\gamma$ the *serial level* of $\mathfrak{M}_\gamma$. The serial P-value is then computed along each maximal chain as described earlier. This means that there will be a collection of serial P-values for each model, one from each maximal chain pointing to the model. We use these for strict hypothesis testing by rejecting any model $\mathfrak{M}_\gamma$ for which the maximum of its serial P-values is less than or equal to its serial level. When this is done, we then look at $\mathbb{M}$ again and confirm any model which was rejected but is more complex than some confirmed model. This second step is necessary for consistency, to insure that no model is rejected when one of its submodels is confirmed.

This is the general serial testing procedure. It has a very important property with regard to the probability of rejecting any model containing π.

First, let $\bar{\alpha}$ be the largest of all the serial levels of all the models in $\mathbf{M}$. We call this the *serial level* of $\mathbf{M}$. Now suppose that $\mathbf{M}$ has the property that whenever $\mathfrak{M}_\gamma$ and $\mathfrak{M}_\eta$ are in $\mathbf{M}$, then so is $\mathfrak{M}_\gamma \cap \mathfrak{M}_\eta$. We may then say that the probability of rejecting any model anywhere in $\mathbf{M}$ which contains π is no more than $\bar{\alpha}$. This is the primary justification for the serial test.

As an example of a collection of models, consider a three-factor cross-classification. The models of interest are shown in Figure IV.1, with arrows indicating the tests. Although such diagrams are useful for thinking about the interrelationships among the models, they are only possible in fairly simple cases. The general ANALYSIS table for Figure IV.1 appears in Table IV.2. Here the models are listed and assigned numbers for ease of reference. The df and χ^2 for the test of fit of each model are then listed. The Chi-square statistics here are $n\mathbb{D}(\hat{p}, \hat{q}_i)$ for model i, and the df is $\#\mathcal{Y} - 1 - \#\mathbf{u}_i$ when the model is of the form $\mathfrak{M}[q:\mathbf{u}_i]$. Note that these Chi-square statistics are not the ones which have appeared in simpler versions of the ANALYSIS table given previously. In the "Next" column are all the tests possible when proceeding from the model listed at the left. The test of $\mathfrak{M}[q:\mathbf{u}_{i+1}]$ given $\mathfrak{M}[q:\mathbf{u}_i]$ is based on $n\mathbb{D}(\hat{q}_i, \hat{q}_{i+1}) = n\mathbb{D}(\hat{p}, \hat{q}_{i+1}) - n\mathbb{D}(\hat{p}, \hat{q}_i)$ with df $= \#\mathbf{u}_i - \#\mathbf{u}_{i+1}$. Thus the tests are made by taking differences of the χ^2-statistics and the dfs are likewise gotten by taking differences of the entries in the "df" column. The serial levels $\bar{\alpha}_i$ and largest serial P-values $\bar{P}_{\max}$ are then given for each model. One can quickly determine the results of the serial test by comparing the $\bar{P}_{\max}$-values with the $\bar{\alpha}$-values.

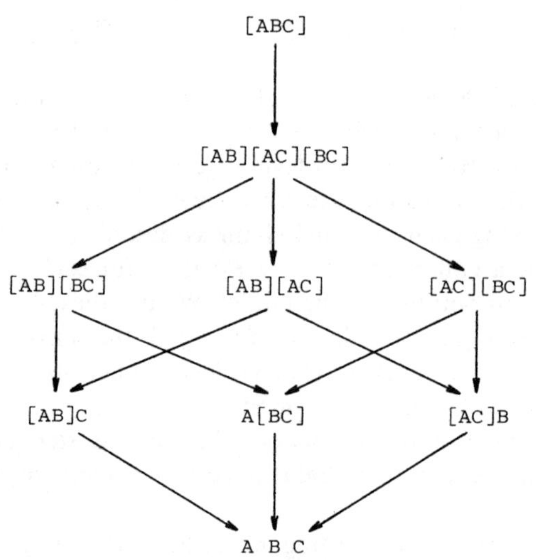

Fig. IV.1 Diagram of the hierarchical models for a three-way table.

Table IV.2

The ANALYSIS *Table for the Hierarchical Models on a Three-Way Table*

	Model	df	χ^2	Next	$\bar{\alpha}$	$\langle \bar{P}_{\max} \rangle$
1	$[ABC]$			2⟨ ⟩		
2	$[AB][BC][AC]$			3⟨ ⟩		
				4⟨ ⟩		
				5⟨ ⟩		
3	$[AB][AC]$			6⟨ ⟩		
				7⟨ ⟩		
4	$[AB][BC]$			6⟨ ⟩		
				8⟨ ⟩		
5	$[AC][BC]$			7⟨ ⟩		
				8⟨ ⟩		
6	$[AB]C$			9⟨ ⟩		
7	$[AC]B$			9⟨ ⟩		
8	$A[BC]$			9⟨ ⟩		
9	$A\ B\ C$					

The final step in the use of the ANALYSIS table is to identify the *maximal rejected* models (rejected models which are not submodels of any other rejected models) and the *minimal confirmed* models (confirmed models all of whose submodels are rejected). It is sometimes useful to append an R to all the rejected models, and an asterisk to all models which are either minimal confirmed or maximal rejected.

3. Classical Log-linear Analysis

Cross-classifications occur quite frequently in practice, and so the historical development of discrete data analysis has been oriented towards models for this kind of data. Exponential models for cross-classifications are called *log-linear models*, and have enough similarity to the linear analysis of factorial experiments that they are often couched in the somewhat old-fashioned terms used in that field. We have not followed the classical approach here, because at an appropriate level of generality it is more complex than the approach we have chosen. Thus one of the main purposes of this section is to establish the points of contact between the two approaches.

Parametrization. One of the first problems with the classical approach is that in general it is more difficult to understand how the models are parametrized. We will present the general framework and show how it relates to the customary expressions for log-linear models.

We begin with a collection of functions $\{\mathbf{u}_s\}$ defined on $\mathcal{Y}$ and taking values in various spaces $\mathbb{R}^i$. For each s, simplify the notation by writing $\mathbb{S}_s = \mathbb{S}(1, \mathbf{u}_s)$, and use $\mathbb{S}_0 = \mathbb{S}(1)$. Let $q \in \mathcal{P}$ be arbitrary; presently we will take q to be uniform. For any $f: \mathcal{Y} \to \mathbb{R}$, let f_s^* be the q-projection of f on $\mathbb{S}_s$ (see Problem II.1).

Whenever $\mathbb{S}_s \subset \mathbb{S}_{s'}$ write $s \preccurlyeq s'$, and if in addition $\mathbb{S}_s \neq \mathbb{S}_{s'}$ then write $s \prec s'$. This establishes a partial order on the set of indices $\{s\}$. The *Möbius function* of $\{s\}$ is a function $c(s', s)$ defined on pairs (s', s) according to the rules

1. if $s \prec s'$, then $c(s', s) = 0$;
2. $c(s, s) = 1$ for all s;
3. for $s' \preccurlyeq s$, $c(s', s)$ is the negative of the sum of all the $c(s'', s)$ for $s' \prec s'' \preccurlyeq s$.

These rules show how to construct c, and we will see examples presently. The fundamental property of c is the *Möbius inversion formula*:

For any functions $F: \{s\} \to \mathbb{R}$ and $G: \{s\} \to \mathbb{R}$, the following statements are equivalent:

1. $F(s) = \Sigma_{0 \preccurlyeq s' \preccurlyeq s} G(s') c(s', s)$ for all s,
2. $G(s) = \Sigma_{0 \preccurlyeq s' \preccurlyeq s} F(s')$ for all s.

We apply this to a $p \in \mathcal{P}$ by defining for each s

$$\lambda_s = \sum_{0 \preccurlyeq s' \preccurlyeq s} (\ln p)_{s'}^* c(s', s)$$

Note that λ_s is a function on $\mathcal{Y}$, because each $(\ln p)_s^*$ is a function on $\mathcal{Y}$. The Möbius Inversion Formula is then

$$\ln p = \sum_s \lambda_s$$

where the summation is over all s, *provided* that $\{s\}$ contains a largest element $\bar{s}$ and $\ln p \in \mathbb{S}_{\bar{s}}$. In a sense $\{\lambda_s\}$ is a parametrization of $\mathcal{E}[q : \mathbf{u}_{\bar{s}}]$, since the last two formulas above give the λ_s uniquely in terms of p, and p uniquely in terms of the λ_s. It is a bit awkward to parametrize a model with collections of functions, but in standard situations things can be simplified.

An Example. Consider a 2^2 table with factors A and B. The index s will indicate which of the indicators u_A and u_B are present, according to the following scheme:

s	$\mathbb{S}_s$
0	$\mathbb{S}(1)$
A	$\mathbb{S}(1, u_A)$
B	$\mathbb{S}(1, u_B)$
A, B	$\mathbb{S}(1, u_A, u_B)$
$[AB]$	$\mathbb{S}(1, u_A, u_B, u_{AB})$

The Möbius function for this collection is shown in Table IV.3. Now using q to stand for the uniform distribution, we compute

$$\lambda_0 = (\ln p)_0^* = \tfrac{1}{4}[\ln p(00) + \ln p(10) + \ln p(01) + \ln p(11)]$$

$$\begin{aligned}\lambda_A = (\ln p)_A^* - \lambda_0 &= \tfrac{1}{2}[\ln p(00) + \ln p(01)](1 - u_A) \\ &\quad + \tfrac{1}{2}[\ln p(10) + \ln p(11)]u_A - \lambda_0 \\ &= \tfrac{1}{4}[\ln p(00) + \ln p(01) - \ln p(10) - \ln p(11)](1 - u_A) \\ &\quad + \tfrac{1}{4}[\ln p(10) + \ln p(11) - \ln p(00) - \ln p(01)]u_A\end{aligned}$$

Similarly,

$$\begin{aligned}\lambda_B = &\tfrac{1}{4}[\ln p(00) + \ln p(10) - \ln p(01) - \ln p(11)](1 - u_B) \\ &+ \tfrac{1}{4}[\ln p(01) + \ln p(11) - \ln p(10) - \ln p(00)]u_B\end{aligned}$$

Table IV.3

The Möbius Function $c(s', s)$ for the Collection of Models on a Two-Way Table

	s				
s'	0	A	B	A, B	$[AB]$
0	1	−1	−1	1	0
A	0	1	0	−1	0
B	0	0	1	−1	0
A, B	0	0	0	1	−1
$[AB]$	0	0	0	0	1

Then

$$\lambda_{A,B} = (\ln p)^*_{A,B} - (\ln p)^*_A - (\ln p)^*_B + (\ln p)^*_0$$

Because A and B are independent under the uniform distribution, the projection $(\ln p)^*_{A,B}$ onto $\mathbb{S}(1, u_A, u_B)$ is exactly $(\ln p)^*_A + (\ln p)^*_B - (\ln p)^*_0$, so that $\lambda_{A,B} = 0$. Finally,

$$\begin{aligned}
\lambda_{[AB]} &= (\ln p)^*_{[AB]} - (\ln p)^*_{A,B} \\
&= \ln p - \tfrac{1}{2}[\ln p(00) + \ln p(01)](1 - u_A) \\
&\quad - \tfrac{1}{2}[\ln p(10) + \ln p(11)]u_A \\
&\quad - \tfrac{1}{2}[\ln p(00) + \ln p(10)](1 - u_B) \\
&\quad - \tfrac{1}{2}[\ln p(01) + \ln p(11)]u_B \\
&\quad + \tfrac{1}{4}[\ln p(00) + \ln p(10) + \ln p(01) + \ln p(11)] \\
&= \tfrac{1}{4}[\ln p(00) - \ln p(10) - \ln p(01) + \ln p(11)] \\
&\quad \times (2u_A - 1)(2u_B - 1)
\end{aligned}$$

There are several features of this parametrization. First, it is clear that λ_0 depends only on $\ln p(00)p(10)p(01)p(11)$, λ_A depends on $\ln\{[p(00)p(01)]/[p(10)p(11)]\}$, λ_B depends on $\ln\{[p(00)p(10)]/[p(01)p(11)]\}$, and $\lambda_{[AB]}$ depends on $\ln\{[p(00)p(11)]/[p(01)p(10)]\}$. Thus there are four underlying parameters, one of which (say λ_0) is redundant. This kind of simplification occurs in higher-dimensional models as well. Secondly, $\lambda_{A,B} = 0$ because $\mathbb{S}_A$ and $\mathbb{S}_B$ had the property that the projection onto $\mathbb{S}_{A,B}$ was the sum of the projections on $\mathbb{S}_A$ and $\mathbb{S}_B$ minus the projection on $\mathbb{S}_0$. This happened because under q any $v \in \mathbb{S}_A$ was independent of any $w \in \mathbb{S}_B$. If q had been chosen differently, then $\lambda_{A,B} \neq 0$ would have been possible. If we were to delete $\mathbb{S}_{A,B}$, then in the partial order $\mathbb{S}_{[AB]}$ would replace $\mathbb{S}_{A,B}$ and thus inherit its part of the Möbius function. This gives

$$\lambda_{[AB]} = (\ln p)^*_{[AB]} - (\ln p)^*_A - (\ln p)^*_B + (\ln p)^*_0$$

which coincides with the customary definition.

Another Example. The preceding example illustrates the general principle that if $\mathbb{S}_{s''}$ is the collection of all functions which can be written as a sum of a function in $\mathbb{S}_s$ and a function in $\mathbb{S}_{s'}$, then $\lambda_{s''} = 0$ provided functions in $\mathbb{S}_s$ are independent of functions in $\mathbb{S}_{s'}$ under q. Thus, one can simplify the collection of models which specify the parametrization. For instance, in a three-way cross-classification with factors A, B, and C, it is enough to specify the models 0, A, B, C, $[AB]C$, $A[BC]$, $[AC]B$, and $[ABC]$. The Möbius function appears in Table IV.4. Once this principle is recognized, extension to higher-dimensional tables is easy.

The customary notation for log-linear models differs from that used here. There are two schemes in use, which we illustrate for the three-way table. Letting ijk stand for $AiBjCk$, one writes

$$\ln p(ijk) = \lambda + \lambda_i^A + \lambda_j^B + \lambda_k^C + \lambda_{ij}^{AB} + \lambda_{ik}^{AC} + \lambda_{jk}^{BC} + \lambda_{ijk}^{ABC}$$

Another scheme uses $u = \lambda$, $u_{A(i)} = \lambda_i^A$, $u_{AB(ij)} = \lambda_{ij}^{AB}$, and so on. Whenever such a parametrization is used it is necessary to mention that many sums are zero; for instance, $\lambda_+^A = 0$, $\lambda_{+j}^{AB} = 0$, $\lambda_{i+}^{AB} = 0$, $\lambda_{i+k}^{ABC} = 0$, and so on. The condition is that any set of λ's summed over any subscript vanishes, and since this is necessary to ensure that the λ's are identifiable, it is called the *identifiability condition*. The relation between the customary notation and our notation is

$$\begin{aligned} \lambda_0(ijk) &= \lambda, & \lambda_{[AB]C}(ijk) &= \lambda_{ij}^{AB} \\ \lambda_A(ijk) &= \lambda_i^A, & \lambda_{A[BC]}(ijk) &= \lambda_{jk}^{BC} \\ \lambda_B(ijk) &= \lambda_j^B, & \lambda_{[AC]B}(ijk) &= \lambda_{ik}^{AC} \\ \lambda_C(ijk) &= \lambda_k^C, & \lambda_{[ABC]}(ijk) &= \lambda_{ijk}^{ABC} \end{aligned}$$

Note that the identifiability condition follows directly from the definitions

Table IV.4

The Möbius Function on the Collection of Models on a Three-Way Table

	0	A	B	C	$[AB]C$	$A[BC]$	$[AC]B$	$[ABC]$
0	1	−1	−1	−1	2	2	2	−4
A	0	1	0	0	−1	−1	−1	2
B	0	0	1	0	−1	−1	−1	2
C	0	0	0	1	−1	−1	−1	2
$[AB]C$	0	0	0	0	1	0	0	−1
$A[BC]$	0	0	0	0	0	1	0	−1
$[AC]B$	0	0	0	0	0	0	1	−1
$[ABC]$	0	0	0	0	0	0	0	1

Table IV.5

Two sets of Functions for Parametrizing Models on the 2^2 Table

y	u_A	u_B	u_{AB}	v_A	v_B	v_{AB}
00	0	0	0	−1	−1	1
01	0	1	0	−1	1	−1
10	1	0	0	1	−1	−1
11	1	1	1	1	1	1

and the Möbius inversion. Establishing the Möbius inversion from the identifiability conditions is harder, and cannot be generalized easily.

The Hierarchy Principle. The classical parametrization makes nonhierarchical models quite confusing, and so it is often said that nonhierarchical log-linear models are not sensible. For instance, consider the nonhierarchical model $A\,AB$ on the 2^2 table. In the customary notation one is tempted to write $\ln p(ij) = \lambda = \lambda_i^A + \lambda_{ij}^{AB}$ as the parametrization, but this is in fact ambiguous. To see why, look at the u- and v-functions in Table IV.5. Either set could be used for a classical parametrization of the hierarchical models, but $\mathbb{S}(1, u_A, u_{AB}) \neq \mathbb{S}(1, v_A, v_{AB})$, so that the model $A\,AB$ itself depends on what set of functions was used. One is better off using the exponential form for nonhierarchical models, and then careful attention should be paid to the interpretation of the parameters.

Collapsibility. The classical approach to parametrization makes the following kind of question seem natural. Suppose we have a cross-classification with cells ij. (Each of i and j could itself represent a cross-classification, but we avoid complicating the notation to reflect this.) We then consider, as before, a parametrization

$$\lambda_s = \sum_{0 \preccurlyeq s' \preccurlyeq s} (\ln p)^*_{s'} c(s', s)$$

We temporarily abuse the notation by writing $\bar{p}(i) = p(i+)$. If $\mathbb{S}_s$ corresponds to a model specified only in terms of i (or one of its subfactors), then let $\mathbb{S}_{\bar{s}}$ stand for the corresponding set of functions defined on the i-marginal table. Thus there is a collection $\{\bar{s}\}$ corresponding to a subset of $\{s\}$. Write

$$\lambda_{\bar{s}} = \sum_{0 \preccurlyeq \bar{s}' \preccurlyeq \bar{s}} (\ln \bar{p})^*_{\bar{s}'} \bar{c}(\bar{s}', \bar{s})$$

where $\bar{c}$ is the Möbius function of $\{\bar{s}\}$. Now the question is, if $\mathbb{S}_{\bar{s}}$ corresponds to $\mathbb{S}_s$, when do we have $\lambda_{\bar{s}} = \lambda_s$? The idea is that λ_s measures something of interest about i in the ij-table, and $\lambda_{\bar{s}}$ measures the same thing in the i-marginal table. If $\lambda_{\bar{s}} = \lambda_s$, then the table is *collapsible along j with respect to* λ_s in the sense that it makes no difference whether one looks at the ij-table or the i-table in studying λ_s. This is a particular case of the general homogeneity problem, which will be considered in Chapter VI, Section D.

Marginal and Partial Association. When faced with a cross-classification and no specific hypotheses to test, it is natural to ask what parameters need to be included in a model to fit the data well. We can set up the ANALYSIS table for the collection of hierarchical models, but for tables involving four or more factors this is quite unwieldy unless there is a reasonable principle for trimming down the collection of models to be considered. The notions of marginal and partial association make it possible to *screen* the data and make a preliminary judgment about the parameters that will be required.

In general, we are interested in whether or not to include a particular parameter in the model. We have used θ for exponential-model parametrizations and λ for classical log-linear models, but since we are dealing only with hierarchical cross-classifications, we can use the letters of the factors to indicate the effect we are investigating, and our conclusion will be the same regardless of the parametrization. In both the marginal and partial approaches, we fit a model containing the effect and one not containing it, and use the customary Chi-square statistic for assessing significance. The only difference between the approaches is in the choice of models.

For ease of exposition, suppose we have factors A, B, C, D, and E. The test for marginal association of A, B, and C is made by collapsing the table to the A, B, C-margin, and then testing $\lambda^{ABC} = 0$ (or $\theta_{ABC} = 0$) in that marginal table. The test for partial association is made by fitting the largest model of third order and then dropping the ABC-term from that model. Thus we have

Marginal association: drop ABC from $[ABC]$ in the A, B, C-marginal table.

Partial association: drop ABC from the model

$$[ABC][ABD][ABE][ACD][ACE][BCD][BCE][CDE][ADE][BDE]$$

Likewise, for testing the $ABCD$ term,

Marginal association: drop $ABCD$ from $[ABCD]$ in the A, B, C, D-marginal table.

Partial association: drop $ABCD$ from the model

$$[ABCD][ABCE][ABDE][ACDE][BCDE]$$

Naturally, there are many other models from which these terms might have been dropped. But the basic idea is that the partial-association test accounts for the presence of the other factors to a reasonable degree, while the marginal-association test ignores the other factors. It would be nice if all other tests had P-values between those of the marginal and partial association tests, but this is not true in general. Nevertheless, these two tests provide a reasonable basis for forming an initial impression of the importance of various effects in situations where a more detailed analysis would be excessively tedious.

Interpretation of Hierarchical Models. If one *plans* the inference of a set of data, one is likely to consider only models that have some substantive interpretations. But if one uses the model-building techniques of the preceding section, it is very easy to arrive at a final fitted model whose interpretation is unclear. There are a couple of tricks for making simpler assertions that follow from a complex model.

The first trick involves *conditionalization*. Recall that $[AB][BC]$ has the interpretation that A and C are conditionally independent given B. We can see this because the latter condition says

$$\frac{p(ijk)}{p(+j+)} = \frac{p(ij+)}{p(+j+)}\,\frac{p(+jk)}{p(+j+)}$$

so that $p(ijk) = p(+j+)p(ij+)p(+jk)$, and this is equivalent to $\ln p \in \mathbb{S}(1, \mathbf{u}_A, \mathbf{u}_B, \mathbf{u}_C, \mathbf{u}_{AB}, \mathbf{u}_{BC})$. We write $AC|B$ for the model asserting A and C independent given B, and we can now write $[AB][BC] = AC|B$. By the same kind of reasoning we can write $[ABC][BD] = [AC]D|B$. It would seem that we should also be able to write $[ABC][BD] = [AB][BD]|C$, but this is not correct. To see why, it is enough to note that $[ABC][BCD] = [AB][BD]|C$ would also be justified, giving the contradiction $[ABC][BD] = [ABC][BCD]$.

In order to be able to make assertions about conditional distributions in higher-way tables, where interpretation is the greatest problem, we will introduce the notation $\uparrow_F$ to denote conditioning on the factor F. Thus $[AB][BC]\uparrow_B AC$ means that "$[AB][BC]$ when conditioned on B gives AC." Note that the model AC in this context refers to the model of conditional distributions of (A, C) given B. We also have $[ABC][BD]\uparrow_B[AC]D$ and $[ABC][BD]\uparrow_C[AB][BD]$. This latter assertion is that every distribution in $[ABC][BD]$ is also in $[AB][BD]|C$. The reversed assertion is not implied, since there are distributions in $[AB][BD]|C$ which are not in $[ABC][BD]$. There is a simple rule for obtaining the form of the conditional models:

Conditioning Rule. To determine $\mathfrak{N}$ in $\mathfrak{M} \uparrow_F \mathfrak{N}$, simply cross out all occurrences of F and its subfactors in the model $\mathfrak{M}$.

Thus, for example, $[ABC][BD] \uparrow_{BC} A\ D$, which could have also been done in two steps,

$$[ABC][BD] \uparrow_C [AB][BD] \uparrow_B A\ D$$

The conditioning rule is useful for determining which assertions follow from a *confirmed* model. When $\mathfrak{M} \uparrow_F \mathfrak{N}$, in general the model $\mathfrak{N}|F$ will be larger that $\mathfrak{M}$, and so confirmation of $\mathfrak{M}$ implies confirmation of $\mathfrak{N}|F$. But rejection of $\mathfrak{M}$ implies nothing. In fact, rejection of $\mathfrak{N}|F$ would imply rejection of $\mathfrak{M}$, but this is not the usual use of the conditioning rule.

The other aid to interpretation is *marginalization*. We write $\mathfrak{M} \downarrow_F \mathfrak{N}$ to mean that any distribution in the model $\mathfrak{M}$, when collapsed over the factor F, gives a distribution in the model $\mathfrak{N}$. Note that in this context $\mathfrak{N}$ refers to a model on the marginal table of all factors not in F. For example, we can establish that $[ABC][BD] \downarrow_A [BC][BD]$, that is, when the table is collapsed along A, one obtains distributions in $[BC][BD]$ on the B, C, D-margin. To see this, let $ijkm$ stand for $AiBjCkDm$ and write

$$\begin{aligned}
p(+jkm) &= \sum_i p(ijkm) \\
&= \sum_i \exp\big[\boldsymbol{\theta}_A'\mathbf{u}_A(i) + \boldsymbol{\theta}_B'\mathbf{u}_B(j) + \boldsymbol{\theta}_C'\mathbf{u}_C(k) + \boldsymbol{\theta}_D'\mathbf{u}_D(m) \\
&\quad + \boldsymbol{\theta}_{AB}'\mathbf{u}_{AB}(ij) + \boldsymbol{\theta}_{AC}'\mathbf{u}_{AC}(ik) + \boldsymbol{\theta}_{BC}'\mathbf{u}_{BC}(jk) \\
&\quad + \boldsymbol{\theta}_{ABC}'\mathbf{u}_{ABC}(ijk) + \boldsymbol{\theta}_{BD}'\mathbf{u}_{BD}(jm) - \delta\big]
\end{aligned}$$

Now note that on the B, C, D-table the function

$$\sum_i \exp\big[\boldsymbol{\theta}_A'\mathbf{u}_A(i) + \boldsymbol{\theta}_{AB}'\mathbf{u}_{AB}(ij) + \boldsymbol{\theta}_{AC}'\mathbf{u}_{AC}(ik) + \boldsymbol{\theta}_{ABC}'\mathbf{u}_{ABC}(ijk)\big]$$

is in $\mathbb{S}(1, \mathbf{u}_B, \mathbf{u}_C, \mathbf{u}_{BC})$, and so it can be written as

$$\exp[\boldsymbol{\zeta}_B'\mathbf{u}_B + \boldsymbol{\zeta}_C'\mathbf{u}_C + \boldsymbol{\zeta}_{BC}'\mathbf{u}_{BC}]$$

Thus,

$$\begin{aligned}
p(+jkm) = \exp\big[&\boldsymbol{\theta}_B'\mathbf{u}_B(j) + \boldsymbol{\theta}_C'\mathbf{u}_C(k) + \boldsymbol{\theta}_D'\mathbf{u}_D(m) + \boldsymbol{\theta}_{BD}'\mathbf{u}_{BD}(jm) \\
&+ \boldsymbol{\zeta}_B'\mathbf{u}_B(j) + \boldsymbol{\zeta}_C'\mathbf{u}_C(k) + \boldsymbol{\theta}_{BC}'\mathbf{u}_{BC}(jk) + \boldsymbol{\zeta}_{BC}'\mathbf{u}_{BC}(jk) - \delta\big]
\end{aligned}$$

which shows that $p(+jkm)$ is in $[BC][BD]$. The general method of obtaining the form of the marginal model is:

Margining Rule. To determine $\mathfrak{N}$ in $\mathfrak{M} \downarrow_F \mathfrak{N}$, first pool all effects (bracketed terms) which have any factor in common with F, then cross out all factors in F from the resulting array.

For example, to solve $[AB][BC][AD] \downarrow_B \mathfrak{N}$, we first pool $[AB]$ and $[BC]$ to form $[ABC]$, and then cross out B to get $\mathfrak{N} = [AC][AD]$. To solve $[ABCD][ACE][DF] \downarrow_{BC} \mathfrak{N}$, pool $[ABCD]$ and $[ACE]$ to form $[ABCDE]$, then cross out BC to get $\mathfrak{N} = [ADE][DF]$. This can also be done in two steps,

$$[ABCD][ACE][DF] \downarrow_B [ACD][ACE][DF] \downarrow_C [ADE][DF]$$

Note that marginalization is also useful only for drawing conclusions from confirmed models. From a given confirmed model it may be of interest to see what the consequences are of both marginalization and conditionalization. An exhaustive analysis of $[ABC][BD]$ would be

$$
\begin{aligned}
[ABC][BD] &\uparrow_A [BC][BD] \uparrow_B CD \\
&\uparrow_B [AC]D \\
&\uparrow_C [AB][BD] \uparrow_B AD \\
&\uparrow_D [ABD] \\
&\downarrow_A [BC][BD] \uparrow_B CD \\
&\downarrow_B [ACD] \\
&\downarrow_C [AB][BD] \uparrow_B AD \\
&\downarrow_D [ABC]
\end{aligned}
$$

so we could confirm $CD|A, B$; $AD|B, C$; $CD|B$; and $AD|C$. Note that these last two models are on the B, C, D- and A, C, D-margins, respectively. Note also that $[ABC][BD] \downarrow_D [ABC]$ means that no conclusions about the A, B, C-margin can be drawn. Another example of this form is $[AB][AC][BC] \downarrow_A [BC]$, that is, the model without a three-factor interaction has no implication for the two-way marginal tables. Note also that $[AB][BC][AC] \uparrow_A [BC]$, so that the important interpretation of the no-three-factor-interaction model is not revealed by either marginalization or conditionalization.

4. Residuals

The discrepancy statistics may not tell the whole story concerning either model fitting or hypothesis testing. For this reason one often wants to look directly at the fitted distribution $\hat{q}$ and consider departures from the model expressed by $\hat{p}(y) - \hat{q}(y)$, and departures from q expressed by $\hat{q}(y) - q(y)$. Both of these quantities are residuals, and for the purpose of comparison we would like to divide each by its standard deviation, or an estimate of its standard deviation.

Proposition 32. Let $\binom{\mathbf{u}}{\mathbf{v}}$ be maximal π-standard, and suppose that $\hat{q}_n$ converges rapidly on $\mathcal{S}(\mathbf{v})$. Then for every y,

1. $\sqrt{n}\,[\hat{p}_n(y) - \hat{q}_n(y)] \to_d N(0, \|\mathbf{v}(y)\|^2\pi(y)^2)$,
2. $\sqrt{n}\,[\hat{q}_n(y) - \pi(y)] \to_d N(0, \|\mathbf{u}(y)\|^2\pi(y)^2)$,

and moreover, every term of the form 1 is asymptotically independent of every term of the form 2.

Proof. Result 1 is immediate from

$$\sqrt{n}\,[\hat{p}(y) - \hat{q}(y)] = \sqrt{n}\,\{M[\mathbf{u}:\hat{p}]'\mathbf{u}(y) - M[\mathbf{u}:\hat{q}]'\mathbf{u}(y)\}\pi(y)$$
$$+\sqrt{n}\,\{M[\mathbf{v}:\hat{p}]'\mathbf{v}(y) - M[\mathbf{v}:\hat{q}]'\mathbf{v}(y)\}\pi(y)$$

since the right-hand side is asymptotically equivalent to $\sqrt{n}\,M[\mathbf{v}:\hat{p}]'\mathbf{v}(y)\pi(y)$ directly from the definition of rapid convergence and the MWT. Likewise, for result 2

$$\sqrt{n}\,[\hat{q}(y) - \pi(y)] = \sqrt{n}\,M[\mathbf{u}:\hat{q}]'\mathbf{u}(y)\pi(y)$$

and the right side is asymptotically equivalent to $\sqrt{n}\,M[\mathbf{u}:\hat{p}]'\mathbf{u}(y)\pi(y)$. □

Recall that both ML and MDI distributions $\hat{q}$ converge rapidly, so that the above proposition pertains to ML inference for exponential models and MDI inference for linear models.

Notice that if we let w denote the indicator of a specific $y \in \mathcal{Y}$, then

$$\frac{1}{\pi}w = 1 + M\left[\frac{1}{\pi}w\mathbf{u}:\pi\right]'\mathbf{u} + M\left[\frac{1}{\pi}w\mathbf{v}:\pi\right]'\mathbf{v}$$

and so evaluating at y we get

$$\|\mathbf{u}(y)\|^2 + \|\mathbf{v}(y)\|^2 = \frac{1-\pi(y)}{\pi(y)}$$

Consequently, an upper bound on the variance term in result 1 is

$$\|\mathbf{v}(y)\|^2 \pi(y)^2 \leqq \pi(y)[1 - \pi(y)]$$

and in result 2 is

$$\|\mathbf{u}(y)\|^2 \pi(y)^2 \leqq \pi(y)[1 - \pi(y)]$$

Thus, as a matter of convenience, one might replace the variances by their estimated upper bounds. We shall use the following terminology:

$$\frac{\sqrt{n}\,[\hat{p}(y) - \hat{q}(y)]}{\sqrt{\hat{p}(y)[1 - \hat{p}(y)]}} = \mathsf{F-RES}$$

$$\frac{\sqrt{n}\,[\hat{p}(y) - \hat{q}(y)]}{\sqrt{\hat{q}(y)[1 - \hat{q}(y)]}} = \mathsf{RES-F}$$

$$\frac{\sqrt{n}\,[\hat{q}(y) - q(y)]}{\sqrt{\hat{q}(y)[1 - \hat{q}(y)]}} = \mathsf{G-RES}$$

$$\frac{\sqrt{n}\,[\hat{q}(y) - q(y)]}{\sqrt{q(y)[1 - q(y)]}} = \mathsf{RES-G}$$

These quantities are printed out by LAMDA under the columns with the above names. These are timid residuals because of the tendency to overestimate their variances. If a more precise analysis is desired, then the variances given in Proposition 32 can be estimated directly, using either $\hat{p}$, $\hat{q}$, or q in place of π.

If $\hat{q}$ fits the data poorly, then we look at **F − RES** rather than **RES − F**, whereas if $\hat{q}$ fits well, we do the reverse. A similar remark applies in the choice between **G − RES** and **RES − G**. LAMDA prints out both sorts of residuals partly for this reason, and also because in an MDI analysis the roles of $\hat{p}$ and q are reversed.

One use of the residuals is as an aid in discovering why a poorly fitting model fits poorly. The distribution of magnitudes and signs of the residuals in the table may give a clue to what sort of model might fit better. Another use is as a check against the possibility of fitting a model which does well overall, but in which a small number of cells show a rather poor fit. Such a situation might arise because of a classification error or data-entry error, or might be a warning that although the model fits well over a broad range of cells, there are regions in the table where it does not fit.

The Uniform Plot. One simple technique for assessing the residuals is the uniform plot. Letting the P-value of the ith residual be denoted P_i, reorder the P-values from smallest to largest, the ith smallest now being denoted $P_{(i)}$. The uniform plot is then a plot of the pairs $(i, P_{(i)})$ for $1 \leqq i \leqq \#\mathcal{Y}$. If $\{P_i\}$ formed a random sample from a uniform distribution, then the mean of $P_{(i)}$ would be $i/(m+1)$ and its variance would be

$$\frac{\dfrac{i}{m+1}\left(1 - \dfrac{i}{m+1}\right)}{(m+1)(m+2)}, \qquad \text{where} \quad m = \#\mathcal{Y}$$

Although the collection of P-values is not such a random sample (due to the linear restraints which define the fitted distribution), the uniform plot is still a useful diagnostic aid. A LAMDA program which may be easily adapted to produce a uniform plot is given in Run IV.B.1.

It is possible to do formal hypothesis testing on the residuals by treating each as a $N(0,1)$ and using the significance level α/m for each test, so that the overall significance level will be α. This is the Bonferroni procedure for testing residuals, and has the property that the probability of falsely detecting any significant residuals is no more than α.

5. Multiple Confidence Intervals for Means

It is sometimes required to estimate the quantities $M[w : \pi]$ for more than one function w. In the case of a single w, an asymptotically valid confidence interval (CI) with confidence coefficient $1 - \alpha$ is given by

$$M[w : \hat{p}] \pm \frac{\xi}{\sqrt{n}} \sqrt{V[w : \hat{p}]}$$

where $F_0(\xi) = 1 - \alpha/2$ (see Problem II.4). If several functions $w_1, w_2, \ldots, w_k$ are of interest, and it is desired to have a simultaneous CI for all of them, then the *Bonferroni CI* for the $M[w_i : \pi]$ is

$$M[w_i : \hat{p}] \pm \frac{\xi}{\sqrt{n}} \sqrt{V[w_i : \hat{p}]} \qquad (1 \leqq i \leqq k)$$

where $F_0(\xi) = 1 - \alpha/2k$. If a model were being fitted along with this procedure, then it would be legitimate to use $\hat{q}$ in place of $\hat{p}$.

It is possible to provide a simultaneous CI for all $M[w : \pi]$ $[w \in \mathcal{S}(1, \mathbf{u})]$. We first observe that from the Cauchy–Schwartz inequality we have for any

$q \in \mathcal{P}$

$$M[w:q] - M[w:\pi] = M\left[w\frac{q-\pi}{q}:q\right] \leqq \sqrt{V[w:q]\mathbb{D}_{-1}(q,\pi)}$$

If we take $\hat{q}$ to be ML in $\mathfrak{S}[q:\mathbf{u}]$, then $M[\mathbf{u}:\hat{q}] = M[\mathbf{u}:\hat{p}]$ and $\hat{q}$ converges rapidly on $\mathbb{S}(\mathbf{v})$ for any π-complement of $\mathbf{u}$, and so we have the asymptotically valid CI

$$M[w:\hat{p}] \pm \frac{\xi}{\sqrt{n}}\sqrt{V[w:\hat{q}]} \qquad [w \in \mathbb{S}(1,\mathbf{u})]$$

where $F_\nu(\xi^2) = 1 - \alpha$, and $\nu + 1$ is the dimension of $\mathbb{S}(1,\mathbf{u})$. If $\mathbf{u}$ has affinely linearly independent components, then $\nu = \#\mathbf{u}$. The form above would be used if a model had been fitted, but if there were no model we would replace $\hat{q}$ by $\hat{p}$ on the right. This is called the *S-method* of constructing a simultaneous CI.

The S-method appears to be quite comprehensive because it produces a simultaneous CI for the infinite collection of w's in an entire space $\mathbb{S}(1,\mathbf{u})$. However, it does so at the cost of providing rather wide CIs. For example, if interest centered on the quantities $M[w_i:\pi]$ $(1 \leqq i \leqq k)$, then one could use either the Bonferroni approach or the S-method applied to $\mathbb{S}(1, w_1, \ldots, w_k)$. The lengths of the CIs are determined by the constants ξ which appear in the definitions. In order to determine which are shorter, enter Table IV.6 with $\nu + 1$ equal to the dimension of $\mathbb{S}(1, w_1, \ldots, w_k)$. If $k < K$ then the Bonferroni intervals are shorter, while if $k \geq K$ the S-method intervals are preferred. It does not take much experience with this table to arrive at the conclusion that the Bonferroni intervals tend to be better.

Table IV.6

Values of K Such that the Bonferroni Intervals Will Be Shorter Than the S-method Intervals for Fewer than K Functions Which Generate a Space of Dimension ν

ν	K
1	6
2	16
3	37
4	85
5	178

B. DATA EXAMPLES

1. 2^3 Table

The data on the gun-registration questionnaire (Table III.24), which was analyzed in Chapter III, Section B.2, will be treated at greater length here. We will use the ANALYSIS diagram of Figure IV.2 with its individual significance levels for the analysis of the data. The LAMDA run which fits the required models appears in Run IV.B.2. The important statistics from the output are $n\mathbb{D}_1(\hat{p}, \hat{q}_i)$, which appear as $D(P, QP)$. The results are displayed in Table IV.7. The $\bar{P}$-values associated with the models are the largest of the serial P-values, as described in Section A.2. For example, to compute $\bar{P}$ for $[FY]G$ we need to consider all $P_i^{w_i}$ for all arrows along chains pointing to $[FY]G$. These are

$$[FGY] \overset{.8994}{\rightarrow} [FG][GY][FY] \overset{.2822}{\rightarrow} [FY][GY] \overset{.6375}{\rightarrow} [FY]G$$

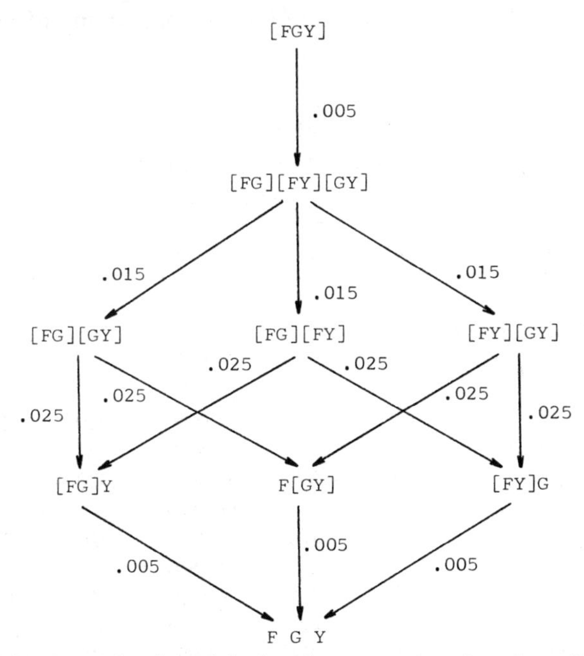

Fig. IV.2 The hierarchical models for the gun-registration data. The individual significance levels appear next to the arrows.

and

$$[FGY] \overset{.8994}{\rightarrow} [FG][GY][FY] \overset{.6759}{\rightarrow} [FG][FY] \overset{.2378}{\rightarrow} [FY]G$$

Now for the upper chain, min = min{.8994, .2822, .6375} = .2822, and so

$$\bar{P} = 1 - (1 - .2822^{1/.1887})(1 - .005)(1 - .2822^{1/.2722})$$
$$= .016$$

where the exponents are the $1/w_i$, and $w_i = -1/\ln \alpha_i$. For the lower chain, min = .2378 and

$$\bar{P} = 1 - (1 - .2378^{1/.1887})(1 - .2378^{1/.2381})(1 - .005)$$
$$= .008$$

The higher of these is $\bar{P}_{max} = .016$.

In Table IV.7 the rejected models are marked R, and the maximal rejected and minimal confirmed models are marked with an asterisk. We are permitted to say that no matter where π is, the probability of rejecting any

Table IV.7

Analysis of the Data from the Gun-registration Questionnaire

	Model	df	χ^2	Next	$\bar{\alpha}$	$\langle \bar{P} \rangle$
1	$[FGY]$	—	—	2⟨.570⟩	—	—
2	$[FG][GY][FY]$	1	0.322	3⟨.193⟩	.005	⟨.570⟩
				4⟨.902⟩		
				5⟨.005⟩		
3	$[FG][FY]$	2	2.015	7⟨.005⟩	.020	⟨.294⟩
				8⟨.839⟩		
4	$[FG][GY]$	2	0.377	6⟨.005⟩	.020	⟨.845⟩
				8⟨.190⟩		
5	$[FY][GY]$	2	8.109	6⟨.839⟩	.020	⟨.006⟩ R*
				7⟨.190⟩		
6	$F[GY]$	3	8.150	9⟨.190⟩	.045	⟨.016⟩ R
7	$[FY]G$	3	9.828	9⟨.839⟩	.045	⟨.016⟩ R
8	$[FG]Y$	3	2.056	9⟨.005⟩	.045	⟨.461⟩ *
9	FGY	4	9.870		.050	⟨.049⟩ R

Table IV.8

Parameter Estimates for the Gun-registration Data

i	G	Y	F	FG
$\hat{\theta}_i$	1.0342	0.3147	0.1500	−0.2709
t_i	14.81	7.08	1.83	−2.79

model which contains π is no more than .05. Since the maximal rejected model is $[FY][GY]$, we may conclude that the form of the question (F) and opinion on gun registration (G) are not independent given the year (Y). The minimal confirmed model is $[FG]Y$, so that we may conclude that the joint factor (F, G) is independent of Y, that is, the joint distribution of (F, G) is the same within each of the two years. The parameter estimates appear in Table IV.8, and one may verify that θ_{FG} is the ln odds ratio common to the two Y-subtables corresponding to $Y0$ and $Y1$. Thus we have a common negative relation between F and G independently of the year, and the sign of this relationship is reasonable because the higher level ($F1$) of F denotes a question slanted against gun registration while the higher level ($G1$) of G denotes favoring gun registration.

The relevant standardized residuals are in the output under **RES-F**, and are reproduced in Table IV.9. Of course, all these values are small when referred to the standard normal distribution, but the pattern of signs is striking. Changing $-$ to 0 and $+$ to 1 in Table IV.9, we discover the function $(1 - u_Y)(1 - u_G) + u_{GY}$. Thus the pattern of signs points in the direction of including u_{GY} in the model. This would yield $[FG][GY]$, which we see from the ANALYSIS table would fit the data better, though not statistically significantly better.

If we had been model-fitting instead of hypothesis-testing, then $[FG][GY]$ and $[FG]Y$ would both be candidates for the final fitted model, due to their

Table IV.9

Standardized Residuals in the Fitted Model [FG] Y

	$Y0$		$Y1$	
	$F0$	$F1$	$F0$	$F1$
$G0$	.826	.417	−.714	−.361
$G1$	−.659	−.153	.584	.135

high serial P-values. In part, the decision between them depends on how one regards the P-value of the test $[FG][GY] \to [FG]Y$, which is .190. However, it is also worthwhile to take the interpretability of the models into account, and since $[FG][GY]$ says that F and Y are independent given G, and it is awkward to think of G as an appropriate factor to condition on, we might settle on $[FG]Y$ as the final fitted model. It is convenient that the final fitted model and minimal confirmed model coincide in this case, but of course this will not happen in general.

2. 2 × 3 × 4 Table

The data in Table IV.10 are from a survey with follow-up in which voters were first asked to express their political preference (P), and then upon follow-up their actual voting was ascertained. The voters were further classified according to their income level (I). In analyzing this data we use the customary factorial indicators $\mathbf{u}_I$, $\mathbf{u}_P$, and u_V. Since the factor I could be considered an interval-level measurement, we introduce the u-function

Table IV.10

The Political Preference Data

	$I0$		$I1$		$I2$		$I3$	
	$V0$	$V1$	$V0$	$V1$	$V0$	$V1$	$V0$	$V1$
$P0$	112	7	83	13	76	8	86	11
$P1$	67	37	75	45	67	57	67	63
$P2$	5	35	3	28	9	35	14	71

P = Political Preference
- $P0$ = Democrat
- $P1$ = Independent
- $P2$ = Republican

V = Actual Vote
- $V0$ = Democrat
- $V1$ = Republican

I = Income Level
- $I0$ = < 10,000 per year
- $I1$ = 10,000–15,000
- $I2$ = 15,000–20,000
- $I3$ = > 20,000

which is linear in I, $u_J(ijk) = i$, where ijk stands for $IiVjPk$. We consider the hierarchical submodels of $[IVP]$ and their linear counterparts in $[JVP]$. The LAMDA program for fitting all these models is in Run IV.B.3.

From the output one immediately notices that the chi-square associated with $[IV][IP]$ is 343.605 with 8 df. Before proceeding we may note that this enormous chi-square value will certainly lead to the rejection of $[IV][IP]$ and all its submodels. Thus we may simplify the ANALYSIS table by ignoring these models. Since $[IV][IP]$ asserts that voting and voting preference are independent within income levels, and it would be regarded as phenomenal if this model fitted the data, we have cut down the possible models in a reasonable way.

We now compute the serial P-values as before, assuming all individual significance levels are equal. In order to have an overall .05 significance level, each individual level must be .0125. The results appear in Table IV.11 and show that $[JP][VP]$ is the minimal confirmed model, which says that income and voting are independent given political preference. If one were model-fitting, then probably $[IP][VP]$ would be the fitted model, and of course this differs from $[JP][VP]$ only in that the income effect is not required to be linear.

The **RES-F** residuals from $[JP][VP]$ appear in Table IV.12. These values all appear acceptably small, and there is no obvious pattern among the

Table IV.11

Analysis of the Political-Preference Data

	Model	df	χ^2	Next	$\bar{\alpha}$	$\langle \bar{P} \rangle$
1	$[IVP]$	—	—	2⟨.275⟩	—	—
2	$[IV][VP][IP]$	6	7.509	3⟨.108⟩	.0125	⟨.275⟩
				4⟨.007⟩		
				5⟨.229⟩		
3	$[JV][VP][JP]$	14	20.604	6⟨.011⟩	.0250	⟨.204⟩
				7⟨.043⟩		
4	$[IV][VP]$	12	25.195	6⟨.350⟩	.0250	⟨.014⟩ R*
				8⟨.000⟩		
5	$[IP][VP]$	9	11.813	7⟨.045⟩	.0250	⟨.406⟩
				8⟨.000⟩		
6	$[JV][VP]$	16	29.630	9⟨.000⟩	.0375	⟨.033⟩ R
7	$[JP][VP]$	15	24.699	9⟨.000⟩	.0375	⟨.129⟩ *
8	$I[VP]$	15	43.160	9⟨.070⟩	.0375	⟨.000⟩ R
9	$J[VP]$	17	48.469		.05	⟨.000⟩ R

Table IV.12

Standardized Residuals from the Model [JP][VP]

	I0		I1		I2		I3	
	V0	V1	V0	V1	V0	V1	V0	V1
P0	1.25	−1.20	−1.03	0.92	−1.07	−0.44	0.82	0.81
P1	0.65	−1.28	1.08	−0.54	−0.52	0.69	−1.10	0.99
P2	0.17	1.95	−1.32	−1.09	0.16	−1.72	0.73	1.05

signs. In Run IV.B.4 we see the LAMDA program which computes the uniform plot of these residuals, which appears in Figure IV.3. The *P*-values are about right at the left of the scale, but are a bit small in the range from about the 10th to the 22nd. We must remember in looking at this plot that the residuals are subject to eight constraints, and it is not clear how these might affect the plot. If we return to model fitting, we might consider the more elaborate model $[IP][VP]$, whose **RES-F** uniform plot appears in Figure IV.4. These *P*-values appear to behave more like a uniform sample than the preceding ones. After viewing the uniform plots it is reasonable to stay with $[JP][VP]$ as the minimal confirmed model and $[IP][VP]$ as the final fitted model.

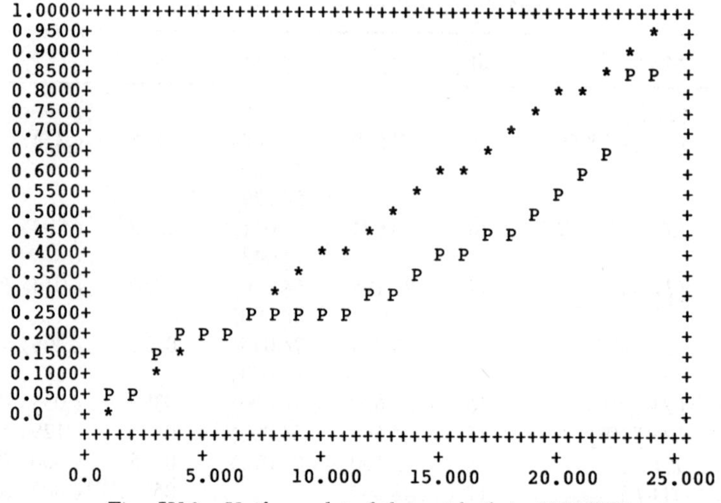

Fig. IV.3 Uniform plot of the residuals in [JP][VP].

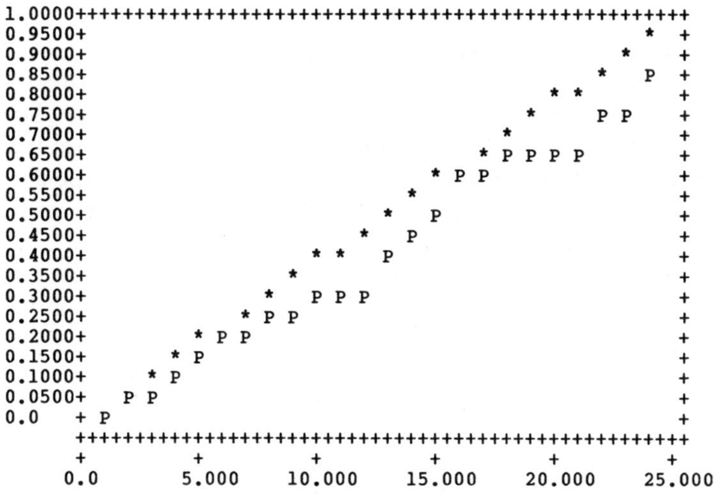

Fig. IV.4 Uniform plot of the residuals in [IP][VP].

3. Residuals for Model Fitting

We found in Chapter III, Section B.2 that the linear model did not describe the smoking behavior of male arteriosclerosis subjects and controls (Table III.32). The **RES − G** residuals from that fitted model appear in Table IV.13. Apparently the seriously large residuals are in the $A1D0$ and $A2D1$ columns. This suggests augmenting the fitted model from $\mathcal{L}[q: v, \mathbf{u}_A, u_D, \mathbf{u}_{AD}]$ to $\mathcal{L}[q: v, \mathbf{u}_A, u_D, \mathbf{u}_{AD}, vu_{A1D0}, vu_{A2D1}]$, since the larger model permits different linear smoking terms for the above two columns.

For convenience in denoting the models, let S stand for the linear smoking term associated with v, so that the $S[AD]$ model was fitted in Chapter III, Section B.2 and S $SA1D0$ $SA2D1[AD]$ is the model being fitted here. We accomplish this by rerunning Run III.B.13 except that at

Table IV.13

Residuals from the Model S[AD] Fitted to the Smoking and Arteriosclerosis Data

	$A0$		$A1$		$A2$	
	$D0$	$D1$	$D0$	$D1$	$D0$	$D1$
$S0$	1.29	1.86	3.49	−0.06	−0.92	−1.61
$S1$	0.51	−0.38	−0.30	0.94	1.49	1.94
$S2$	−1.32	0.36	−3.22	−0.07	0.95	3.30

line 32 we remove variable 11 ($SA1D0$) and variable 14 ($SA2D1$). We combine this with the previous analysis in the simple ANALYSIS of Table IV.14. The new model fits the data well, but the high P-value of .717 does not have an interpretation in terms of significance levels, because the model was suggested by the data. The **G-RES** uniform plot is in Figure IV.5. The P-values here tend to be too high, which is an indication that the data have been over-fitted. Of course, this is precisely the risk one takes in using residuals as a guide to model fitting.

Although LAMDA does not provide the estimates of the parameters in the case of MDI fitting, we may deduce from the fitted distribution in Table IV.15 that

$$\hat{\alpha}_S = (.0545 - .0189) \times 18 = (.090 - .0545) \times 18 = .640$$

$$\hat{\alpha}_{SA1D0} = (.0866 - .0133) \times 18 - \hat{\alpha}_S$$

$$= (.1600 - .0866) \times 18 - \hat{\alpha}_S = .680$$

$$\hat{\alpha}_{SA2D1} = (.0147 - .0172) \times 18 - \hat{\alpha}_S$$

$$= (.0122 - .0147) \times 18 - \hat{\alpha}_S = -.685$$

Thus, in a descriptive sense we have about the same distribution of smoking over all but two of the (A, D) groups, with $A1D0$ smoking more and $A2D1$ smoking less.

4. 2×3^2 with Linear Effects

Continuing the data of the preceding example, from the hypothesis-testing viewpoint we might have been concerned at the outset with the effects of age (A) and disease status (D) on smoking. It would then be appropriate to consider the nested linear models involving the terms S, A, D, SA, SD, AD, SAD in that order. The parameter of greatest interest would be α_{SD},

Table IV.14

Analysis of the Smoking and Arteriosclerosis Data

Model	df	χ^2	$\langle P \rangle$
$S\ SA1D0\ SA2D1[AD]$	9	6.245	$\langle .717 \rangle$
$S[AD]$	2	36.416	$\langle .000 \rangle$
0	6	91.534	$\langle .000 \rangle$

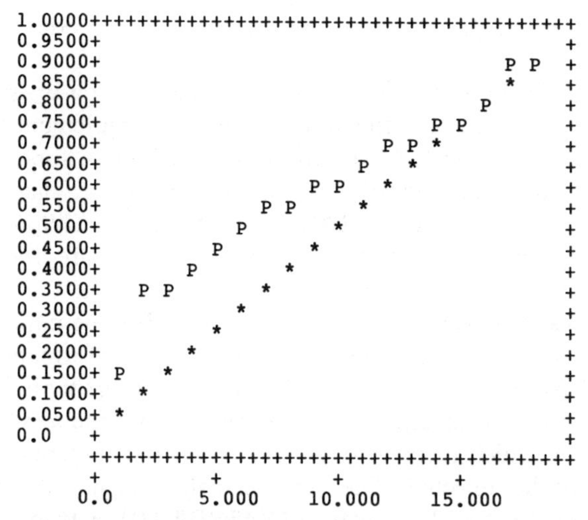

Fig. IV.5 Uniform plot of the residuals for the smoking and arteriosclerosis data.

since it represents a main effect due to disease on the smoking gradient. The chain of $\mathfrak{L}$-models we will examine and their individual significance levels are then

Model	Level
$[SAD]$	.005
$[SA][SD][AD]$	.010
$[SD][SA]$	.030
$D[SA]$	.005

and the test $[SD][SA] \rightarrow D[SA]$ is the one of interest. Naturally we hope

Table IV.15

Fitted Distribution to the Smoking and Arteriosclerosis Data

	$A0$		$A1$		$A2$	
	$D0$	$D1$	$D0$	$D1$	$D0$	$D1$
$S0$	.0189	.0575	.0133	.0094	.0496	.0172
$S1$	.0545	.0854	.0866	.0259	.0661	.0147
$S2$	.0900	.1134	.1600	.0425	.0827	.0122

that the model $[SD][SA]$ will fit, so that we can procede to test the SD effect.

In order to fit these models we will need the function v^2, symbolized by S^2, and then the additional functions needed to complement those already defined are S^2, S^2D, S^2A, S^2AD. In Run IV.B.5 these functions are constructed, and then standardized in the order given above. Then, for example, $[SD][SA]$ is fitted by doing **LXM** with the terms AD, SAD, S^2, S^2D, S^2A, S^2AD being **SEL**ected. The result is that we are forced to reject $[SD][SA]$ in favor of $[SA][SD][AD]$, which shows that the AD terms are highly significant. Since we had not considered AD to be important prior to the analysis, our ANALYSIS table no longer provides information about SD. We should have begun with a more elaborate ANALYSIS table.

In order to try to salvage something, we consider the analysis displayed in Table IV.16, which includes our previous analysis except for the models we know will be rejected. The serial P-values again lead us to reject $[SA][AD]$ and $[SD][AD]$ in favor of $[SA][SD][AD]$, with overall level .05. We now have a test of the SD effects which shows them to be significant.

Part of the LAMDA output appears in Table IV.17, and the uniform plot (**G-RES**) is in Figure IV.6. There appears to be a good overall fit, except possibly for the 9th cell in which there are too many observed $S2A1D0$ (heavy-smoking, middle-aged) cases. One expects that this cell has a fair amount to do with the better fit of $[SAD]$, and one possible decision at this point would be to use this latter model for fitting and describing the data. However, we will proceed with the $[SA][SD][AD]$ model.

Since LAMDA is designed to do ML analysis, it does not display the MDI parameter estimates or estimates of their standard deviations. From Problem III.1, we can compute these from the fact that $\hat{q} = (\hat{\alpha}_0 + \hat{\boldsymbol{\alpha}}'\mathbf{u})q$ implies

Table IV.16

	Model	df	χ^2	Next	$\bar{\alpha}$	$\langle \bar{P} \rangle$
1		—	—	2⟨.658⟩	—	—
2	$[SAD]$	6	4.156	3⟨.015⟩	.005	⟨.658⟩
3	$[SA][SD][AD]$	8	12.492	4⟨.000⟩ 5⟨.018⟩ 6⟨.000⟩	.015	⟨.023⟩ *
4	$[SA][SD]$	10	33.286		.045	⟨.000⟩ R*
5	$[SA][AD]$	9	18.052	7⟨.000⟩	.045	⟨.025⟩ R*
6	$[SD][AD]$	10	29.533	7⟨.001⟩	.045	⟨.000⟩ R*
7	$S[AD]$	11	42.655		.050	⟨.000⟩ R

Table IV.17

Part of the LAMDA *Output from Fitting [SA][SD][AD] to the Smoking and Arteriosclerosis Data*

POS	GEN	OBS	FIT	RES	F-RES	G-RES	RES-F	RES-G
1	4.000	11.167	2.979	8.187	2.521	-0.596	4.779	-0.515
2	9.000	11.167	12.286	-1.119	-0.345	0.967	-0.330	1.121
3	20.000	11.167	21.591	-10.425	-3.210	0.363	-2.375	0.375
4	9.000	11.167	12.198	-1.032	-0.318	0.945	-0.305	1.091
5	20.000	11.167	17.321	-6.154	-1.895	-0.673	-1.547	-0.631
6	24.000	11.167	22.443	-11.277	-3.472	-0.349	-2.525	-0.339
7	3.000	11.167	4.598	6.569	2.023	0.754	3.099	0.929
8	14.000	11.167	14.256	-3.090	-0.951	0.070	-0.849	0.071
9	34.000	11.167	23.915	-12.748	-3.925	-2.197	-2.777	-1.898
10	2.000	11.167	1.107	10.060	3.098	-0.852	9.589	-0.635
11	4.000	11.167	6.582	4.585	1.412	1.023	1.817	1.304
12	10.000	11.167	12.057	-0.890	-0.274	0.611	-0.264	0.667
13	14.000	11.167	10.311	0.855	0.263	-1.179	0.273	-1.022
14	10.000	11.167	13.460	-2.294	-0.706	0.976	-0.647	1.123
15	15.000	11.167	16.609	-5.442	-1.676	0.412	-1.394	0.432
16	4.000	11.167	4.130	7.036	2.167	0.065	3.498	0.066
17	2.000	11.167	3.095	8.071	2.485	0.627	4.623	0.778
18	3.000	11.167	2.061	9.106	2.804	-0.658	6.376	-0.546

$M[\mathbf{u}:\hat{q}] = M[\mathbf{u}:q] + C[\mathbf{u}:q]\hat{\boldsymbol{\alpha}}$, so that

$$\hat{\boldsymbol{\alpha}} = C[\mathbf{u}:q]^{-1}(M[\mathbf{u}:\hat{q}] - M[\mathbf{u}:q])$$

Run IV.B.6 computes the quantities on the right, and also $C[\mathbf{u}:\hat{q}]$, since the asymptotic covariance of $\sqrt{n}(\hat{\boldsymbol{\alpha}} - \boldsymbol{\alpha})$ is estimated by $C[\mathbf{u}:q]^{-1}C[\mathbf{u}:\hat{q}]$ $C[\mathbf{u}:q]^{-1}$. This gives the results in Table IV.18.

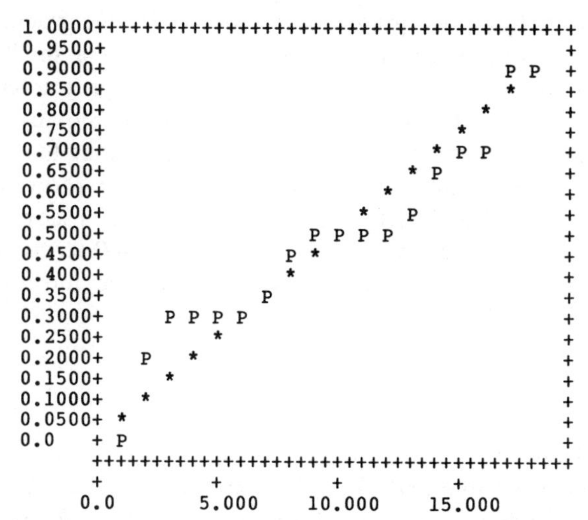

Fig. IV.6 Uniform plot of the residuals (G − RES) *from Table IV.17.*

Table IV.18

Parameter estimates in [SA][SD][AD]

i	$\hat{\theta}_i$	$\sqrt{V[\hat{\theta}_i]}$	t_i
D	0.828	.283	2.93
S	0.834	.174	4.79
$A1$	0.147	.266	0.55
$A2$	0.659	.282	2.34
SD	−0.375	.171	−2.19
$SA1$	0.032	.225	0.13
$SA2$	−0.552	.211	−2.62
$DA1$	−1.141	.358	−3.19
$DA2$	−1.382	.338	−4.09

The differences between two $\hat{q}$-probabilities as we move down a column are

$$\hat{q}(S1AiDj) - \hat{q}(S0AiDj) = \hat{q}(S2AiDj) - \hat{q}(S1AiDj)$$

and reflect the smoking gradient in the $AiDj$ group. The coefficients $SA1$ and $SA2$ indicate how these gradients are affected by age group; the middle group $A1$ has a larger gradient than $A0$, while $A2$ has a smaller gradient. Likewise, SD shows that $D1$ (controls) have a smaller gradient than $D0$ (cases). Thus we have found a relationship between smoking and disease status in a model which permits effects for age, so that the age effect is separated from the smoking effect to the extent possible. Moreover, we have confirmed at level .05 a model which says that there is a lack of three-way interaction among the factors S, A, and D (because SAD terms are omitted), so that SA and SD measure the additive effects of age and disease on the smoking gradient.

5. A Flow Model

The data in Table IV.19 consist of counts of instances in which one person (the influencer) has had an effect on the opinion of another (the influencee). The ages of the individuals involved were recorded, and also the context of the influence: whether both persons were in the same family or not.

The purpose of this study was to determine how the flow of influence relates to age categories. Define the functions

$$u_{S1}(FiBjAk) = \begin{cases} 1 & \text{if } jk \text{ is 01 or 10} \\ 0 & \text{otherwise} \end{cases}$$

$$u_{S2}(FiBjAk) = \begin{cases} 1 & \text{if } jk \text{ is 02 or 20} \\ 0 & \text{otherwise} \end{cases}$$

$$u_{T1} = u_{A1} - u_{B1}, \qquad u_{T2} = u_{A2} - u_{B2}$$

which are associated with a paired-comparison model. The concept of "strength" measured by the γ-parameters of Chapter III, Section B.2 becomes "flow of influence" in the present context. Since we are not interested in flows within age categories, we will reduce $\mathcal{Y}$ by discarding all the diagonal cells, *FiBjAj*. We then have the following models and their interpretations:

$[FS][FT]$	The paired-comparison model holds separately in each *F*-table.
$[FS]T$	The "flow parameters" are the same in both *F*-tables.
$[FS]FT1\ FT2$	The *F*0-table is symmetric, while the *F*1-table satisfies the paired-comparison model.
$[FS](1-F)T1\,(1-F)T2$	The *F*1-table is symmetric, while the *F*0-table satisfies the paired-comparison model.
$[FS]$	Both *F*-tables are symmetric.
FST	The two *F*-tables satisfy the paired-comparison model homogeneously.

The results of Run IV.B.7 are displayed in diagrammatic form in Figure IV.7. Using an overall level of .05, divided equally among the tests, we arrive at the minimal confirmed models FST and $[FS]FT1\ FT2$. Since $[FS]$ is rejected, we conclude that there is evidence of a flow of influence, but it is not entirely clear whether FST or $[FS]FT1\ FT2$ should be reported as the final model. The **F-RES** residuals for these two models show reasonable values for $[FS]FT1\ FT2$ and one exceptional (-3.98) residual for FST. On the basis of this, and the rather higher serial *P*-value of $[FS]FT1\ FT2$, we choose the latter model. Thus we find no flow of influence outside the family, but a flow inside the family, and one may

Table IV.19

Data on the Flow of Influence

	*F*0			*F*1		
	*A*0	*A*1	*A*2	*A*0	*A*1	*A*2
*B*0	14	2	1	3	6	2
*B*1	6	19	6	15	43	9
*B*2	1	12	20	15	14	30

F = Family Context
*F*0 = Outside Family
*F*1 = Inside Family
A = Age of Influencee
*A*0 = 15–24 Years
*A*1 = 25–44 Years
*A*2 = 45 + Years
B = Age of Influencer
*B*0 = 15–24 Years
*B*1 = 25–44 Years
*B*2 = 45 + Years

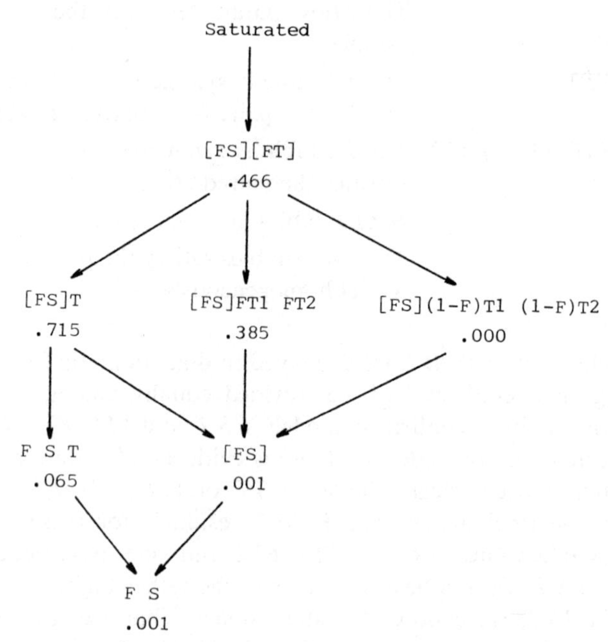

Fig. IV.7 Display of the results of the analysis of the "flow of influence" data. Serial P-values appear beneath the models.

compute the flow parameters (and their t-values) to be

$$\hat{\gamma}_1 = -.5444\ (-2.56)$$

$$\hat{\gamma}_2 = -.8312\ (-3.54)$$

$$\hat{\gamma}_2 - \hat{\gamma}_1 = -.2868\ (-1.22)$$

Thus we find significant flows from both $A1$ and $A2$ to $A0$, and a nonsignificant flow from $A2$ to $A1$.

It must be said that because we could not reject FST, which embodies the same model in both F-tables, the data are not sufficiently decisive to rule out the possibility that flow is the same within and outside the family.

6. A Four-Way Table

The data in Table IV.20 involve five factors and 120 cells. If one had a specific set of questions about data this complex, then the methods introduced earlier could be used in a fairly straightforward way. On the other hand, if we approach this data with the idea of doing a general analysis, the first difficulty which arises is that there is a very large number of reasonable models for a five-way table. It is neither practical nor desirable to fit all hierarchical models to this data. What is required is a principle for narrowing the collection of models that need to be examined.

Collapsing. One obvious approach is to eliminate one of the factors by collapsing the table. The "total" column of Table IV.20 gives the A, S, C, V-margin of the full table, and if we use these frequencies as the basis for analysis, then we have reduced the size of the problem greatly. However, this preliminary collapsing must be regarded as tentative, and the final analysis would be incomplete without some assessment of its justification. We will proceed with the four-way table here, and return to the collapsibility issue in Chapter VI, Section D.

Interpretability. Even with only four factors, the number of hierarchical models is rather large. One possible approach is to restrict consideration to some class of hierarchical models which have a particular interpretation. A natural choice here is the class of models which exhibit conditional independence of some factors given values of the others. In Table IV.21 is a listing of all such models for a four-way table, together with all their intersections.

We use this table as a worksheet, and begin at the top by fitting models 2–7 (Run IV.B.8; see also the comments in Chapter VI, Section A.2 on this

Table IV.20

A Five-Way Table and One of Its Four-Way Margins

				U0	U1	U2	U3	U4	Total
A0	S0	C0	V0	2	2	2	1	1	8
			V1	5	3	9	2	8	27
		C1	V0	21	18	20	11	24	94
			V1	53	58	64	31	69	275
	S1	C0	V0	0	0	1	0	0	1
			V1	3	6	3	2	5	19
		C1	V0	5	5	5	6	9	30
			V1	87	60	58	62	68	335
A1	S0	C0	V0	11	8	4	11	6	40
			V1	58	42	37	42	66	245
		C1	V0	4	1	2	2	9	18
			V1	40	39	42	36	60	217
	S1	C0	V0	0	2	0	0	1	3
			V1	13	10	17	13	12	65
		C1	V0	1	1	4	2	3	11
			V1	47	41	36	37	58	219
A2	S0	C0	V0	12	16	6	8	12	54
			V1	142	113	127	91	193	666
		C1	V0	1	4	1	2	7	15
			V1	54	26	23	18	52	173
	S1	C0	V0	2	3	0	0	1	6
			V1	19	22	31	26	54	152
		C1	V0	0	0	0	0	0	0
			V1	24	23	20	15	45	127

A = Age
*A*0 = 16–35 Years
*A*1 = 36–55 Years
*A*2 = Over 56 Years
S = Sex
*S*0 = Male
*S*1 = Female
C = Cardiovascular Disease
*C*0 = Present
*C*1 = Absent
V = Driving Violations
*V*0 = Present
*V*1 = Absent
U = Urban Area
*U*0 = Rural Area
*U*1–*U*4 = Urban Areas

Table IV.21

The Conditional Independence Models and Their Intersections, on a Four-Way Table

No.	Name	df	Chi-square	Next	$\bar{\alpha}$	$\bar{P}max$
1	[ABCD]	-	-	2,3,4,5,6,7	-	-
2	[ABC][ABD]			8,9,10,11		
3	[ABC][ACD]			8,12,13,14		
4	[ABC][BCD]			9,12,15,16		
5	[ABD][ACD]			10,13,17,18		
6	[ABD][BCD]			11,15,17,19		
7	[ACD][BCD]			14,16,18,19		
8	[AD][ABC]			20,24,28,30		
9	[BD][ABC]			20,25,29,35		
10	[AC][ABD]			21,24,29,33		
11	[BC][ABD]			21,25,28,37		
12	[CD][ABC]			20,26,31,34		
13	[AB][ACD]			22,24,31,32		
14	[BC][ACD]			22,26,30,39		
15	[AB][BCD]			23,25,34,36		
16	[AC][BCD]			23,26,35,38		
17	[CD][ABD]			21,27,32,36		
18	[BD][ACD]			22,27,33,38		
19	[AD][BCD]			23,27,37,39		
20	[ABC]D			49,50,51,57,59,60		
21	[ABD]C			46,47,48,56,58,60		
22	[ACD]B			43,44,45,55,58,59		
23	[BCD]A			40,41,42,55,56,57		
24	[AB][AC][AD]			43,46,49		
25	[AB][BC][BD]			40,47,50		
26	[AC][BC][CD]			41,44,51		
27	[AD][BD][CD]			42,45,48		
28	[AB][AD][BC]			46,50,54		
29	[AB][AC][BD]			47,49,53		
30	[AC][AD][BC]			43,51,54		
31	[AC][AB][CD]			44,49,52		
32	[AD][AB][CD]			45,46,52		
33	[AD][AC][BD]			43,48,53		
34	[BC][AB][CD]			41,50,52		
35	[BC][BD][AC]			40,51,53		
36	[BD][AB][CD]			42,47,52		
37	[BD][AD][BC]			40,48,54		
38	[CD][AC][BD]			42,44,53		
39	[CD][BC][AD]			41,45,54		
40	A[BC][BD]			56,57		
41	A[BC][CD]			55,57		
42	A[BD][CD]			55,56		
43	B[AC][AD]			58,59		
44	B[AC][CD]			55,59		
45	B[AD][CD]			55,58		
46	C[AB][AD]			58.60		
47	C[AB][BD]			56,60		
48	C[AD][BD]			56,58		
49	D[AB][AC]			59,60		
50	D[AB][BC]			57,60		
51	D[AC][BC]			57,59		
52	[AB][CD]			55,60		
53	[AC][BD]			56,59		
54	[AD][BC]			57,58		
55	AB[CD]			61		
56	AC[BD]			61		
57	AD[BC]			61		
58	BC[AD]			61		
59	BD[AC]			61		
60	CD[AB]			61		
61	ABCD			-		

Table IV.22

Partial Analysis of the Four-Way Table

No.	Name	df	Chi-square	Next
1	[ASCV]	-	-	2<.175>3<.341>4<.452>5<.080>6<.052>7<.108>
2	[ASC][ASV]	3	4.954	8<.000>9<.000>10<.000>11<.000>
3	[ASC][ACV]	5	5.661	8<.000>12<.000>13<.000>14<.000>
4	[ASC][SCV]	6	5.749	9<.000>12<.000>15<.000>16<.000>
5	[ASV][ACV]	5	9.833	10<.000>13<.000>17<.000>18<.000>
6	[ASV][SCV]	6	12.46	11<.000>15<.000>17<.000>19<.000>
7	[ACV][SCV]	6	10.42	14<.000>16<.000>18<.000>19<.000>
8	[AV][ASC]	9	66.64	20,24,28,30
9	[SV][ASC]	10	88.77	20,25,29,35
10	[AC][ASV]	9	154.1	21,24,29,33
11	[SC][ASV]	10	854.3	21,25,28,37
12	[CV][ASC]	10	118.5	20,26,31,34
13	[AS][ACV]	9	203.4	22,24,31,32
14	[SC][ACV]	10	77.02	22,26,30,39
15	[AS][SCV]	12	919.5	23,25,34,36
16	[AC][SCV]	12	92.89	23,26,35,38
17	[CV][ASV]	10	1108.	21,27,32,36
18	[SV][ACV]	10	300.9	22,27,33,38
19	[AV][SCV]	12	994.9	23,27,37,39

run) and then models 8–19. The results appear in Table IV.22. Now since 11 terms are to be dropped in going from the saturated model $[ASCV]$ to the independence model $A\ S\ C\ V$, if we assign a level of .005 to every test, then the overall serial level will be .055. This gives us the partial analysis of Table IV.22, and it is clear that every model lower than 7 will be rejected. We are thus left with models 2–7 as being minimal confirmed, and this is not a very decisive set of affairs. If it is required to find a "final" model, then $[ASC][SCV]$ is the leading candidate because of its high serial P-value, but it must be admitted that the interpretability approach has not yielded a satisfactory analysis.

Marginal and Partial Association. An alternative approach is to screen the data with the objective of determining factors which are probably significant. The P-values associated with the marginal and partial association tests appear in Table IV.23. One might conclude that the terms ASC, AS, AC, AV, SC, and SV should be included, giving the model $[ASC][AV][SV]$. Whether CV should also be included is problematical.

Although this fitted model is very easily determined by the use of marginal and partial association tests, it cannot be used to draw any inferential conclusions about the true distribution.

Forward and Backward Methods. Another *ad hoc* procedure is to begin with a reasonable hierarchical model, and then test that model against each hierarchical model containing exactly one additional term. Keep adding the significant terms and repeating this test until no more terms are significant.

Table IV.23

P-values of Marginal and Partial Association Tests

Term	Marginal	Partial
$ASCV$	⟨.375⟩	⟨.375⟩
SCV	⟨.465⟩	⟨.138⟩
ACV	⟨.295⟩	⟨.247⟩
ASV	⟨.419⟩	⟨.383⟩
ASC	⟨.028⟩	⟨.019⟩
CV	⟨.148⟩	⟨.053⟩
SV	⟨.000⟩	⟨.000⟩
SC	⟨.000⟩	⟨.000⟩
AV	⟨.000⟩	⟨.000⟩
AC	⟨.000⟩	⟨.000⟩
AS	⟨.000⟩	⟨.000⟩

Then test all models having exactly one fewer term. Keep dropping terms until retained terms are significant. Then return to tests for adding terms, and so on. If this process ever stops, one has found a model all components of which would appear significant if they were dropped, and to which no significant term can be added—provided one stays within the framework of hierarchical models and one-step tests.

In the example at hand, we begin with $[ASC][AV][SV]$, and test for the addition of terms CV ⟨.375⟩ and ASV ⟨.384⟩. Since neither term needs to be added, we consider deleting ASC ⟨.019⟩, AV ⟨.000⟩, and SV ⟨.000⟩. Depending on how one feels about a P-value of .019, one could now stay with the original model or proceed to test $[AS][AC][SC][AV][SV]$ against the deletions AS ⟨.000⟩, AC ⟨.000⟩, SC ⟨.000⟩, AV ⟨.000⟩, and SV ⟨.000⟩. Finding no possible deletions, one considers the addition of CV ⟨.148⟩, ASC ⟨.019⟩, and ASV ⟨.384⟩. It is now clear that one must make up one's mind about a P-value of .019 and choose one of the two models.

Conclusion. When faced with a table having four or more factors one needs to consider the tradeoffs involved in (1) the labor of fitting a large number of models, with its security of knowing that no plausible model had gone uninvestigated, (2) the maintenance of a bound on the probability of rejecting a true model, and (3) the desire to finish with an interpretable, defensible fitted model. It is always tempting to collapse the table, thus ignoring some factors. In some cases this is necessary because of the paucity of data, but in others it is highly undesirable because a large number of interactions are of real interest. We will review the collapsibility issue in Chapter VI, Section D.

One further avenue in the analysis of higher-dimensional tables involves the division of the factors into "explanatory" and "response" factors. This important technique is the subject of Chapter V.

PROBLEMS

IV.1. Analyze the following data:

	$W0$		$W1$		$W2$		$W3$	
	$A0$	$A1$	$A0$	$A1$	$A0$	$A1$	$A0$	$A1$
$I0$	12	4	9	9	8	19	18	52
$I1$	13	12	9	12	9	23	14	57
$I2$	71	27	38	17	16	17	18	34

W = Occupation
- $W0$ = Commerce
- $W1$ = Finance and Records
- $W2$ = Manufacturing
- $W3$ = Building and Maintenance

A = Political Affiliation
- $A0$ = Republican
- $A1$ = Democrat

I = Income
- $I0$ = Less than 4000
- $I1$ = 4000–6000
- $I2$ = More than 6000

IV.2. Analyze the following data:

	$A0$	$A1$	$A2$	$A3$
$S0$	44	52	45	67
$S1$	23	22	21	13

A = Anxiety Level
- $A0$ = High
- $A1$ = Mid to High
- $A2$ = Low to Mid
- $A3$ = Low

S = Scholarship
- $S0$ = Low
- $S1$ = High

IV.3. Consider the following data on eyesight:

	$L0$	$L1$	$L2$	$L3$
$R0$	492	179	82	36
$R1$	205	1772	362	117
$R2$	78	432	1512	234
$R3$	66	124	266	1520

R = Right Eyesight, graded worst (0) to best (3)
L = Left Eyesight, graded worst (0) to best (3)

Formulate the symmetry model $[p(ij) = p(ji)]$ and the marginal-homogeneity model $[p(i+) = p(+i)]$ as linear models. Fit both models and present the ANALYSIS table. Compute the parameter estimates and their estimated standard deviations.

IV.4. The data below show the occupation distribution of an original sample and three follow-up samples:

	$F0$	$F1$	$F2$	$F3$
$S0$	71	53	62	33
$S1$	167	59	74	46
$S2$	151	31	19	10
$S3$	404	77	34	10

F = Sample
$F0$ = Original
Fi = ith Follow-up
S = Occupational Status
$S0$ = Business and Professional
$S1$ = High White Collar
$S2$ = Low White Collar
$S3$ = Wage Worker

The issue is whether the follow-up samples differ from the original sample, and if so, whether there is a bias in one direction or the other on the occupational scale. On the table $\{SiFj : 0 \leqq i \leqq 3, 1 \leqq j \leqq 3\}$, use the generating distribution which renders S and F independent, and reproduces the observed F-margin and the original sample ($F0$) distribution in the S-margin. In addition to the usual indicators of S and F, use also the factor L defined by $u_L(SiFj) = i$. Fit the models $S[FL]$, SFL, FS, FL, and F; give the ANALYSIS table; interpret the parameters and the results.

IV.5. The data in Table IV.24 are counts of the number of boys in families of size eight. A Binomial model was fitted and the statistician observed from the residuals that including the function

$$v(y) = \begin{cases} 1 & \text{if } y = 0, 2, 4, 6, 8 \\ 0 & \text{if } y = 1, 3, 5, 7 \end{cases}$$

should improve the fit. He did so, and concluded that the fit was improved. Repeat this analysis, and discover what the statistician left out of his analysis. Can you find a function in place of v which is more satisfactory?

IV.6. The data in Table IV.25 give the number of dolphin sightings from an airplane by different teams, using different methods, on the morning or afternoon of three different days. Since only one airplane was available, only one team–method combination could be used each time, which accounts for the fixed empty cells in the table. The pattern of fixed empty cells has been balanced, but it is still only possible to fit certain models. Analyze these data, beginning with $[MT][MD][MA]$ and its submodels, remembering that assessment of the methods (M) is of principal importance.

IV.7. 1. Consider models $\mathfrak{M}_1$, $\mathfrak{M}_2$, and $\mathfrak{M}_{12} = \mathfrak{M}_1 \cap \mathfrak{M}_2$. Let $\hat{q}_1$, $\hat{q}_2$, and $\hat{q}_{12}$ be MDEs in these models, and consider the ANALYSIS table of tests Saturated $\to \mathfrak{M}_1$, Saturated $\to \mathfrak{M}_2$, $\mathfrak{M}_1 \to \mathfrak{M}_{12}$, and $\mathfrak{M}_2 \to \mathfrak{M}_{12}$, with all tests being carried out at the same level. Show that if

$$\mathbb{D}(\hat{p}, \hat{q}_{12}) = \mathbb{D}(\hat{p}, \hat{q}_1) + \mathbb{D}(\hat{p}, \hat{q}_2)$$

Table IV.24

y = Number of Boys	$N(y)$
0	215
1	1485
2	5331
3	10649
4	14959
5	11929
6	6678
7	2092
8	342

Table IV.25

	A0			A1		
	D0	D1	D2	D0	D1	D2
				T0		
M0	104	—	—	—	—	90
M1	—	4	—	20	—	—
M2	—	—	58	—	56	—
				T1		
M0	—	16	—	0	—	—
M1	—	—	63	—	12	—
M2	7	—	—	—	—	14
				T2		
M0	—	—	3	—	3	—
M1	1	—	—	—	—	30
M2	—	4	—	9	—	—

T = Team
A = Time
$A0$ = Morning
$A1$ = Afternoon
D = Day
M = Method

then the serial P-values for $\mathfrak{M}_{12}$ obtained along the chains Saturated → $\mathfrak{M}_1$ → $\mathfrak{M}_{12}$ and Saturated → $\mathfrak{M}_2$ → $\mathfrak{M}_{12}$ coincide.

2. Generalize this to the case of three models.
3. Show that in an (A, B, C)-table the following are examples of part 1:

$$\mathfrak{M}_1 = A[BC], \qquad \mathfrak{M}_2 = B\,C, \qquad \mathfrak{M}_{12} = A\,B\,C$$

and

$$\mathfrak{M}_1 = A\,B, \qquad \mathfrak{M}_2 = [AB][BC], \qquad \mathfrak{M}_{12} = A[BC].$$

IV.8. Table IV.26 gives the fate of manuscripts submitted to a publisher in the nineteenth century. The research hypothesis underlying these

Table IV.26

	$T0$		$T1$		$T2$	
	$S0$	$S1$	$S0$	$S1$	$S0$	$S1$
			$F0$			
$D0$	2	7	4	51	1	27
$D1$	9	54	9	61	18	82
$D2$	6	39	26	94	7	28
			$F1$			
$D0$	0	0	1	0	1	0
$D1$	33	24	52	17	60	61
$D2$	13	4	29	14	15	4

F = Type
- $F0$ = Nonfiction
- $F1$ = Fiction (novels)

T = Time Period

S = Sex
- $S0$ = Female
- $S1$ = Male

D = Disposition
- $D0$ = Accepted
- $D1$ = Sent to reader
- $D2$ = Rejected

data is that the novel increased in stature and prestige with respect to nonfiction over the time period from $T0$ to $T2$, and as a consequence women authors were squeezed out of the fiction market and into the nonfiction market. Analyze the data with this hypothesis in mind.

IV.9. In a crossover design each subject is given a placebo and the treatment, but some are given the placebo first, some the treatment first. Let C denote the placebo response ($C1$ = favorable) and T denote the treatment response:

	Placebo First		Treatment First	
	$C0$	$C1$	$C0$	$C1$
$T0$	6	3	5	22
$T1$	16	7	2	1

Table IV.27

u_M	0	0	1	1
	0	0	1	1
u_T	0	0	0	0
	1	1	1	1
u_C	0	1	0	1
	0	1	0	1
u_F	0	1	0	0
	0	1	1	1
u_R	0	0	0	1
	1	0	0	0
u_L	0	1	0	0
	0	0	1	0

Of course the main question is whether the treatment is better than the placebo, but an important preliminary step is to try to see whether the order of administration makes a difference. Consider the exponential model generated by the uniform distribution and determined by the functions given in Table IV.27 (written in the same pattern as the table above). Interpret the parameters in this model, and explain how it embodies the notion that the order of administration has no effect. What parameter measures the treatment effect, and how can it be estimated? Analyze the data.

IV.10. 1. Consider a two-way cross-classification with factors A and B. Let $\mathbf{v}_A, \mathbf{u}_A, \mathbf{u}_B, \mathbf{u}_{AB}$ be functions depending only on the subscript factors. Prove that if $\mathfrak{E}[q_A : \mathbf{u}_A]$ is the saturated model on the A-margin, $\hat{q}$ is the ML distribution in $\mathfrak{E}[q : \mathbf{u}_A, \mathbf{u}_B, \mathbf{u}_{AB}]$, and $\hat{q}_A$ is the ML distribution in $\mathfrak{E}[q_A : \mathbf{v}_A]$ on the A-margin, then the ML distribution $\hat{q}_0$ in $\mathfrak{E}[q_A : \mathbf{v}_A] \cap \mathfrak{E}[q : \mathbf{u}_A, \mathbf{u}_B, \mathbf{u}_{AB}]$ is

$$\hat{q}_0(ij) = \hat{q}_A(i)\hat{q}_i(j)$$

2. Use this result to express the ML distribution which lies in $[AB][BC][AC]$ and $A\,B$ (on the A, B-margin) in terms of the ML distribution in $[AB][BC][AC]$.

IV.11. Let $\mathbf{u}' = (\mathbf{u}_1', \mathbf{u}_2')$. Let $\hat{q}_1$ be the ML distribution in $\mathfrak{E}[q : \mathbf{u}_1]$, and $\hat{q}_2$ be the ML distribution in $\mathfrak{E}[q : \mathbf{u}]$, both determined by the same empirical distribution, $\hat{p}$. Show the following:

1. $\hat{q}_2$ is the ML distribution in $\mathcal{E}[\hat{q}_1 : \mathbf{u}]$ based on the data $\hat{p}$.
2. $\hat{q}_1$ is the ML distribution in $\mathcal{E}[q : \mathbf{u}_1]$ based on the data $\hat{q}_2$.

REMARKS

Kullback (1968) initiated the use of a precursor of our ANALYSIS table. Further use and extensions appear in Kullback, Kupperman, and Ku (1962), Ireland, Ku, and Kullback (1969), Ku, Varner, and Kullback (1971), Kullback (1971), Ku and Kullback (1974), Gokhale and Johnson (1978), and especially Gokhale and Kullback (1978). Of course, the ANALYSIS table works because of the additive decompositions of Propositions 30 and 31. The general analysis table and its accompanying serial test are new, and provide a resolution of some difficulties which will now be described.

Interference

Let $\mathfrak{M}_0$ stand for the saturated model, and let the intersection of models $\mathfrak{M}_1$ and $\mathfrak{M}_2$ be denoted $\mathfrak{M}_{12}$. Now consider the simple collection of tests $\mathfrak{M}_0 \to \mathfrak{M}_1$, $\mathfrak{M}_0 \to \mathfrak{M}_2$, $\mathfrak{M}_1 \to \mathfrak{M}_{12}$, and $\mathfrak{M}_2 \to \mathfrak{M}_{12}$. We know that the serial P-values for $\mathfrak{M}_{12}$ will generally differ along the two chains pointing to $\mathfrak{M}_{12}$. We might say, then, that there is *interference* in the setup; inference along one chain interferes with inference along the other chain, because they do not yield the same evidence about $\mathfrak{M}_{12}$. However, as shown in Problem IV.7, there are circumstances in which this does not happen. When the two serial P-values for $\mathfrak{M}_{12}$ coincide, then we say there is *noninterference*.

In cases of noninterference it is often true that the tests $\mathfrak{M}_0 \to \mathfrak{M}_1$ and $\mathfrak{M}_2 \to \mathfrak{M}_{12}$ are in a sense the same test, and similarly for $\mathfrak{M}_0 \to \mathfrak{M}_2$ and $\mathfrak{M}_1 \to \mathfrak{M}_{12}$. When there is noninterference the chi-squares for these pairs of tests actually coincide. Probably the simplest interesting example [Kullback (1968), Chapter 8, Table 3.3] is on a three-way cross-classification with $\mathfrak{M}_1 = A[BC]$ and $\mathfrak{M}_2 = B\,C$, so that $\mathfrak{M}_{12} = A\,B\,C$. According to Problem IV.7 there is no interference in this situation. The chi-square for testing $\mathfrak{M}_0 \to \mathfrak{M}_2$ is the usual chi-square for independence in a two-way table, and is identical to the chi-square for $\mathfrak{M}_1 \to \mathfrak{M}_{12}$. For another example [Kullback (1968), Chapter 8, Tables 3.5 and 6.1] take $\mathfrak{M}_1 = [AB][BC]$, $\mathfrak{M}_2 = A\,B$, and $\mathfrak{M}_{12} = A[BC]$. Goodman (1969) also emphasized the noninterference of this example, and provided a correct test of $\mathfrak{M}_0 \to [AB][BC][AC]$ to extend the analysis. Many of the examples cited by Kullback and his associates involve one or more marginal distributions, since these often enter into noninterfering patterns with various independence and conditional independence models.

In Ku, Varner, and Kullback (1971), Table 5, we see the chain

$$[AB][AC][AC][BC][BD][CD] \rightarrow [AB][AC][AD][BC][BD] \rightarrow$$

$$[AB][AC][AD][BC] \rightarrow [AB][AC][AD] \rightarrow [AB][AC]D \rightarrow [AB]CD,$$

and Table 6 shows a chain involving three-factor effects. It is clear that there are other chains which begin and end at the same models, but it is also clear that these chains will interfere with one another, which perhaps explains why most writers have preferred to avoid dealing with the multiplicity of possible chains. An illuminating discussion of this dilemma can be found in Darroch and Silvey (1963).

It is common practice now to list a collection of models in a table which superficially resembles our ANALYSIS table, and then to sort through the goodness-of-fit chi-squares in an ostensibly reasonable fashion before settling on a "final model" [Bishop, Fienberg and Holland (1975), Fienberg (1980)]. Goodman (1971c) suggests beginning with a reasonably fitting model, and then testing successively for the addition and deletion of various terms. This stepwise procedure amounts to wandering around in our ANALYSIS table, trying to stay in the class of reasonably fitting models. When the table is simple, there is no reason to do this, but in complex tables it may be attractive despite the fact that a model determined by such a stepwise procedure has no demonstrable inferential properties. Of course, the problem stems from the fact that there is interference throughout most ANALYSIS tables, so that there is no uniquely defined chi-square for each effect. If there were noninterference the problem would essentially vanish, and at least one author [Haberman (1974), p. 122ff.] recommends behaving as if this were the case.

The marginal and partial association tests of Brown (1976) [see also Benedetti and Brown (1978)] are also oriented towards dealing with the interference problem, and offer a reasonable way of proceeding when the size of the problem overwhelms the attempt to produce a precise analysis. Another procedure, due to Aitkin (1979), attempts to produce "minimal adequate sets" of terms to include in a model, but it has quite different power properties than the serial test has.

Table IV.21 was used in Aickin (1982) to carry out a simultaneous test of all hierarchical models on a four-way table, by means of a "look-ahead" procedure.

Other Topics

Arrival at the classical parametrization of a cross-classification has been a slow process. Important papers are Bartlett (1935), Roy and Kastenbaum (1956), Birch (1963) and Darroch (1974). The role of the Möbius inversion in parametrizing log-linear models has not been recognized previously, although Whittemore (1978) evidently used it implicitly. See this paper and also Bishop (1971) for discussion of the "collapsibility" issue.

Proposition 32 is a good example of a result which is computationally tedious (at least, if LAMDA is not available) but conceptually simple. Explicit expressions for the variance of residuals in particular situations can be found in Haberman (1973) and (1978).

The use of simultaneous inference for means goes back at least as far as Goodman (1965a); see also Bhapkar and Somes (1976) and Wackerly and Dietrich (1976).

An improvement on the Bonferroni procedure for testing the significance of residuals has been provided recently by Holm (1979). Using Holm's method, one compares the P-value of the most extreme residual with α/m, the P-value of the second most extreme with $\alpha/(m - 1)$, the P-value of the third most extreme with $\alpha/(m - 2)$, and so on. The procedure continues so long as the residuals are found to be too large; no further testing is done after the first time a P-value is found to be greater than the corresponding level. Thus Holm's method differs from Bonferroni in that the levels rise as more and more residuals are rejected, and the procedure is sequential. However, the overall probability of falsely detecting an extreme residual is still bounded by α with Holm's method.

CHAPTER V

Covariate Inference

Models which permit estimation of the effects of covariates on experimental results are well known in the analysis of "continuous" variables under the names regression, analysis of variance, and analysis of covariance. These are all special cases of the General Linear Model. In this chapter we investigate models which perform the same function for discrete data.

The discrete case poses some problems because a well-fitting discrete model usually requires several parameters. We will see that a covariate model permits these parameters to depend in a linear fashion on functions of the covariates. Thus, the result of the analysis is typically a set of regressionlike equations. Since each parameter in the model is going to appear in a regressionlike equation, it is more important than ever to construct models whose parameters are interpretable.

We begin in the following section by showing how the response parameters may be linearly related to covariates, with special attention to linear exponential models. ML inference for these models follows the same pattern we have seen in earlier chapters, but since the proofs are much harder, we only present the results. The technique of conditional inference is discussed, since it is very important for analyzing complex data sets. Finally, the LAMDA command `LXM` is extended to cover linear exponential models, and a number of actual data sets are analyzed in Section B.

A. BILINEAR MODELS

1. Responses and Covariates

The statistical analysis described in preceding chapters has been based on the assumption that each experimental unit entering the experiment or survey sorts itself into one of the cells of the table $\mathcal{Y}$ according to a probability distribution π. We think of the experimental unit as having produced a *response*, and of π as being the true distribution of responses. The distribution π remains the same from one experimental unit to the next, and so we are treating each experimental unit just like the others. This is often highly unrealistic, since in most experiments one knows before the response occurs some characteristics which ought to influence or be associated with it. These characteristics, usually known before the experimental unit responds, are called *covariates*.

We will continue the convention of letting y stand for an element of $\mathcal{Y}$, and we now introduce the use of x to stand for a value of one or more covariates. The true distribution upon which we wish to work some inference should now be written as $\pi_x(y)$, the true probability of a response y on the part of an experimental unit characterized by the covariate value x.

In a particular experiment or survey the values $x_1, x_2, \ldots, x_n$ of the covariates of the n experimental units are fixed numbers known before the responses are made, and consequently there is no inferential interest in relations among the various components of the x_i; nor is it even meaningful to ask questions about the true distribution of these quantities, since they are not random variables but fixed values. All interest centers on relations among responses expressed in the structure of the u-functions, and in relationships between response factors and covariates.

We can get at this sort of inference by posing the *bilinear model* as an extension of the linear model. Suppose that $\{q_x^{\boldsymbol{\theta}}\}$ is a model; for each value of $\boldsymbol{\theta}$ in some subset of $\mathbb{R}^r$ we have a uniquely defined set of distributions $q_x^{\boldsymbol{\theta}}(y)$ $(y \in \mathcal{Y})$, one for each value of x in some specified set of possible covariates. In order for the model to be linear we will suppose that $\boldsymbol{\theta}$ is the parameter of a linear or exponential model, as described in Chapter III, Section A.1. In order for it to be bilinear we assume further that there is a function $\mathbf{t}$ defined over the set of covariates, assuming values in $\mathbb{R}^s$, and such that $q_x^{\boldsymbol{\theta}}$ is the element of the linear or exponential model for which $\boldsymbol{\theta} = \mathbf{B}'\mathbf{t}(x)$. Here $\mathbf{B}$ is an $s \times r$ matrix of parameters.

For example, one could begin with the exponential model

$$q^{\boldsymbol{\theta}}(y) = e^{\boldsymbol{\theta}'\mathbf{u}(y) - \delta(\boldsymbol{\theta})} q(y)$$

We will call the bilinear form of this model a *linear exponential model*:

$$q_x^{\boldsymbol{\theta}}(y) = q^{\mathbf{B}'\mathbf{t}(x)}(y) = e^{\mathbf{t}(x)'\mathbf{B}\mathbf{u}(y) - \delta(\mathbf{t}(x)'\mathbf{B})} q(y)$$

Note that the parameter $\boldsymbol{\theta}$ is nonessential once we have seen the relation between the exponential and linear exponential models, and so we will generally use $q_x^{\mathbf{B}}$ to stand for the model probabilities.

Similarly, from the linear model

$$q^{\boldsymbol{\alpha}}(y) = [1 + \boldsymbol{\alpha}'\mathbf{u}(y)]q(y)$$

we can construct the bilinear model

$$q_x^{\boldsymbol{\alpha}}(y) = q^{\mathbf{B}'\mathbf{t}(x)}(y) = [1 + \mathbf{t}(x)'\mathbf{B}\mathbf{u}(y)]q(y)$$

However, the linear exponential model is the only one in use currently, and so we will concentrate on its exposition.

Examples. The simplest sort of linear exponential model is called *logistic regression*. In this case $\mathcal{Y} = \{0, 1\}$, $u(y) = y$ is essentially the only u-function possible, the t-functions are arbitrary, and the parameter matrix can be written as a vector $\mathbf{b}$. Thus the model could be written

$$q_x^{\mathbf{b}}(1) = \frac{e^{\mathbf{t}(x)'\mathbf{b}}q(1)}{q(0) + e^{\mathbf{t}(x)'\mathbf{b}}q(1)}$$

Another way of writing it is

$$\ln\frac{q_x^{\mathbf{b}}(1)}{q_x^{\mathbf{b}}(0)} = \mathbf{t}(x)'\mathbf{b} + \ln\frac{q(1)}{q(0)}$$

which shows explicitly that the ln odds favoring a 1-response at covariate level x is the ln odds for the generating distribution q plus a factor linear in $\mathbf{t}(x)$.

The vector $\mathbf{t}(x)$ plays the same role here as it would in an ordinary linear model. For example, if x consists of four variables x_1, x_2, x_3, x_4, with x_1 and x_2 being measured on a "continuous" scale while x_3 and x_4 assume only the values 0 or 1, then some possible choices for $\mathbf{t}$ are:

$\mathbf{t}(x)'$	Name
$(1, x_1)$	Simple linear regression
$(1, x_1, x_2)$	Bivariate regression
$(1, x_1, x_2, x_1^2, x_2^2, x_1x_2)$	Polynomial regression
$(1, x_3, x_4)$	Analysis of variance
$(1, x_3, x_4, x_3x_4)$	
$(1, x_1, x_1x_3)$	Analysis of covariance
$(1, x_1, x_1x_3, x_1x_4)$	

At the next level of complexity, if $\mathcal{Y} = \{ij\}$ is a 2×2 contingency table, then we could consider the linear exponential model obtained from the saturated exponential model, in which

$$\mathbf{B} = \begin{bmatrix} \mathbf{b}_A & \mathbf{b}_B & \mathbf{b}_{AB} \end{bmatrix}$$

$$q_x^{\mathbf{B}} = e^{\mathbf{t}(x)'\mathbf{b}_A u_A + \mathbf{t}(x)'\mathbf{b}_B u_B + \mathbf{t}(x)'\mathbf{b}_{AB} u_{AB} - \delta} q$$

Then

$$\ln \frac{q_x^{\mathbf{B}}(11)\, q_x^{\mathbf{B}}(00)}{q_x^{\mathbf{B}}(01)\, q_x^{\mathbf{B}}(10)} = \mathbf{t}(x)'\mathbf{b}_{AB} + \ln \frac{q(11)\, q(00)}{q(01)\, q(10)}$$

so that the ln odds ratio is expressed in terms linear in $\mathbf{t}(x)$, and consequently we may relate the interdependence of A and B to the covariate x through the structure function $\mathbf{t}$. Of course, all the t-functions illustrated in the preceding paragraph would be appropriate here as well.

In the case of the independence model for the 2×2 table, we have

$$\mathbf{B} = \begin{bmatrix} \mathbf{b}_A & \mathbf{b}_B \end{bmatrix}$$

and with the uniform generating distribution

$$q_x^{\mathbf{B}}(1+) = \frac{e^{\mathbf{t}(x)'\mathbf{b}_A}}{1 + e^{\mathbf{t}(x)'\mathbf{b}_A}}$$

$$q_x^{\mathbf{B}}(+1) = \frac{e^{\mathbf{t}(x)'\mathbf{b}_B}}{1 + e^{\mathbf{t}(x)'\mathbf{b}_B}}$$

so that the marginal A and B distributions can be studied in terms of their dependence on x through $\mathbf{t}(x)$. Of course, this is identical to two independent logistic regressions.

For the general rectangular table, we would have

$$\mathbf{B} = \begin{bmatrix} \mathbf{B}_A & \mathbf{B}_B & \mathbf{B}_{AB} \end{bmatrix}$$

$$q_x^{\mathbf{B}} = e^{\mathbf{t}(x)'\mathbf{B}_A \mathbf{u}_A + \mathbf{t}(x)'\mathbf{B}_B \mathbf{u}_B + \mathbf{t}(x)'\mathbf{B}_{AB} \mathbf{u}_{AB} - \delta} q$$

and $\mathbf{t}(x)'\mathbf{B}_{AB}$ would represent the A, B-interactions written linearly in terms of $\mathbf{t}$. Higher-dimensional tables are similarly parametrized.

Models on an A, B-table could be embellished in a variety of ways. For example, let q_A be a Binomial distribution on the A-margin, and q_B be a

Truncated Inverse Binomial distribution on the B-margin; take $q(ij) = q_A(i)q_B(j)$, $u_A(ij) = i$, $u_B(ij) = j$, and then

$$\mathbf{B} = \begin{bmatrix} \mathbf{b}_A & \mathbf{b}_B & \mathbf{b}_{AB} \end{bmatrix}$$

$$q_x^{\mathbf{B}} = e^{\mathbf{t}(x)'\mathbf{b}_A u_A + \mathbf{t}(x)'\mathbf{b}_B u_B + \mathbf{t}(x)'\mathbf{b}_{AB} u_A u_B - \delta} q$$

so that $\mathbf{t}(x)'\mathbf{b}_{\text{AB}}$ represents a kind of interdependence appropriate for relating a Binomial and a Truncated Inverse Binomial for experimental units characterized by covariate x.

It should be apparent from these examples that a linear exponential model can be thought of as a combination of two pieces. The response part of the model is determined by the response set $\mathcal{Y}$, the u-function $\mathbf{u}$, and the generating distribution q. The covariate part is determined by a t-function $\mathbf{t}$, which may consist of factorial indicators, linear terms, or any functions which make sense in terms of the components of the covariates. If one understands the model $\mathcal{E}[q:\mathbf{u}]$ in the sense that one has some interpretation for its parameters, and if one understands how $\mathbf{t}$ permits covariate-dependent variation, then as a consequence one understands the conjunction of the two in a linear exponential model.

Covariate Model Notation. It will be convenient to extend the notation developed in Chapter III for the purpose of conciseness in referring to various models. To illustrate this, suppose that A, B, C, D, and E are factors determining a five-way table, and let $\mathbf{u}_A, \mathbf{u}_B$, and so on denote the corresponding factorial indicators when we think of them as u-functions, and $\mathbf{t}_A, \mathbf{t}_B$, and so on be the same functions thought of as t-functions. The linear exponential model obtained from $\mathcal{E}[q:\mathbf{u}_A, \mathbf{u}_B]$ and with t-function $(1, \mathbf{t}'_D, \mathbf{t}'_E)$ will be denoted $\mathcal{E}[q:\mathbf{u}_A, \mathbf{u}_B | 1, \mathbf{t}_D, \mathbf{t}_E]$, and abbreviated to $AB|DE$ when convenient. Notice the important fact that the u-function must be such that the function 1 is not in $\mathcal{S}(\mathbf{u})$, but in general the t-functions will have 1 in $\mathcal{S}(\mathbf{t})$. Thus, although we only write DE for the covariate, it is implicit that the t-function is really $(1, \mathbf{t}'_D, \mathbf{t}'_E)$. If we did not adhere to this convention, there would be no guarantee that the model would permit a distribution free of x but different from q.

Using this notation, we see that $[AB][BC]|DE$ is the model under which, when D and E are taken into account, A and C are conditionally independent given B, and all the parameters are linearly related to levels of D and E. $[AB][BC]|[DE]$ has the same response part, but now the parameters are no longer linear expressions in $\mathbf{t}_D$ and $\mathbf{t}_E$ separately, but can be written as any function of the joint (D, E)-factor, since $\mathcal{S}(1, \mathbf{t}_D, \mathbf{t}_E, \mathbf{t}_{DE})$ contains all functions on the (D, E)-table. The model $[AB]C|[DE]$ stipulates that C is independent of the joint factor (A, B), provided the (D, E)-factor is accounted for. It is worthwhile to recognize that this model does not imply that C is independent of (A, B) without taking (D, E) into account. In fact, $[AB]C$ is a submodel of $[AB]C|[DE]$ as a direct consequence of the general observation that $\mathfrak{S}[q:\mathbf{u}|1] = \mathfrak{S}[q:\mathbf{u}]$, so that $\mathfrak{S}[q:\mathbf{u}] \subset \mathfrak{S}[q:\mathbf{u}|1, \mathbf{t}]$ for any $\mathbf{t}$.

2. ML Inference

Although inference for covariate models parallels that of covariate-free models to a great extent, there are some more or less problematical differences. As indicated earlier, we will only study linear exponential models.

A measure of the discordance between a particular distribution q and the data $\hat{p}$ is the *unlikelihood*, defined by

$$U(\hat{p}, q) = -2 \sum_{i=1}^{n} \ln\left[q_{x_i}(y_i)\right]$$

where $\{(y_i, x_i) : 1 \leqq i \leqq n\}$ is the sample upon which $\hat{p}$ is based. In the covariate-free case we have

$$U(\hat{p}, q) = -2n \sum_{y} \hat{p}(y) \ln q(y)$$

so that $n\mathbb{D}_1(\hat{p}, q) = U(\hat{p}, q) - U(\hat{p}, \hat{p})$. If $\mathfrak{S}[q:\mathbf{u}_2] \subset \mathfrak{S}[q:\mathbf{u}_1]$ and $\hat{q}_2$ and $\hat{q}_1$ are the ML distributions in these models, then we have

$$\begin{aligned} n\mathbb{D}_1(\hat{q}_1, \hat{q}_2) &= n\mathbb{D}_1(\hat{p}, \hat{q}_2) - n\mathbb{D}_1(\hat{p}, \hat{q}_1) \\ &= U(\hat{p}, \hat{q}_2) - U(\hat{p}, \hat{q}_1) \end{aligned}$$

Thus all of our preceding inference based on $\mathbb{D}_1$-MDEs could have been expressed in terms of U rather than $\mathbb{D}_1$. When it comes to covariate analysis, there is no counterpart to $\mathbb{D}_1$ and we must use U.

The ML distribution $\hat{q}$ in $\mathfrak{S}[q:\mathbf{u}|\mathbf{t}]$ is that distribution which minimizes $U(\hat{p}, \hat{q})$. Thus, ML estimation for covariate models generalizes ML inference for covariate-free models. The proofs of the properties of ML analysis in the context of covariate models are rather more difficult than before, and so we omit them.

Assertion 1. Assume that the parameter in $\mathfrak{S}[q:\mathbf{u}|\mathbf{t}]$ is identifiable. Then the ML distribution $\hat{q}$ is unique if it exists, and it satisfies

$$\sum_{i=1}^{n} \mathbf{u}(y_i)\mathbf{t}(x_i)' = \sum_{i=1}^{n} M\left[\mathbf{u}:\hat{q}_{x_i}\right]\mathbf{t}(x_i)'$$

where $\{(y_i, x_i): 1 \leqq i \leqq n\}$ is the sample.

On account of this result, we will always tacitly assume that the parameter matrix is identifiable. The next assertion is that ML inference is consistent.

Assertion 2. Under regularity conditions on $\{\mathbf{t}(x_i): 1 \leqq i < \infty\}$, the ML distributions $\hat{q}_n$ exist eventually as $n \to \infty$, and we have $\hat{q}_n \to_p \pi$ if π lies in the model. If π has parameter matrix $\mathbf{B}$ and $\hat{q}_n$ has parameter matrix $\hat{\mathbf{B}}_n$, then $\hat{\mathbf{B}}_n \to_p \mathbf{B}$.

For the next assertion, write $\mathbf{B} = [\mathbf{b}_1 \ \mathbf{b}_2 \ \cdots \ \mathbf{b}_r]$ and then define $\boldsymbol{\beta}' = (\mathbf{b}_1', \mathbf{b}_2', \ldots, \mathbf{b}_r')$. Let $\mathbf{u} \otimes \mathbf{t}$ denote the vector

$$\mathbf{u} \otimes \mathbf{t} = \begin{pmatrix} u_1\mathbf{t} \\ u_2\mathbf{t} \\ \vdots \\ u_r\mathbf{t} \end{pmatrix}$$

The reason for doing this is to make the next assertion easier to state, and is based on the fact that $\mathbf{t}'\mathbf{B}\mathbf{u} = \boldsymbol{\beta}'\mathbf{u} \otimes \mathbf{t}$.

Assertion 3. Under regularity conditions on $\{\mathbf{t}(x_i): 1 \leqq i < \infty\}$, if π has parameter $\boldsymbol{\beta}$ in the model $\mathfrak{S}[q:\mathbf{u}|\mathbf{t}]$ and $\hat{\boldsymbol{\beta}}_n$ is the ML estimate, then

$$\sqrt{n}\,(\hat{\boldsymbol{\beta}}_n - \boldsymbol{\beta}) \to_d N\left(\mathbf{0}, \left\{\lim \frac{1}{n} \sum_i C\left[\mathbf{u} \otimes \mathbf{t}(x_i) : \pi_{x_i}\right]\right\}^{-1}\right)$$

This assertion permits us to estimate $\boldsymbol{\beta}$ by $\hat{\boldsymbol{\beta}}_n$ and to estimate the covariance matrix of $\hat{\boldsymbol{\beta}}_n$ by

$$\left(\sum_{i=1}^{n} C\left[\mathbf{u} \otimes \mathbf{t}(x_i) : \hat{q}_{x_i}\right]\right)^{-1}$$

For the next assertion, let $\{\mathbf{u}_i\}$ be such that all the components of $\mathbf{u}_i$ are also components of $\mathbf{u}_{i-1}$, and similarly for $\{\mathbf{t}_i\}$. Let the subscripts $i_1 j_2$, $i_2 j_2$, and so on be such that the models

$$\mathfrak{M}_k = \mathcal{E}\left[q : \mathbf{u}_{i_k} | \mathbf{t}_{j_k}\right]$$

satisfy $\mathfrak{M}_k \subset \mathfrak{M}_{k-1}$. Finally, let

$$\nu_k = \#\left(\mathbf{u}_{i_{k-1}} \otimes \mathbf{t}_{j_{k-1}}\right) - \#\left(\mathbf{u}_{i_k} \otimes \mathbf{t}_{j_k}\right)$$

so that ν_k is the number of parameters in $\mathfrak{M}_{k-1}$ minus the number of parameters in $\mathfrak{M}_k$.

Assertion 4. If the ML distribution $\hat{q}_1$ in $\mathfrak{M}_1$ exists, then so do the ML distributions $\hat{q}_i$ in $\mathfrak{M}_i$. Moreover, under regularity conditions on $\{\mathbf{t}(x_i) : 1 \leqq i < \infty\}$, if $\pi \in \mathfrak{M}_k$ then

$$U(\hat{p}, \hat{q}_j) - U(\hat{p}, \hat{q}_{j-1}) \rightarrow_d \chi^2(\nu_j) \qquad (2 \leqq j \leqq k)$$

and these statistics are asymptotically independent.

This important result permits the concept of an ANALYSIS table and all the associated inferential machinery to be extended immediately to cover covariate models.

3. Conditional Inference

Some linear exponential models can be obtained by conditioning exponential models. In general, let $\mathcal{E}[q : \mathbf{u}_1, \mathbf{u}_2]$ be an exponential model, and consider the conditional distribution of y given $\mathbf{u}_2(y)$ determined by a distribution in the model. Partition $\boldsymbol{\theta}$ as $(\boldsymbol{\theta}_1', \boldsymbol{\theta}_2')$ to correspond to $(\mathbf{u}_1', \mathbf{u}_2')$, and then compute

$$q^{\boldsymbol{\theta}}(y | \mathbf{u}_2(y)) = \frac{\exp\left[\boldsymbol{\theta}_1'\mathbf{u}_1(y) + \boldsymbol{\theta}_2'\mathbf{u}_2(y)\right] q(y)}{\Sigma_z \exp\left[\boldsymbol{\theta}_1'\mathbf{u}_1(z) + \boldsymbol{\theta}_2'\mathbf{u}_2(z)\right] q(z)}$$

$$= \frac{\exp\left[\boldsymbol{\theta}_1'\mathbf{u}_1(y)\right] q(y)}{\Sigma_z \exp\left[\boldsymbol{\theta}_1'\mathbf{u}_1(z)\right] q(z)}$$

where the summation is over $\mathcal{Z} = \{z \in \mathcal{Y} : \mathbf{u}_2(z) = \mathbf{u}_2(y)\}$. This shows that $q^{\boldsymbol{\theta}}(y | \mathbf{u}_2)$ is the element of $\mathcal{E}[q : \mathbf{u}_1]$ defined on $\mathcal{Z}$ with parameter $\boldsymbol{\theta}_1$. The importance of this result is that $\boldsymbol{\theta}_2$ has disappeared in the process of conditioning.

For a more specific example, suppose that $\mathbf{u}_A$ and $\mathbf{u}_B$ pertain to two factors in a two-way table, $\mathcal{Y} = \{ij\}$. An exponential model which permits interaction between A and B is $\mathcal{E}[q : \mathbf{u}_A, \mathbf{u}_B, \mathbf{u}_A \otimes \mathbf{u}_B]$. Note that if $\mathbf{u}_A$ and $\mathbf{u}_B$ were factorial indicators, then $\mathbf{u}_A \otimes \mathbf{u}_B$ is a more precise way of writing the second-order factorial indicator which we have denoted $\mathbf{u}_{AB}$ (see Problems II.9 and II.12). The conditional model for y given $\mathbf{u}_B$ is

$$q^{\boldsymbol{\theta}}(y|\mathbf{u}_B(y)) = \frac{\exp[\boldsymbol{\theta}_A'\mathbf{u}_A(y) + \boldsymbol{\theta}_{AB}'\mathbf{u}_A \otimes \mathbf{u}_B(y)]q(y)}{\Sigma_z \exp[\boldsymbol{\theta}_A'\mathbf{u}_A(z) + \boldsymbol{\theta}_{AB}'\mathbf{u}_A \otimes \mathbf{u}_B(z)]q(z)}$$

This can be viewed as the linear exponential model $\mathcal{E}[q : \mathbf{u}_A | 1, \mathbf{u}_B]$ by defining the parameter matrix $\mathbf{B}$ so that $\boldsymbol{\beta}' = (\boldsymbol{\theta}_A', \boldsymbol{\theta}_{AB}')$, since

$$\begin{pmatrix} 1 \\ \mathbf{u}_B \end{pmatrix}' \mathbf{B}\mathbf{u}_A = \boldsymbol{\beta}'\mathbf{u}_A \otimes \begin{pmatrix} 1 \\ \mathbf{u}_B \end{pmatrix} = \boldsymbol{\theta}_A'\mathbf{u}_A + \boldsymbol{\theta}_{AB}'\mathbf{u}_A \otimes \mathbf{u}_B$$

Another way of looking at this is to consider $\mathcal{Y} = \mathcal{Y}_1 \times \mathcal{Y}_2 = \{ij\}$, write $\mathbf{u}_A(ij) = \mathbf{u}_A(i)$, $\mathbf{u}_B(ij) = \mathbf{u}_B(j)$, and then see that the above model can be written

$$q^{\mathbf{B}}(i|j) = \frac{\exp[(1, \mathbf{u}_B(j)')\mathbf{B}\mathbf{u}_A(i)]q(i+)}{\Sigma_k \exp[(1, \mathbf{u}_B(j)')\mathbf{B}\mathbf{u}_A(k)]q(k+)}$$

where the summation is over all k in $\mathcal{Y}_1$. Thus we see that the conditional model for A given B, obtained from $\mathcal{E}[q : \mathbf{u}_A, \mathbf{u}_B, \mathbf{u}_A \otimes \mathbf{u}_B]$, is the linear exponential model $\mathcal{E}[q : \mathbf{u}_A | 1, \mathbf{u}_B]$ defined on $\mathcal{Y}_1$ with covariate values in $\mathcal{Y}_2$. In the process of conditioning $\boldsymbol{\theta}_B$ has dropped from the model.

For a concrete example, let A and B define a 2×2 table, and let C be a factor with levels $\{0, 1, \ldots, c\}$. Staying within the class of hierarchical models, according to the conditioning rule (Chapter IV, Section A.3) we have

$$[ABC] \uparrow_C [AB]$$

$$[AC][BC] \uparrow_C A\,B$$

$$[AB]C \uparrow_C [AB]$$

and the models on the right can be thought of as linear exponential models. In the first and second cases the conditional distributions of (A, B) depend on the value of C, while in the third case they are free of C.

In accordance with the discussion in Chapter V, Section A.1, by conditioning we are conceptually changing one of the factors from a response to a covariate. Inference carried out on the covariate model rather than the original model is called *conditional inference*. In the example of the preceding paragraph, conditional inference would be especially appropriate if the sampling plan were believed to be faulty with regard to the C factor, but valid on subsamples within C, because then the parameters concerning the (A, B) distribution within C ($\boldsymbol{\theta}_A$, $\boldsymbol{\theta}_B$, and $\boldsymbol{\theta}_{AB}$) and the relationship between (A, B) and C ($\boldsymbol{\theta}_{AC}$ and $\boldsymbol{\theta}_{BC}$) would be properly estimated, while the parameter $\boldsymbol{\theta}_C$, relating to C alone, would be improperly estimated, so that the latter parameter would have been justifiably removed from the model.

Exact Conditional Tests. The range of usefulness of conditioning is broader than just the changing of exponential models into linear exponential models. It is sometimes possible to obtain a test based on a conditional distribution which is *exact*, in the sense that it does not depend on an asymptotic approximation.

In order to develop such a test, we need to shift our point of view somewhat. Imagine a contingency table $\mathcal{Y} = \{y_1, y_2, \ldots, y_r\}$ in which the elements y_i have been arranged in a specific order. Let $\mathcal{Y}^*$ denote the derived contingency table consisting of all integer vectors $\mathbf{n}$ having r components, and such that $n_i \geqq 0$ for all i and $\mathbf{1}'\mathbf{n} = n$. Thus $\mathcal{Y}^*$ is the set of all possible observations of a sample of size n in the original table $\mathcal{Y}$, with n_i being the frequency in cell y_i. One can then view the experiment as n independent observations in a model

$$q^{\boldsymbol{\theta}}(y) = e^{\boldsymbol{\theta}'\mathbf{u}(y) - \delta} q(y) \qquad (y \in \mathcal{Y})$$

or as a single observation in the model

$$q_*^{\boldsymbol{\theta}}(\mathbf{n}) = \exp[\boldsymbol{\theta}'\mathbf{u}^*(\mathbf{n}) - n\delta]\, q^*(\mathbf{n}) \qquad (\mathbf{n} \in \mathcal{Y}^*)$$

where

$$\mathbf{u}^*(\mathbf{n}) = \sum_i n_i \mathbf{u}(y_i) \quad \text{and} \quad q^*(\mathbf{n}) = \binom{n}{\mathbf{n}} \prod_i q(y_i)^{n_i}$$

and $\binom{n}{\mathbf{n}}$ is the multinomial coefficient giving the number of ways the n original observations could be distributed in $\mathcal{Y}$ in order to produce frequency n_i in cell y_i. This latter model is $\mathcal{E}[q^* : \mathbf{u}^*]$ on $\mathcal{Y}^*$.

We now apply the conditioning argument to $\mathcal{Y}^*$. Partition

$$\mathbf{u}^* = \begin{pmatrix} \mathbf{u}_1^* \\ \mathbf{u}_2^* \end{pmatrix}$$

and compute

$$q_*^{\theta}(\mathbf{n}|\mathbf{u}_2^*(\mathbf{n})) = \frac{\exp[\boldsymbol{\theta}_1'\mathbf{u}_1^*(\mathbf{n})]\,q^*(\mathbf{n})}{\Sigma_{\mathbf{m}}\exp[\boldsymbol{\theta}_1'\mathbf{u}_1^*(\mathbf{m})]\,q^*(\mathbf{m})}$$

where the summation is over all $\mathbf{m} \in \mathcal{Y}^*$ for which $\mathbf{u}_2^*(\mathbf{m}) = \mathbf{u}_2^*(\mathbf{n})$. Thus we arrive at a collection of conditional distributions in $\mathcal{E}[q^* : \mathbf{u}_1^*]$, and the exact conditional test is based on the distribution for which $\mathbf{u}_2^*(\mathbf{n})$ was observed.

McNemar's Test. One of the simplest exact conditional tests is McNemar's test for marginal homogeneity in a 2^2 table. Let q be uniform, and let $\mathbf{u}$ be defined by Table V.1. Note that u_1 and (u_2, u_3) are q-complements. Thus the model of marginal homogeneity is

$$q^{\boldsymbol{\alpha}} = (\alpha_0 + \alpha_2 u_2 + \alpha_3 u_3)q$$

In terms of $\mathcal{Y}^*$ this becomes

$$q_*^{\boldsymbol{\alpha}}(\mathbf{n}) = (\alpha_0 + \alpha_2 + \alpha_3)^{n(00)}\alpha_0^{n(01)+n(10)}(\alpha_0 + \alpha_2 - \alpha_3)^{n(11)}q^*(\mathbf{n})$$

where

$$q^*(\mathbf{n}) = \binom{n}{\mathbf{n}}\left(\frac{1}{4}\right)^n.$$

Thus

$$q_*^{\boldsymbol{\alpha}}(\mathbf{n}|n(00), n(11)) = \frac{\alpha_0^{n(01)+n(10)}\binom{n}{\mathbf{n}}}{\Sigma_{\mathbf{m}}\alpha_0^{m(01)+m(10)}\binom{n}{\mathbf{m}}}$$

where the sum is over $\mathbf{m}$ for which $m(11) = n(11)$ and $m(00) = n(00)$. But

Table V.1

Functions on the 2^2 Table Associated with McNemar's Test

u_1		u_2		u_3	
0	1	1	0	1	0
−1	0	0	1	0	−1

then $m(01) + m(10) = n(01) + n(10)$ and so

$$q_*^{\alpha}(\mathbf{n}|n(00), n(11)) = \frac{\binom{n}{\mathbf{n}}}{\Sigma_{\mathbf{m}}\binom{n}{\mathbf{m}}}$$

$$= \binom{n(01) + n(10)}{n(01)}\left(\frac{1}{2}\right)^{n(01)+n(10)}$$

Therefore, conditional on the diagonal values, the distribution of $n(01)$ is Binomial with $n(01) + n(10)$ trials and success probability $\frac{1}{2}$. In this case both of the parameters α_2 and α_3 have been conditioned out of the model. The marginal homogeneity hypothesis is now rejected on the basis of a P-value computed from the above Binomial distribution.

When $n(01) + n(10)$ is large, one can approximate McNemar's test by treating $(|n(01) - n(10)| - 1)^2/[n(01) + n(10)]$ as a $\chi^2(1)$ observation. This is an example of an *approximate conditional test*.

Fisher's Test. Probably the most widely known exact conditional test is Fisher's test for independence in a 2^2 table. Continue the notation of the preceding section, and write

$$q^{\boldsymbol{\theta}}(ij) = \exp[\theta_A u_A(i) + \theta_B u_B(j) + \theta_{AB} u_{AB}(ij) - \delta]q(ij)$$

where q leaves rows and columns independent, and

$$\mathbf{u}(ij) = \begin{pmatrix} u_A(i) \\ u_B(j) \\ u_{AB}(ij) \end{pmatrix} = \begin{pmatrix} i \\ j \\ ij \end{pmatrix}$$

Then

$$\mathbf{u}^*(\mathbf{n}) = \begin{pmatrix} n(1+) \\ n(+1) \\ n(11) \end{pmatrix}$$

and

$$q_*^{\boldsymbol{\theta}}(\mathbf{n}) = \exp[\boldsymbol{\theta}'\mathbf{u}^*(\mathbf{n}) - n\delta]q^*(\mathbf{n})$$

where

$$q^*(\mathbf{n}) = \binom{n}{\mathbf{n}} q(00)^{n(00)} q(01)^{n(01)} q(10)^{n(10)} q(11)^{n(11)}$$

$$= \binom{n}{\mathbf{n}} q(1+)^{n(1+)} q(0+)^{n(0+)} q(+1)^{n(+1)} q(+0)^{n(+0)}$$

Conditioning, we obtain

$$q_*^{\boldsymbol{\theta}}(\mathbf{n}|n(+1))$$

$$= \frac{\exp[\theta_A n(1+) + \theta_{AB} n(11)] \binom{n}{\mathbf{n}} q(1+)^{n(1+)} q(0+)^{n(0+)}}{\Sigma_{\mathbf{m}} \exp[\theta_A m(1+) + \theta_{AB} m(11)] \binom{n}{\mathbf{m}} q(1+)^{m(1+)} q(0+)^{m(0+)}}$$

where the sum is over $\mathbf{m}$ for which $m(+1) = n(+1)$. Now if $\theta_A = 0$ and $\theta_{AB} = 0$, then this reduces to

$$q_*^{\mathbf{0}}(\mathbf{n}|n(+1)) = \binom{n(+0)}{n(00)} \binom{n(+1)}{n(11)} q(1+)^{n(1+)} q(0+)^{n(0+)}$$

so that conditional on $n(+1)$, the distribution of $n(00)$ and $n(11)$ is that of independent Binomials with $n(+0)$ and $n(+1)$ trials and success probabilities $q(0+)$ and $q(1+)$, respectively. A test of $\theta_A = 0$ and $\theta_{AB} = 0$ would be based on this distribution.

If in the above model we further condition on $n(1+)$, we get

$$q_*^{\boldsymbol{\theta}}(\mathbf{n}|n(+1), n(1+)) = \frac{\exp[\theta_{AB} n(11)] \binom{n}{\mathbf{n}}}{\Sigma_{\mathbf{m}} \exp[\theta_{AB} m(11)] \binom{n}{\mathbf{m}}}$$

where the sum is over all $\mathbf{m}$ having $m(+1) = n(+1)$ and $m(1+) = n(1+)$. When $\theta_{AB} = 0$ this becomes

$$q_*^{\mathbf{0}}(\mathbf{n}|n(+1), n(1+)) = \frac{\binom{n(+0)}{n(00)} \binom{n(+1)}{n(01)}}{\binom{n}{n(0+)}}$$

Thus, conditional on the margins of the table, $n(00)$ is Hypergeometric, and the test of $\theta_{AB} = 0$ is based on this distribution. This test of independence is sometimes called the *Fisher–Irwin test*.

Elimination of Nuisance Parameters. The conditioning principle involves changing the model one is using for inference, going from the full model $\mathfrak{S}[q:\mathbf{u}_1,\mathbf{u}_2]$ to a conditional model $\mathfrak{S}[q:\mathbf{u}_1]$ dependent on the observed value of $\mathbf{u}_2$. Although the conditional model is valid, not all statisticians agree that it is appropriate. On the other hand, some statisticians insist that only the conditional model is appropriate.

Whatever position one chooses to take, there is one situation in which the logic of conditional inference is overpowering. Begin with the covariate model

$$q_x^{\boldsymbol{\theta}}(y) = \exp[\boldsymbol{\theta}_0'\mathbf{u}_0(y) + \boldsymbol{\theta}_x'\mathbf{u}_1(y) - \delta]q(y)$$

The symbol $\boldsymbol{\theta}_x$ is intended to represent the fact that $\boldsymbol{\theta}_x$ may depend in any way whatsoever on x. In particular, we could take x to be the number of the experimental unit, so $x = 1, 2, \ldots, n$. Now when the sample reaches size n there are parameters $\boldsymbol{\theta}_0, \boldsymbol{\theta}_1, \ldots, \boldsymbol{\theta}_n$, and so the number of parameters increases with the sample size. Conventional ML inference does not cover this situation, and in fact there are many cases where it can be shown to fail, in the sense that misleading inferences are likely and the situation does not improve with larger samples.

When the parameters $\boldsymbol{\theta}_x$ are of no interest, they simply confound the analysis of $\boldsymbol{\theta}_0$, and so they are called *nuisance parameters*. As we have seen several times now, upon conditioning on $\mathbf{u}_1$ we obtain

$$q_x^{\boldsymbol{\theta}}(y|\mathbf{u}_1(y)) = \exp[\boldsymbol{\theta}_0'\mathbf{u}_0(y) - \delta]q(y)$$

a model with a fixed parameter $\boldsymbol{\theta}_0$ and free of x. Note that we could have begun with $\boldsymbol{\theta}_0 = \mathbf{B}'\mathbf{t}(x)$, in which case we would have obtained a linear exponential model, conditional on the value of $\mathbf{u}_1(y)$. ML inference is valid for these latter models, and so conditioning is effective in dealing with the problem of nuisance parameters.

4. LAMDA Notes

For a ML fit to a linear exponential model, it is necessary to have the t-functions in the C-file in addition to all the quantities necessary for fitting an exponential model. Whereas each row in the C-file corresponded to one possible response y when fitting an exponential model, in the case of a linear exponential model each row corresponds to one or more experimental units. The `WGT` variable (variable 1 by default) is a count of the number of cases (experimental units) the row represents. The values of the u-functions and t-functions for those cases then occupy other variables.

For example, suppose that row i represents c_i cases, each of which had response y_i and covariate x_i. The row would then appear as follows:

Var1	Var2	$\cdots$	Var$(r+1)$	Var$(r+2)$	Var$(r+3)$	$\cdots$	Var$(r+s+2)$
c_i	$u_1(y_i)$		$u_r(y_i)$	1	$t_1(x_i)$		$t_s(x_i)$

The command to fit the corresponding model $\mathcal{E}[q:\mathbf{u}|\mathbf{t}]$ would be

```
aSEL 2 3 4 ··· (r + s + 2)

rLXM
```

where $a = r + s + 1$, the number of integers following **SEL**. Thus a represents the total of all u- and t-functions (including the function 1, if it is one of the t-functions), while r is the number of u-functions. The order of the specification of the variables in the **SEL** command is not important, except that all the u-functions must precede all of the t-functions. Note that the constant 1 can be indicated in the **SEL** command by using the special variable number 0.

There is one further way in which LAMDA treats the covariate and covariate-free cases differently. Consider the segments

1. ```
   4SEL 2 3 4 5
   3LXM
   ```

2. ```
   3SEL 2 3 4
   3LXM
   ```

in which variable 5 is the constant 1. Since $\mathcal{E}[q:\mathbf{u}|1] = \mathcal{E}[q:\mathbf{u}]$, one might expect the two analyses to be identical, but this may not be the case. Recall that in segment 2 each row is presumed to stand for a cell in the table, but in segment 1 each row corresponds to one or more cases. Thus in segment 1 LAMDA does not know in advance how many cells there are in the table, and it determines this by counting how many distinct values $\mathbf{u}(y_i)$ occur in the C-file, and then pooling rows with identical entries (of the u- and t-functions) together. Consequently, two identical rows in segment 2 will be kept separate as two cells with coincident probabilities, but in segment 1 they will be pooled to form one cell. Apart from this, however, both commands produce the same result when there are no identical rows.

The generating distribution is assumed uniform by **LXM** unless a **GEN** command is used. Since the number of cells and number of rows in the C-file will nearly always be different in the covariate case, the **G = V** command is inappropriate.

The COEFFICIENTS printed out by LAMDA are in the form of the parameter matrix **B**. The columns thus correspond to the u-functions and the rows to the t-functions, in the order they were specified in the **SEL** command.

To see how the COVARIANCE MATRIX relates to the parameters, write

$$\mathbf{B} = \begin{bmatrix} \mathbf{b}_1 & \mathbf{b}_2 & \cdots & \mathbf{b}_r \end{bmatrix}$$

that is, think of **B** as a collection of columns, and then write

$$\boldsymbol{\beta} = \begin{pmatrix} \mathbf{b}_1 \\ \mathbf{b}_2 \\ \vdots \\ \mathbf{b}_r \end{pmatrix}$$

The covariance matrix is then the estimated covariance matrix of $\hat{\boldsymbol{\beta}}$. Stated another way, the row and column of the covariance matrix corresponding to the parameter b_{ij} (in the matrix **B**) would be $(j-1)s+i$, where **B** is $s \times r$.

LAMDA does not provide much diagnostic help with finer questions about residuals, since it reports only the fit over the $\mathscr{Y}$-table, ignoring covariates. We will see in some of the data examples how such a finer analysis can be carried out.

B. DATA EXAMPLES

1. Logistic Regression: Indicators of a Single Factor

Look again at the data on the polio epidemic among Eskimos in Table III.11. Assume now that the disease status of every member of the population is known. It is then natural to consider the probability of contracting polio as a function of age, and so we consider the response space $\{0, 1\}$ and the age categories 0 through 6. Thus we have $y \in \{0, 1\}$ and $x \in \{0, 1, \ldots, 6\}$ for each subject.

To begin model-fitting, let u be the indicator of 1 and $t_i(x)$ be the indicator of $x = i$, and consider the model $\mathscr{E}[q: u \mid 1, t_1, \ldots, t_6]$ where q is uniform. Under the model,

$$q_x^{\mathbf{b}}(1) = \frac{1}{1 + e^{-b_0 - b_x}}$$

and since each age group is permitted its own parameter, the model is saturated. The LAMDA run appears in Run V.B.1, and part of the output in

Figure V.1. Note that $D(P, QP)$ cannot serve as a measure of fit, because here we have fitted the data exactly but $D(P, QP)$ takes the apparently meaningless value -19.751. It is usually the case for covariate models that $D(P, QP)$ is negative; one can think of $D(P, QP) = 0$ indicating that a covariate-free model fits the data perfectly, so that the addition of any covariates will make $D(P, QP)$ go negative.

Examination of the fitted coefficients suggests that $b_1 = b_2 = b_4 = b_5 = b_6$. In order to fit this model, we alter the preceding run as in Run V.B.2. Note that variable 4 has been redefined to be the indicator of $\{1, 2, 4, 5, 6\}$. From the second output we obtain the unlikelihood $U = 261.211$. Thus the fit is hardly changed at all, since $261.211 - 260.926 = 0.285$ is treated as a χ^2 for the omission of four parameters. Of course, there is no question of hypothesis testing here, since the simpler model was suggested by the initial analysis.

We can now write the final fitted model as

$$\ln \frac{\hat{q}_x(1)}{\hat{q}_x(0)} = -3.2387 + 1.9990(t_1 + t_2 + t_4 + t_5 + t_6) + 2.9285 t_3$$

```
*** RUN V.B.1 ***
=== LXM ===
U(P,QP)     =       260.925
DELTA(C) =   0.00015665
DELTA(U) =   0.00033569
 NO OF CASES =     275.000
 6 ITERATIONS

- - - A N A L Y S I S - - -
D(P,QP)     =       -19.751
D(QP,Q)     =       120.306
D(P,Q)      =       100.555

- - - COEFFICIENTS - - -

        -3.2387
         2.0424
         2.0992
         2.9285
         1.8524
         2.0347
         1.8524

- - - T-VALUES - - -

        -4.4932
         2.5945
         2.5372
         3.5589
         2.1712
         2.5678
         2.1116
```

Fig. V.1 LAMDA *output of the logistic regression of polio on age.*

and so we come to the same conclusion as in Chapter III, Section B.1, that there are similar attack probabilities in all but two age groups, infants having lower and teenagers higher attack rates.

2. Logistic Regression: Two Linear Factors

A common type of experiment for which logistic regression is appropriate involves a response of survival or nonsurvival under certain controlled conditions. The data in Table V.2 display the numbers of surviving and nonsurviving irradiated mice, classified according to their log concentration of granulocytes (G) and lymphocytes (L). In the raw data, no doubt the exact values of G and L were recorded for each mouse, but unfortunately the mice were grouped into categories before the analysis. The values of G and L are thus the midpoints of the intervals into which the mice were classified.

It is reasonable to fit linear and second-order terms involving G and L in a logistic regression, since the probability of survival could be assumed to be a simple continuous increasing function of G and L. In Run V.B.3 we see

Table V.2

Survival Ratios S1 / S0 for Irradiated Mice

	L					
G	2.7	2.9	3.1	3.3	3.5	3.7
1.625	1/2	0/2	—	—	—	—
1.875	—	—	1/2	0/2	—	—
2.125	—	1/2	—	—	—	—
2.375	1/0	0/3	1/3	2/2	1/2	—
2.625	—	0/2	2/1	5/2	—	—
2.875	—	—	—	0/1	3/1	1/0
3.125	—	—	—	2/0	2/0	2/0
3.375	—	2/0	—	—	2/0	4/0

S = Vital Status
$S0$ = Nonsurvival
$S1$ = Survival
G = Log Granulocyte Concentration
1.625(.25)3.375
L = Log Lymphocyte Concentration
2.7(.2)3.7

that the data are coded

$$\text{Variable 1} = G \text{ with values } 0(1)7$$

$$\text{Variable 2} = L \text{ with values } 0(1)5$$

$$\text{Variable 3} = S \text{ with values } 0, 1$$

$$\text{Variable 4} = \text{Frequency}$$

and that the program transforms G and L to their true values in lines 35 and 36. In the transformed C-file we have

Variable Number	Variable Name
1	Frequency
2	S
3	G
4	L
5	G^2
6	L^2
7	GL

The three models fitted by this run are $S|G, L, G^2, L^2, GL$; $S|G, L, G^2, L^2$, and $S|G, L$. We will also consider the models $S|G$ and $S|L$, as well as the model S in which survival is unrelated to G and L. From the ANALYSIS Table V.3, either model fitting or hypothesis testing would lead to $S|G$. The fitted distribution is

$$\ln \frac{\hat{q}_x(1)}{\hat{q}_x(0)} = -7.3023 + 2.9519 t_G$$

Since S does not fit the data, we would reject the hypothesis that the coefficient of t_G was zero. However, despite the fact that 2.9519 is statistically significant, there is still the question of its practical significance. To assess this, we can display the fitted probability of survival $\hat{q}_x(1)$ as a function of $x = G$. A natural question which arises here is the following: for a given level of G, what is a confidence interval for the probability of survival? From the LAMDA output the estimated covariance matrix of $(\hat{b}_0, \hat{b}_G)$ is

$$\begin{bmatrix} 4.52787 & -1.75975 \\ -1.75975 & 0.69898 \end{bmatrix}$$

From this we can compute the estimated variance of $\hat{b}_0 + \hat{b}_G t_G$ as

$$\hat{\sigma}^2 = 4.52787 - 2 \times 1.75975 t_G + 0.69898 t_G^2$$

and thus a $1 - \alpha$ CI for $b_0 + b_G t_G$ is

$$\hat{b}_0 + \hat{b}_G t_G \pm \xi \hat{\sigma}$$

where $F_0(\xi) = 1 - \alpha/2$. The implied $1 - \alpha$ CI for $\pi_x(1)$ is then

$$\left[\frac{1}{1 + \exp[-\hat{b}_0 - \hat{b}_G t_G + \xi\hat{\sigma}]}, \frac{1}{1 + \exp[-\hat{b}_0 - \hat{b}_G t_G - \xi\hat{\sigma}]} \right]$$

By a modification of an argument due to Scheffé, if we take ξ so that $F_2(\xi^2) = 1 - \alpha$, then the above CI is valid for all values of x simultaneously, and thus provides a $1 - \alpha$ CI for the entire curve $\pi_x(1)$. In Run V.B.4 the fitted probabilities with upper and lower confidence bounds for $\xi = 1.177$ $(1 - \alpha = .5)$ and $\xi = 2.448$ $(1 - \alpha = .95)$ are computed and plotted. From Figure V.2 one can assess both the statistical and practical significance of the results.

In order to examine the residuals for this model we should consider the **RES - F**-type residuals

$$\frac{\sqrt{N_x(+)}\,[\hat{p}_x(y) - \hat{q}_x(y)]}{\sqrt{\hat{q}_x(1)\hat{q}_x(0)}}$$

for each observed x and $y \in \{0, 1\}$. Due to the requirements $\hat{p}_x(0) + \hat{p}_x(1) = 1$, $\hat{q}_x(0) + \hat{q}_x(1) = 1$, we see that the residuals satisfy constraints so severe that half are the negatives of the other half. Thus we should only produce a uniform plot, for, say, the cases where $y = 1$. This is shown in Figure V.3 (see Run V.B.5) and indicates no troublesome irregularities.

Table V.3

Analysis of the Survival Data on Irradiated Mice

	Model	df	χ^2	Next	$\bar{\alpha}$	$\langle \bar{P} \rangle$
1	$S\|G, L, G^2, L^2, GL$	—	—	2 ⟨.452⟩	—	—
2	$S\|G, L, G^2, L^2$	1	0.36	3 ⟨.172⟩	.01	⟨.452⟩
3	$S\|G, L$	3	3.89	4 ⟨.721⟩ 5 ⟨.001⟩	.02	⟨.314⟩
4	$S\|G$	4	4.24	6 ⟨.000⟩	.03	⟨.432⟩ *
5	$S\|L$	4	14.40	6 ⟨.000⟩	.03	⟨.003⟩ R*
6	S	5	24.72		.04	⟨.000⟩ R

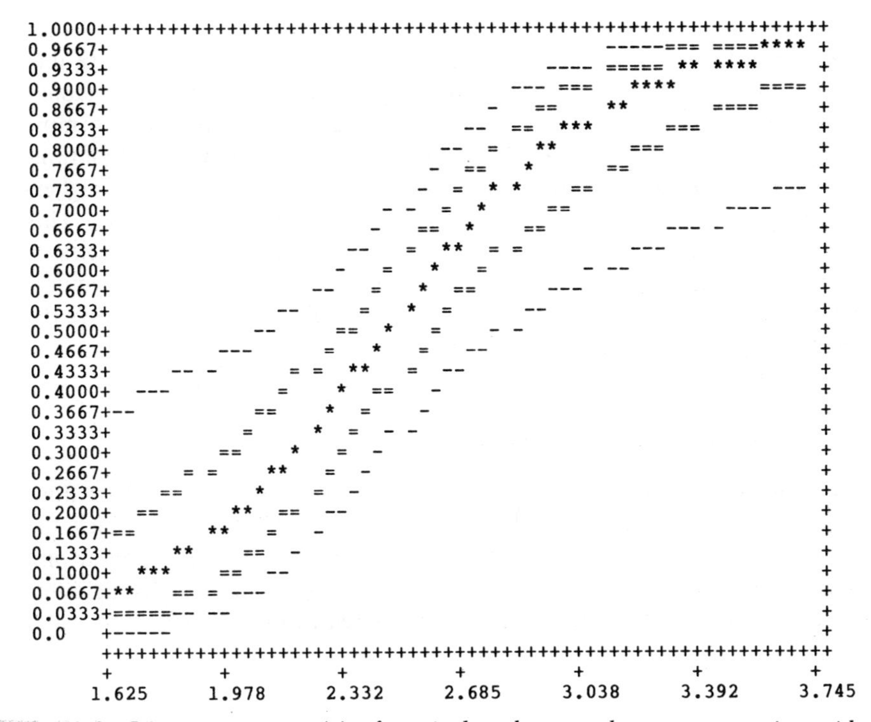

Fig. V.2 Logistic regression (*) of survival on log granulocyte concentration, with simultaneous .5 (=) and .95 (−) confidence regions.

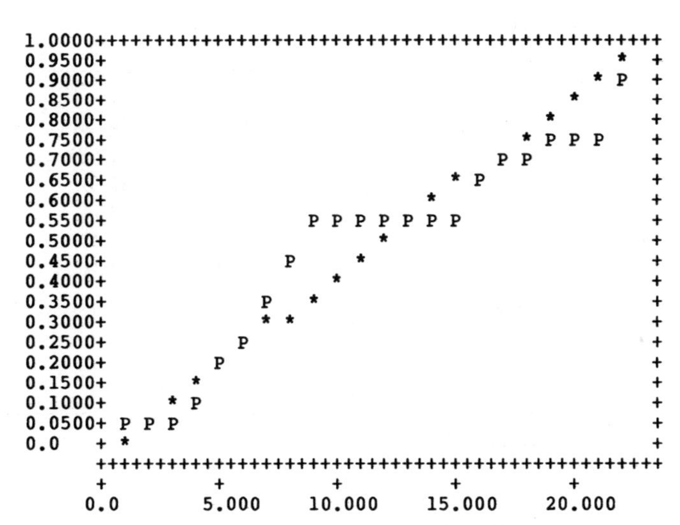

Fig. V.3 Uniform plot of half the residuals from the logistic regression.

In the data examples of preceding chapters there was always an absolute measure of fit due to the fact that a saturated model could be fitted. In this case G and L happened to be groupings obtained from "continuous" measures, and it is not clear what should be a candidate for the most complex model to be considered. When covariates are retained as "continuous" this creates a problem, because one can continue to add ever higher-degree polynomial terms until finally the data are fitted exactly. This dilemma is not entirely resolved by the uniform plot; it is nonetheless a useful, if imperfect, indicator of possible residual variation which requires attention, and perhaps the fitting of a more complex model. In this case the residual plot suggests that our model fits well.

3. Logistic Regression: Two Quadratic Factors

As another example of an experiment which results in survival or nonsurvival for the test animal, consider the data in Table V.4, in which the nonsurviving and surviving ($S1$ and $S0$) numbers of insect larvae are reported at different dosages of phosphine (P) gas and at different durations of exposure (D). The quadratic logistic regression fitted by Run V.B.6

Table V.4

Nonsurvival Ratios (S1 / S0) of Larvae

	D					
P	7	10	14	20	28	40
0.10	—	—	—	—	—	27/66
0.20	—	—	—	11/35	24/21	30/13
0.26	—	—	1/27	7/32	21/20	26/7
0.35	—	4/38	16/24	27/5	32/6	39/1
0.49	6/35	19/19	23/15	30/7	40/4	—
0.73	16/29	21/20	29/16	40/6	41/4	—
1.04	16/24	22/23	34/11	35/3	34/3	—
1.43	26/12	23/20	33/6	41/3	—	—
1.96	27/18	34/6	41/3	37/1	—	—
2.81	21/19	31/9	39/5	—	—	—
4.00	31/9	35/8	37/5	—	—	—

S = Vital Status
$S0$ = Survived
$S1$ = Not Survived
P = Phosphine-Gas Concentration
D = Duration of Exposure

```
*** RUN V.B.6 ***
=== LXM ===
U(P,QP)    =     1961.712
DELTA(C) =  0.00004222
DELTA(U) =  0.0
 NO OF CASES =  1765.000
 6 ITERATIONS

- - - A N A L Y S I S - - -
D(P,QP)    =    -301.015
D(QP,Q)    =     485.097
D(P,Q)     =     184.083

- - - COEFFICIENTS - - -

      -1.3688
       0.4609
       0.2917
      -0.6824
      -0.2218
       1.2736

- - - T-VALUES - - -

      -2.8174
       1.1384
       0.7206
      -0.9356
      -3.4263
       6.1832
```

Fig. V.4 LAMDA *output of the logistic regression on two factors.*

is $S|P, D, P^2, D^2, PD$. From the results in Figure V.4 we see that some of the coefficients do not appear to be significantly different from 0. One might be reluctant to drop them from the model, though, because when the quadratic terms P^2 and PD are present it seems natural to include the linear terms as well. This might be justified in the present case on the grounds that retention of the apparently nonsignificant terms P, D, and D^2 involves only 3 df in the face of abundant data. Moreover, we might be especially interested in obtaining estimates of nonsurvival probabilities which are relatively free from bias, and the inclusion of some extra terms would tend to remove bias (at the expense of decreasing the precision of parameter estimates). Therefore, we will not delete any parameters.

Run V.B.7 computes the fitted probabilities for selected levels of P and D. One needs to be careful with the output (Figure V.5) because probabilities are computed for some combinations of P and D not present in the experiment. Values like $(P, D) = (1, 14)$ create no problem, because they are surrounded by actual experimental points (P, D), but values like $(P, D) = (2, 42)$ are remote from experimental points, and they should probably not be trusted. The dotted boundaries enclose probabilities occurring in the range of experimental x-values.

An intriguing feature of this table of fitted probabilities is the fact that in the first column ($D = 7$) the nonsurvival probabilities first rise and then fall

P \ D	7.0000	14.0000	21.0000	28.0000	35.0000	42.0000
0.1000	0.2762	0.3425	0.3995	0.4428	0.4703	0.4812
0.2000	0.2990	0.3890	0.4706	0.5372	0.5864	0.6183
0.3000	0.3220	0.4366	0.5419	0.6281	0.6928	0.7379
0.4000	0.3448	0.4843	0.6104	0.7097	0.7812	0.8297
0.5000	0.3674	0.5311	0.6739	0.7790	0.8491	0.8936
0.6000	0.3895	0.5763	0.7306	0.8350	0.8982	0.9351
0.7000	0.4109	0.6192	0.7800	0.8785	0.9324	0.9609
0.8000	0.4317	0.6594	0.8219	0.9114	0.9555	0.9767
0.9000	0.4515	0.6964	0.8567	0.9358	0.9708	0.9861
1.0000	0.4705	0.7302	0.8852	0.9536	0.9808	0.9917
1.1000	0.4884	0.7607	0.9083	0.9665	0.9874	0.9950
1.2000	0.5052	0.7880	0.9268	0.9758	0.9917	0.9970
1.3000	0.5209	0.8123	0.9415	0.9825	0.9946	0.9982
1.4000	0.5355	0.8338	0.9533	0.9873	0.9964	0.9989
1.5000	0.5489	0.8527	0.9626	0.9908	0.9976	0.9993
1.6000	0.5611	0.8693	0.9700	0.9932	0.9984	0.9996
1.7000	0.5722	0.8838	0.9759	0.9951	0.9989	0.9998
1.8000	0.5821	0.8965	0.9805	0.9964	0.9993	0.9998
1.9000	0.5909	0.9075	0.9842	0.9973	0.9995	0.9999
2.0000	0.5986	0.9172	0.9872	0.9980	0.9997	0.9999
2.1000	0.6051	0.9256	0.9895	0.9985	0.9998	1.0000
2.2000	0.6106	0.9330	0.9914	0.9989	0.9998	1.0000
2.3000	0.6150	0.9394	0.9929	0.9992	0.9999	1.0000
2.4000	0.6183	0.9450	0.9942	0.9994	0.9999	1.0000
2.5000	0.6206	0.9499	0.9952	0.9995	1.0000	1.0000
2.6000	0.6218	0.9542	0.9960	0.9996	1.0000	1.0000
2.7000	0.6220	0.9580	0.9966	0.9997	1.0000	1.0000
2.8000	0.6212	0.9613	0.9972	0.9998	1.0000	1.0000
2.9000	0.6192	0.9642	0.9976	0.9998	1.0000	1.0000
3.0000	0.6163	0.9668	0.9980	0.9999	1.0000	1.0000
3.1000	0.6123	0.9690	0.9983	0.9999	1.0000	1.0000
3.2000	0.6072	0.9710	0.9985	0.9999	1.0000	1.0000
3.3000	0.6010	0.9727	0.9987	0.9999	1.0000	1.0000
3.4000	0.5937	0.9742	0.9989	1.0000	1.0000	1.0000
3.5000	0.5853	0.9756	0.9991	1.0000	1.0000	1.0000
3.6000	0.5758	0.9767	0.9992	1.0000	1.0000	1.0000
3.7000	0.5651	0.9777	0.9993	1.0000	1.0000	1.0000
3.8000	0.5532	0.9786	0.9994	1.0000	1.0000	1.0000
3.9000	0.5402	0.9794	0.9994	1.0000	1.0000	1.0000
4.0000	0.5261	0.9800	0.9995	1.0000	1.0000	1.0000

Fig. V.5 Fitted probabilities of nonsurvival as a function of P (vertical axis) and D (horizontal axis). Values between the dotted lines are points within the range of observed (P, D)-values.

with increasing dose, achieving a maximum in the neighborhood of $P = 2.7$. This stands in contrast to the other columns, and suggests reworking the table so that D goes from 7 to 14 in steps of 1, in order to see for what values of D this phenomenon seems to persist. If we had omitted the P^2, D^2, and PD terms from the model, then of course such curvature would have been prohibited by the very form of the model.

A finer version of the same table would permit one to sketch the contours of LD50 (the P, D-combinations at which 50% mortality is predicted) or LD90, and so on. A further step would be to permit the third-order terms P^3, P^2D, PD^2, and D^3 to enter the model to see whether a better fit might be obtained. To see that something like this is necessary, note that in the

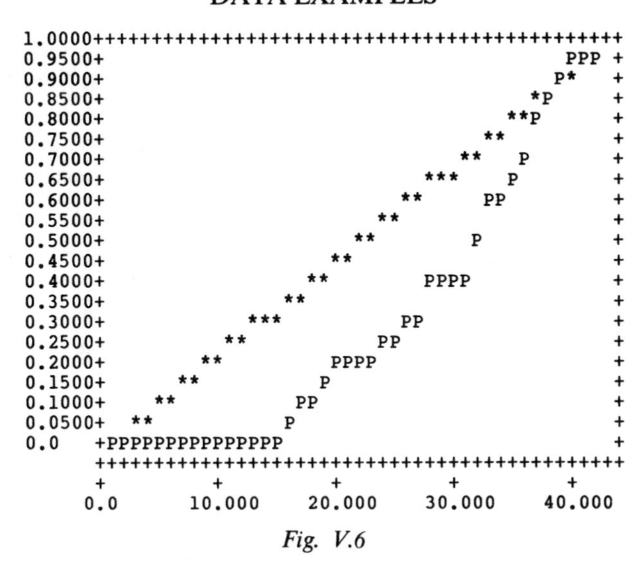

Fig. V.6

uniform plot of **RES − F** residuals (Run V.B.8; see Figure V.6) there are far too many large residuals for comfort with the fitted model. These residuals are shown in Table V.5, and it is clear that the largest ones tend to occur at the boundary of observed x-values, particularly for small values of P.

Since log dose rather than dose itself is often used for this kind of data, it might be wise to try replacing P by $\ln P$, and perhaps also D by $\ln D$. If

Table V.5

Standardized Residuals for the Larvae Data

	D					
P	7	10	14	20	28	40
0.10	—	—	—	—	—	−3.66
0.20	—	—	—	−3.00	0.05	1.17
0.26	—	—	−4.10	−4.01	−1.04	1.29
0.35	—	−3.89	−0.77	3.22	2.25	3.02
0.49	−2.91	0.81	0.97	2.05	2.15	—
0.73	−0.84	0.00	0.18	1.54	0.46	—
1.04	−0.98	−1.54	0.20	0.78	−1.24	—
1.43	1.79	−2.23	0.11	−0.45	—	—
1.96	0.06	1.13	0.43	−0.38	—	—
2.81	−1.25	−1.16	−2.60	—	—	—
4.00	3.16	−0.67	−4.59	—	—	—

such a model fitted better, this would indicate that the original model specification was in error. However, if this did not improve the fit, it might be that there is a rather sharp change in the probability of nonsurvival along the boundary of x-values, so that no smooth model will be able to fit the data successfully. Repetition of the experiment at extreme levels of P would shed light on this question.

4. Logistic Regression: Four Binary Factors

The data in Table V.6 are from a sample of high-school boys. The point of the study was to relate college expectations (C) to the other factors,

Table V.6

Data on College Expectations

	$E0$				$E1$			
	$A0$		$A1$		$A0$		$A1$	
	$P0$	$P1$	$P0$	$P1$	$P0$	$P1$	$P0$	$P1$
				$S0$				
$C0$	2	2	10	23	2	4	2	16
$C1$	38	55	21	22	6	12	4	2
				$S1$				
$C0$	8	12	45	91	4	23	28	76
$C1$	44	61	37	30	9	23	10	6

S = Social Status
$S0$ = High
$S1$ = Low
E = Parental Encouragement about Athletics
$E0$ = High
$E1$ = Low
A = Academic Performance
$A0$ = High
$A1$ = Low
P = Participation in Athletics
$P0$ = Some
$P1$ = None
C = College Expectations
$C0$ = None
$C1$ = Some

participation in athletics (P) being the variable of primary interest to the investigators who collected the data. This suggests the model $C|[SEAP]$ and its submodels, a choice which is reinforced by the fact that the investigators specified that sampling with regard to the (S, E, A, P) factors was not random.

The LAMDA program in Run V.B.9 fits the models $C|[SEA][SEP][EAP][SAP]$, $C|[SE][SA][SP][EA][EP][AP]$, and $C|S\,E\,A\,P$, in that order. The t-functions of these models contain all third-, all second-, and all first-order terms, respectively. The coefficients of the fitted models appear in Table V.7, and the analysis in Table V.8. We thus confirm the model $C|S\,E\,A\,P$ in which the four factors each have main effects on C and there are no interactions. We could attempt to fit models with one or more factors dropped (such as $C|S\,E\,A$), but it is clear from the t-statistics in Table V.7 that none of these will be confirmed.

The signs of the coefficients in $C|S\,E\,A\,P$ are consistent with what one might have expected, high social status, parental encouragement, academic performance, and participation in sports all being associated with a higher probability of having college expectations.

From the model-fitting point of view, one would want to examine the chi-square 8.538 to see whether any of its six components (1 df each) were remarkably larger than the others. A complete analysis would consist of fitting all possible models between $C|S\,E\,A\,P$ and $C|[SE][SA][SP][EA]$

Table V.7

Fitted Parameters and t-values for the College-Expectations Data

i	b_i	t_i	b_i	t_i	b_i	t_i
1	2.7628	4.3417	2.9271	5.6145	3.2565	11.3632
S	−1.0015	−1.3833	−1.2684	−2.3765	−1.0439	−5.0090
E	−1.3960	−1.4695	−1.2500	−2.0396	−1.3096	−6.2595
A	−1.9649	−2.7197	−2.1069	−4.0645	−2.4250	−11.8784
P	0.7657	0.8396	0.1278	0.2499	−0.8385	−4.2961
SE	0.3136	0.3006	0.2648	0.5290		
SA	−0.0105	−0.0129	0.2439	0.4999		
SP	−0.9379	−0.9605	−0.0663	−0.1512		
EA	1.0212	0.9498	0.2761	0.6249		
EP	−1.1552	−1.0700	−0.7531	−1.7092		
AP	−1.6414	−1.6852	−0.9858	−2.3201		
SEA	−0.7039	−0.6600				
SEP	0.6811	0.6443				
SAP	−0.2761	−0.3010				
EAP	0.9347	0.9020				

$[EP][SP]$, of which there are 30, but it is easier and quicker to rely on the t-statistics associated with the coefficients in the latter model, from Table V.7. Clearly the term one should be concerned about here is AP, and since P was the factor of interest initially in the study, there needs to be some investigation. The component of the logistic regression due to A and P is

$$-2.11t_{A1} + 0.13t_{P1} - 0.99t_{A1P1}$$

with the first and last coefficients differing statistically from 0, the middle one not. This says that at the $A0$ level (high academic performance) participation in athletics (P) has virtually no effect on college expectations, while at the $A1$ level (low academic performance) athletics (P) has the effect of raising college expectations. Thus one might posit that athletics has a positive effect on college expectations only among low academic achievers.

5. $2^2|2^2$

The data in Table III.26 are concerned with the factors of Marijuana Use (M), Religion (R), Geographical Region (G) and Family Status (F). In Chapter III, Section B.2 we found that the model $[MGF][RGF][RM]$, which stipulates a common ln odds ratio between M and R in each of the four (G, F)-subtables, does not fit the data very well. It is natural to ask how the ln odds ratio varies with G and F, and so we look at $[MR]|[GF]$ and its submodels. The program for doing this is Run V.B.10, and the results are in Table V.9. Note that $[MR]|[GF]$ is obtained by conditioning the saturated model $[MRGF]$, and so no covariate model depending on (G, F) alone can fit the data any better than $[MR]|[GF]$. From the hypothesis-testing point of view $[MR]|G\,F$ is the minimal confirmed model, but $[MR]|[GF]$ would be the fitted model according to a model-fitting strategy.

Table V.8

Analysis of the College-Expectations Data

Model	df	χ^2	$\langle P\rangle$	$\langle \bar{P}\rangle$
$C\|[SEA][SEP][EAP][SAP]$	1	0.657	⟨.417⟩	⟨.417⟩
$C\|[SE][SA][SP][EA][EP][AP]$	4	1.483	⟨.590⟩	⟨.660⟩
$C\|S\,E\,A\,P$	6	8.538	⟨.200⟩	⟨.488⟩

Table V.9
Analysis of the Marijuana Data

	Model	df	χ^2	Next	$\bar{\alpha}$	$\langle \bar{P} \rangle$
1	$[MR] \| [GF]$	—	—	2⟨.000⟩ 3⟨.038⟩	—	—
2	$MR \| [GF]$	4	42.72		.01	⟨.000⟩
3	$[MR] \| GF$	3	8.41	4⟨.000⟩ 5⟨.000⟩	.01	⟨.038⟩ *
4	$[MR] \| G$	6	64.94		.02	⟨.000⟩ R*
5	$[MR] \| F$	6	25.13		.02	⟨.000⟩ R*

Part of the LAMDA output for $[MR]||[GF]$ is shown in Figure V.7. It is useful to think of the parameter matrix as being of the following form:

	M	R	MR
1	.	.	.
G	.	.	.
F	.	.	.
GF	.	.	.

```
M,R,MR/G,F,GF
=== LXM ===
U(P,QP)    =     1690.890
DELTA(C) =   0.00000814
DELTA(U) =   0.00024414
 NO OF CASES =    772.000
 6 ITERATIONS

- - - A N A L Y S I S - - -
D(P,QP)    =      -80.063
D(QP,Q)    =      449.547
D(P,Q)     =      369.484

- - - COEFFICIENTS - - -

     -0.3959      -0.3403       1.6525
      0.3959       1.4389      -1.1953
      2.0053       1.4389      -1.2545
     -0.9349      -2.1892       1.8062

- - - T-VALUES - - -

     -1.8107      -1.5823       5.7526
      1.0785       4.4571      -2.7198
      2.9968       2.0543      -1.6082
     -1.1626      -2.6341       1.9069
```

Fig. V.7 LAMDA *output for the marijuana data.*

Thus, we can write the response parameters $\boldsymbol{\theta}_R$, $\boldsymbol{\theta}_M$, and $\boldsymbol{\theta}_{MR}$ in terms of covariates as

$$\begin{pmatrix} \boldsymbol{\theta}_M \\ \boldsymbol{\theta}_R \\ \boldsymbol{\theta}_{MR} \end{pmatrix} = \begin{bmatrix} -0.3959 & 0.3959 & \mathbf{2.0053} & -0.9349 \\ -0.3404 & \mathbf{1.4389} & \mathbf{1.4389} & \mathbf{-2.1892} \\ \mathbf{1.6525} & \mathbf{-1.1953} & -1.2545 & 1.8062 \end{bmatrix} \begin{pmatrix} 1 \\ t_G \\ t_F \\ t_{GF} \end{pmatrix}$$

where the boldface entries are those which are more than two standard deviations from 0. We are primarily interested in $\boldsymbol{\theta}_{MR}$, and we recall that $M1$ is non-use of marijuana and $R1$ is religious, so that a positive relation between M and R might have been posited. Looking at the equation for $\boldsymbol{\theta}_{MR}$ as a function of G and F, we see that in $F0$ starting at $G0$ (San Francisco) and going to $G1$ (Contra Costa) lowers the MR relationship; beginning at $G0$ and then moving from $F0$ to $F1$ also lowers the MR relationship. Thus, when both F and G are at their lower levels, changing either of them has a negative effect on the MR relation. If then the other factor is changed too, the MR relation rises. Therefore, there is an interaction effect of F and G on the interaction of M and R. However, if we were to take the statistical significance of the estimates into account, we might wonder whether in fact G were not the only real effect present here.

We can also interpret these equations for the other parameters. $\boldsymbol{\theta}_M$ pertains to marijuana non-use among individuals at level $R0$; here we see that it is F, not G, which exerts an effect, lower marijuana use occurring in $F1$. Likewise, $\boldsymbol{\theta}_R$ pertains to religiousness among individuals in $M0$; here we have the complex situation that both G and F have effects, and there is an interaction.

Note that to obtain information about variation in marijuana use among $R1$ individuals, it would be necessary to exchange the coding of $R0$ and $R1$, and then refit the model. Similarly, to examine how religiousness varies for $M1$ individuals, we must exchange the coding of $M0$ and $M1$. In all of these cases we would be fitting the same model, $[MR]||[GF]$, but the parameters would be redefined depending on how we want to interpret them.

6. Poisson Regression

The data in Table V.10 pertain to the number of times the same bass was caught in a controlled fishing experiment. The basic null hypothesis is that the fish do not learn to avoid the hook after having been caught once, so that the number of times each fish is caught can be considered to have a Binomial distribution. Since the probability that a given fish will be the one

Table V.10

Number of Times Caught in a Controlled Fishing Experiment

		$C0$	$C1$	$C2$	$C3$	$C4$
$S0$	$L0$	22	52	29	2	2
	$L1$	58	27	12	2	0
$S1$	$L0$	53	31	7	1	1
	$L1$	81	19	1	0	0

C = Number of Times Caught
0(1)4
S = Season
$S0$ = Spring
$S1$ = Summer
L = Lure Type
$L0$ = Live
$L1$ = Artificial

caught is rather small, and a large number of catches were involved, it is reasonable to approximate the null hypothesis with a Poisson distribution.

As generating distribution we take the Truncated Poisson with $\lambda = 1$, and we consider u-functions among the polynomials in y. The program for fitting models up to fourth order is Run V.B.11, and the results are in Table V.11. We conclude that the Poisson model $C|S\,L$ does not fit the data adequately, and adopt the model $C, C^2|S\,L$. For the fitted model we have

$$\ln\frac{\hat{q}_x(y)}{q(y)} = (1.6775 - 1.40t_{S1} - 1.3948t_{L1})\,y$$
$$+ (-0.4343 + 0.2784t_{S1} + 0.292t_{L1})\,y^2$$
$$+ \text{constant}$$

In Run V.B.12 we see a program for plotting this quantity as a function of y for the four combinations of S and L. In the resulting Figure V.8 we see that the $S0L1$ and $S1L0$ groups do not depart so much from the Poisson model, since there is not much curvature in the S and L graphs. In contrast, there are more captures than predicted by the Poisson model in the $S0L0$ group, and fewer than predicted in the $S1L1$ group. Thus we might conclude that there is a learning effect from spring to summer, and that the effect of using artificial rather than live bait is to lower repeated captures.

Table V.11

Analysis of the Fish Data

Model	df	χ^2	$\langle P \rangle$	$\langle \bar{P} \rangle$
$C, C^2, C^3, C^4 \mid S L$	4	4.715	$\langle .317 \rangle$	$\langle .317 \rangle$
$C, C^2, C^3 \mid S L$	3	0.644	$\langle .885 \rangle$	$\langle .534 \rangle$
$C, C^2 \mid S L$	3	7.920	$\langle .047 \rangle$	$\langle .135 \rangle$
$C \mid S L$	3	19.903	$\langle .0002 \rangle$	$\langle .001 \rangle$

7. Triangular Table

The data in Table V.12 relate the smoking behavior of a sample of individuals to their age, sex, and disease status. A subset of these data was treated in Chapter III, Section B.2, and in Chapter IV, Sections B.3 and B.4.

Interesting questions here involve relating the various smoking behaviors to the A, D, and X factors, with particular interest in D. Instead of considering smoking to be a factor with six levels, we will view it as the

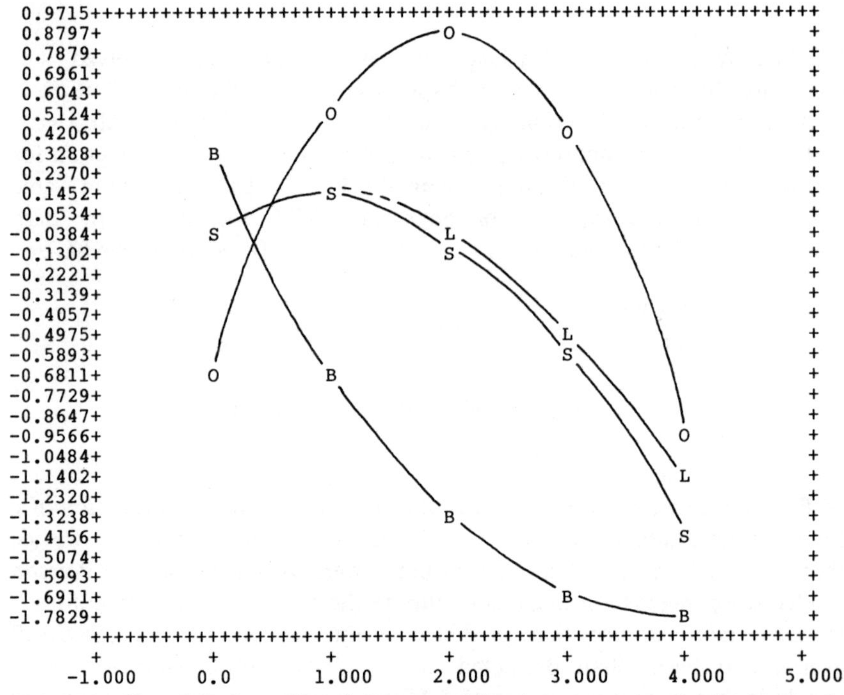

Fig. V.8 Plot of the logarithm of the ratio of fitted to generating distribution for the fish data.

Table V.12

Data on Smoking and Arteriosclerosis

	$A0$		$A1$		$A2$	
	$D0$	$D1$	$D0$	$D1$	$D0$	$D1$
				$X0$		
$S0$	3	23	3	15	12	15
$S1$	2	9	5	8	8	14
$S2$	3	11	1	1	3	2
$S3$	4	9	3	2	14	4
$S4$	9	20	14	4	10	2
$S5$	20	24	34	10	15	3
				$X1$		
$S0$	3	135	18	165	58	126
$S1$	1	13	3	15	7	18
$S2$	0	11	1	8	3	3
$S3$	12	42	21	28	18	11
$S4$	22	45	28	20	10	3
$S5$	14	16	19	15	8	2

X = Sex
- $X0$ = Male
- $X1$ = Female

A = Age
- $A0$ = 35–54
- $A1$ = 55–64
- $A2$ = 65 +

D = Disease Status
- $D0$ = Arteriosclerosis Cases
- $D1$ = Controls

S = Smoking Behavior
- $S0$ = Never Smoked
- $S1$ = Quit > 5 Years Ago
- $S2$ = Quit < 5 Years Ago
- $S3$ = < One Pack/Day
- $S4$ = One Pack/Day
- $S5$ = > One Pack/Day

triangular table (see Chapter I, Section A.1)

$$\begin{matrix} S0 & S1 & S3 \\ & S2 & S4 \\ & & S5 \end{matrix}$$

The u-functions are defined in Table V.13. Thus S stands for ever having smoked, Q stands for ever having quit, P stands for quitting recently, and R represents a gradient related to the rate of current smoking. Run V.B.13 fits the models $SQPR\|[AD][AX][DX]$ and $SQPR|ADX$. Taking the former of these as the most elaborate model to be considered, the chi-square for testing the smaller model is 21.369 $\langle .375 \rangle$ on 20 df, and so we will work with the smaller model. The fitted equations are

$$\begin{pmatrix} \theta_S \\ \theta_Q \\ \theta_P \\ \theta_R \end{pmatrix} = \begin{bmatrix} 1.1129 & \mathbf{-0.7449} & \mathbf{-1.4983} & \mathbf{-1.7416} & -0.4071 \\ -1.5686 & \mathbf{0.9018} & \mathbf{1.7737} & \mathbf{1.5331} & \mathbf{-1.3086} \\ 0.3832 & \mathbf{-1.0989} & \mathbf{-1.6104} & -0.4426 & 0.0390 \\ 0.7882 & 0.0884 & \mathbf{-0.5490} & \mathbf{-0.3679} & \mathbf{-0.7963} \end{bmatrix} \begin{pmatrix} 1 \\ t_{A1} \\ t_{A2} \\ t_D \\ t_X \end{pmatrix}$$

where the boldface estimates have large t-values. Thus we see that cases tended to smoke more than controls, that controls tended to have quit more than cases, and that among current smokers the controls tended to smoke at a lower rate. Disease status does not appear to be related to how long ago the subject quite smoking. All of these findings about the relationship between smoking and disease appear in a model in which age and sex effects are also present, and we have confirmed a model in which there are no interactions among the A-, D-, and X-factors in their impact on smoking.

Run V.B.14 computes the empirical and fitted distributions, and the residuals. At line 25, variables 1–10 are as noted on Run V.B.13. In lines

Table V.13

u-functions for Analyzing the Smoking and Arteriosclerosis Data

	u_S	u_Q	u_P	u_R
$S0$	0	0	0	0
$S1$	1	1	0	0
$S2$	1	1	1	0
$S3$	1	0	0	0
$S4$	1	0	0	1
$S5$	1	0	0	2

Table V.14

	$H0$					$H1$				
	$F0$	$F1$	$F2$	$F3$	$F4$	$F0$	$F1$	$F2$	$F3$	$F4$
$C0$	7	14	8	5	8	7	2	6	0	0
$C1$	13	18	10	8	18	20	26	16	10	21
$C2$	25	7	0	14	9	18	17	13	20	14
	$H2$					$H3$				
	$F0$	$F1$	$F2$	$F3$	$F4$	$F0$	$F1$	$F2$	$F3$	$F4$
$C0$	3	5	10	6	5	19	10	5	6	3
$C1$	21	24	27	10	9	17	21	19	15	15
$C2$	12	16	5	11	14	9	14	11	9	10

C = Covert Status
$C0$ = Not Under a Covert
$C1$ = Under a Covert with Ribbons
$C2$ = Under a Covert Without Ribbons
F = Fish Group
H = Channel
$H0$, $H1$ Covered
$H2$, $H3$ Open

63–94 the value of $\mathbf{t}(x)'\hat{\mathbf{B}}\mathbf{u}(y)$ is computed for each of the 72 cells, so at line 101 variable 1 is the observed count $N_x(y)$ and variable 2 is the corresponding expression $\exp[\mathbf{t}(x)'\hat{\mathbf{B}}\mathbf{u}(y)]$. At line 102 cells having the same value of x are arranged on the same row, and at line 103 the observed frequencies are grouped in front, the fitted expressions $\exp[\mathbf{t}(x)'\hat{\mathbf{B}}\mathbf{u}(y)]$ at the back of the row. At line 105 we have alternating rows of observed counts and fitted expressions. In lines 106–117 each line is reduced to probabilities; thus we have alternating rows of $\hat{p}_x$ and $\hat{q}_x$. Lines 120–125 rearrange these so that variables 2 and 3 are $\hat{p}_x(y)$ and $\hat{q}_x(y)$. The observed $N_x(+)$ are then stored in variable 1 (lines 126–138), and then the residuals are computed in lines 139–148. After changing these to positive values, they may be fed into the uniform plot program, Run IV.B.1. The resulting uniform plot (not shown) suggests that the model fits well.

PROBLEMS

V.1. Five groups of trout were placed in controlled flow channels, half of which were covered with a burlap screen. Each channel had coverts,

Table V.15

		$E0$	$E1$	$E2$	$E3$	$E4$
$S0$	$T0$	0	11	21	38	13
	$T1$	32	12	4	10	0
	$T2$	17	29	37	17	0
$S1$	$T0$	14	17	7	12	6
	$T1$	26	7	8	31	4
	$T2$	46	8	5	9	1
$S2$	$T0$	24	15	13	29	27
	$T1$	91	7	8	7	0
	$T2$	27	12	6	26	7
$S3$	$T0$	10	1	14	34	8
	$T1$	8	9	18	40	2
	$T2$	1	28	34	46	0
$S4$	$T0$	0	10	21	48	6
	$T1$	1	8	15	51	5
	$T2$	32	3	9	14	2
$S5$	$T0$	13	6	7	26	7
	$T1$	65	12	1	3	0
	$T2$	35	22	3	4	0
$S6$	$T0$	0	2	9	33	42
	$T1$	0	10	19	12	0
	$T2$	—	—	—	—	—
$S7$	$T0$	6	9	9	12	7
	$T1$	11	19	39	33	0
	$T2$	37	10	2	6	1
$S8$	$T0$	0	4	16	48	34
	$T1$	2	0	20	60	17
	$T2$	3	8	9	36	6
$S9$	$T0$	0	17	29	67	7
	$T1$	0	18	12	48	6
	$T2$	2	25	16	12	2

E = Exposure Period in which Mortality Occurred
- $E0$ = After First Exposure
- $E1$ = After Second Exposure
- $E2$ = After Third Exposure
- $E3$ = Within 24 Hours after Fourth Exposure
- $E4$ = 24–48 Hours after Fourth Exposure

T = Temperature
- $T0$ = 10°C
- $T1$ = 20°C
- $T2$ = 30°C

S = Species of Fish

half with plastic ribbons, the others without. Sightings were made at fixed times to determine the positions of the fish. The data are given in Table V.14. Analyze these data, beginning with the model $C||[FH]$.

V.2. Ten species of fish were exposed to monochloramine on four successive occasions, and mortality was noted as in Table V.15. Analyze these data, beginning with the model $E|TS$ and an appropriate generating distribution.

V.3. Squirrel-bait carrier was applied around several squirrel habitats according to four different methods. Rodents were then trapped, and those which had taken the bait were identified by the presence of a chemical marker. This was done for a period of six months. Analyze the data in Table V.16, beginning with the model $M||[BTS]$.

Table V.16

		$B0$		$B1$		$B2$		$B3$	
		$M0$	$M1$	$M0$	$M1$	$M0$	$M1$	$M0$	$M1$
$T0$	$S0$	4	6	0	10	1	11	8	8
	$S1$	8	12	3	16	2	18	6	22
$T1$	$S0$	7	20	6	28	5	15	3	11
	$S1$	15	29	14	41	17	19	7	16
$T2$	$S0$	9	11	4	13	6	7	8	9
	$S1$	14	10	7	14	11	12	9	13
$T3$	$S0$	7	5	3	4	4	3	4	1
	$S1$	19	17	11	16	11	16	10	7
$T4$	$S0$	4	6	11	28	3	26	5	29
	$S1$	2	8	20	26	3	9	1	20
$T5$	$S0$	17	33	8	34	6	23	8	26
	$S1$	10	24	4	5	0	1	2	8

B = Bait Method
T = Month
S = Sex
$S0$ = Male
$S1$ = Female
M = Marking
$M0$ = Unmarked
$M1$ = Marked

V.4. Two separate pools of 100 bass were maintained over an 87-day period. At various intervals the pools were fished, one with live bait and the other with artificial lures, fish being returned after capture. At each fishing the amount of time (hours) spent was recorded, along with the catch. At the end of the experiment the pools were drained to determine the number of fish in each; there were 94 in the

Table V.17

D	*L*	*T*	*C*1
0	0	2	19
0	1	2	25
3	0	1.67	3
3	1	1.25	3
17	0	2.91	39
17	1	4	22
24	0	2	14
24	1	1.42	5
32	0	6.33	15
32	1	2.5	0
37	0	4.25	7
37	1	1.67	2
52	0	1.5	1
52	1	1.5	10
59	0	2	9
59	1	2	6
66	0	2	15
66	1	2	1
68	0	8	12
68	1	5.5	0
80	0	2	12
80	1	2	3
87	0	2	3
87	1	2	1

C = Catch Status
 *C*0 = Not Caught
 *C*1 = Caught
D = Day (from beginning of experiment)
 0 to 87
L = Lure
 *L*0 = Live
 *L*1 = Artificial
T = Time Spent Fishing (Hours)

live-bait pool and 100 in the artificial-bait pool. The data are shown in Table V.17. Fit a model expressing C in terms of L, D, and T. It is of particular interest to estimate any learning effect by which the fish tend to avoid the hook later in the season.

V.5. According to the "cube law" of political science, in a two-party system the proportion of contested legislative seats S won by one of the parties satisfies

$$S = \frac{V^3}{3V^2 - 3V + 1}$$

where V is the proportion of the popular vote won by that party. Reformulate this law as a logistic regression and test it against the American record shown in Table V.18. Generalize the model to the case of three or more parties.

V.6. Consider the following data:

	$S0$				$S1$			
	$X0$		$X1$		$X0$		$X1$	
	$T0$	$T1$	$T0$	$T1$	$T0$	$T1$	$T0$	$T1$
$E0$	74	5	84	3	80	10	79	3
$E1$	12	4	14	2	14	4	14	1

S = Strain of Mouse
X = Sex
$X0$ = Female
$X1$ = Male
T = Tumor
$T0$ = Absent
$T1$ = Present
E = Experimental Condition
$E0$ = Control
$E1$ = Treatment

Analyze these data with a view towards writing the ln odds ratio between T and E as a function of S and X.

Table V.18

Year	Democrat	Republican	Proportion of Vote for Democrats
1928	166	267	.452
1930	215	214	.497
1932	313	117	.567
1934	315	103	.562
1936	325	89	.585
1938	258	164	.523
1940	261	162	.531
1942	223	208	.501
1944	243	190	.518
1946	188	245	.479
1948	263	171	.536
1950	235	199	.517
1952	213	221	.496
1954	232	203	.528

V.7. Explain how the logistic regression model

$$\ln \frac{q_x(1)}{q_x(0)} = \alpha + \beta x$$

with the constraint $\alpha = \alpha_0$, can be fitted using LAMDA.

V.8. Consider a two-way table with factors A and B, and assume that A is ordinal so that $v(ij) = i$ is a meaningful function. Letting $\mathbf{u}_A$ and $\mathbf{u}_B$ denote the usual factorial indicators, determine the form of the $A|B$ model obtained from $\mathcal{E}[q : \mathbf{u}_A, \mathbf{u}_B, \mathbf{u}_B v]$, where q is uniform.

1. Is the conditional model for $A|B$ a linear exponential model?
2. Does the conditional model for $A|B$ remain the same when we change the labels for the B factor [that is, replace j by $\sigma(j)$, where σ is a permutation of the levels of B]?
3. Explain the relationship between the conditional model for $A|B$ and $\mathcal{E}[q : v|1, \mathbf{u}_B]$.

V.9. According to Piagetian theory, children progress from being nonconservers (that is, not perceiving that a conservation principle, like conservation of mass, operates in a special situation) through a transitional phase and eventually become conservers. The following are data on some school children:

	A							
	60	72	84	96	108	121	133	144
$S0$	10	7	4	1	2	15	3	0
$S1$	0	3	4	3	0	3	2	2
$S2$	0	0	2	6	8	10	5	8

A = Age (Months)
S = Stage
$S0$ = Nonconserver
$S1$ = Transitional
$S2$ = Conserver

Fit the model $S|A, A^2, A^3$, where S is represented by the function $u(SiAj) = i$, and use the marginal S-distribution for the generating distribution. Let $\mathbf{t}' = (1, t_A, t_A^2, t_A^3)$. Plot $\mathbf{t}'\hat{\mathbf{b}}$ as a function of age, and discover why the educators were concerned about these data. Use this fitted model as a simple hypothesis for testing with the follow-up data:

	A					
	92	103	113	121	131	144
$S0$	3	1	0	1	1	0
$S1$	1	1	0	1	3	0
$S2$	2	4	6	4	2	6

(Use the new S-marginal for the generating distribution, and $\hat{\mathbf{b}}$ from the first data set.) Is the alarming trend confirmed?

V.10. Suppose that $\{T(y) : y \in \mathcal{Y}\}$ are independent Poisson random variables, with $T(y)$ having parameter $\lambda(y)$. Let $\{N(y) : y \in \mathcal{Y}\}$ be a sample of n independent observations in $\mathcal{Y}$ following the distribution $p \in \mathcal{P}(\mathcal{Y})$. Show that the distribution of $\{N(y)\}$ is the same as the distribution of $\{T(y)\}$ given $T(+)$, provided we make the identifications $n = T(+)$ and $p(y) = \lambda(y)/\lambda(+)$.

V.11. Four quail lines were established; two were controls ($S0$), one was administered a higher dose of aflatoxin ($A1$), and the other a lower dose ($A0$). Five generations were followed to observe the trend of

Table V.19

		A0		A1	
		S0	S1	S0	S1
G0	M0	39	119	21	73
	M1	26	20	53	48
G1	M0	25	113	10	91
	M1	66	66	77	66
G2	M0	25	118	3	105
	M1	94	34	97	58
G3	M0	15	114	15	162
	M1	54	15	54	23
G4	M0	19	164	12	162
	M1	56	11	58	18

A = Aflatoxin Group
$A0$ = 2.5 mg/kg
$A1$ = 3.0 mg/kg
S = Status
$S0$ = Control
$S1$ = Treated
M = Mortality
$M0$ = Survived
$M1$ = Not survived
G = Generation
$G0$ to $G4$

genetic resistance to aflatoxin. In the data in Table V.19 the $A0S0$ group is evidently to be thought of as the control for the $A0S1$ group, and similarly for the $A1S0$ and $A1S1$ groups. Analyze the (M, S)-interaction, beginning with the model $[MS]\|[AG]$.

V.12. Show that if the set of possible covariates x is finite, and we let $N_x(y)$ denote the number of observations of $(y_i, x_i) = (y, x)$ in the sample, then provided the left-hand side is well defined we have

$$U(\hat{p}, q) - U(\hat{p}, \hat{p}) = \sum_x N_x(+)\mathbb{D}_1(\hat{p}_x, q_x)$$

REMARKS

The linear exponential models were introduced by Mantel (1966) and by Bock (1969), although simpler examples had been given before, notably by Dyke and Patterson (1952). Logistic regression has attracted the most attention, as attested by the numerous papers in the references on this topic. Berkson has advocated the "minimum logit chi-square" method of fitting this model [Berkson (1949), (1955)], and Cox (1958) has given a conditional test for the slope parameter.

There has been a tendency in the literature to restrict applications of covariate models to cases where the covariate is discrete. It is important to recognize that this is not a limitation of linear exponential models; there is no problem treating the situation in which each covariate value appears exactly once in the sample (assuming an identifiable parameter).

There has been some argument among statisticians concerning conditional inference; the words used in the literature are "sufficiency" and "ancillarity," but this literature is not referenced here. The paper of Andersen (1970) is important for the viewpoint taken here.

The Assertions in this chapter can be found in my thesis (Linear Exponential Models and Discrete Prediction, 1976, University of Washington; available from University Microfilm Service, Ann Arbor, Michigan). The proofs are not included here because of their length and difficulty. The regularity conditions on the covariates mentioned in Assertions 2, 3, and 4 amount to saying that they do not drift out to infinity too much, and they do not tend to a configuration which makes the parameter unidentifiable. These are weak conditions which should not make practical difficulties.

Computation

Although LAMDA is specifically oriented towards the linear exponential model presented here, there are a few other programs for the analysis of contingency tables. The brief descriptions given below are not intended to be definitive.

BMDP

Biomedical Computer Programs, P-Series, is a large statistical package which has several subprograms for discrete data. The most useful is BMDP3F, which fits hierarchical log-linear models to cross-classifications. Unlike many other programs, it can fit a large number of models to the data on a single run. For two- and three-way tables, all hierarchical models can be fitted with a single instruction. For any (reasonable-size) cross-classification the marginal and partial association tests for all terms can also be requested with a single instruction. Another very useful feature in model building is that one can request the fitting of all hierarchical models which differ from a given model by only one term. Display of full and marginal tables and their fitted values is easy to produce. However, the generating distribution of all models is assumed to be uniform.

The primary reference is *BMDP-79* (distributed by the Health Science Computing Facility, UCLA). The documentation is excellent, and has very good examples of input and output. Even though it is necessary to become familiar with some of the more general BMDP control language in order to use P3F, the detailed explanation and illustrations in the P3F section make it possible to use the program with minimal understanding of the general package. Of course, the substantial statistical and data-manipulation capabilities of BMDP are available for use in conjunction with P3F.

The 1979 version of BMDP also contains a program, PLR, for performing logistic regression, with a stepwise procedure for model building.

[Since the preceding comments were written, the facilities of P3F have been enhanced and combined with features of other BMDP routines to produce P4F. Aside from the addition of printing options and choices of different statistics which may be computed, the main improvements in P4F are (1) fixed empty cells may be specified, (2) various residuals may be computed and displayed, and (3) the parameter estimates and their estimated standard deviations are printed; these are the log-linear parameters described in Chapter IV, Section A.3.]

ECTA

Everyman's Contingency Table Analyzer is a program for fitting log-linear models to cross-classifications. The input is according to a rigid format consisting entirely of numerical strings, some of which are column-sensitive. However, the specification of the models to be fitted is very easy, as it should be where only hierarchical models are concerned. It is possible to use different generating distributions, and to deal with fixed empty cells.

The primary reference is the *Description for Users* (distributed by the Department of Statistics, University of Chicago). The documentation is poor, and contains no examples of either input or output. The attitude of the distributer is evidently that the articles by Goodman [see Goodman (1978) for a collection] complement and complete the program description.

ECTA is approximately twice as large as LAMDA, in terms of executable FORTRAN card images.

GLIM

Generalized Linear Interactive Modelling is a sophisticated package for fitting linear models to continuous and discrete data. For the analysis of discrete data it uses the fact that Poisson models on contingency tables can be conditioned to become multinomial models (see Problem V.10). Thus GLIM can fit all models of type $\mathfrak{S}[q:\mathbf{u}]$ and $\mathfrak{L}[q:\mathbf{u}]$, and uses ML rather than MDI inference for $\mathfrak{L}$-models. The special case of the binomial is distinguished in GLIM, so that logistic regression is particularly easy. It is probably possible to fit the more general covariate models $\mathfrak{S}[q:\mathbf{u}|\mathbf{t}]$ (and $\mathfrak{L}[q:\mathbf{u}|\mathbf{t}]$ as well), but this would appear to require substantial facility with the control language when the t-functions contain "continuous" components.

The specification of customary hierarchical models in GLIM is very convenient, and does not require that indicator variables be created for factors. One always has the option, however, of using created factorial indicators (as in LAMDA) so that nonhierarchical or other special models can be used. Data manipulations are quite easy in GLIM, and display and plotting are also available. Parameter estimates and their standard deviations and residuals can all be displayed.

The primary reference source is *The Glim System Release 3 Manual* (distributed by the Numerical Algorithms Group). The documentation is quite good as far as it goes, but the manual by itself is inadequate as an introduction to the models. One should also read the more prominent articles referenced in the manual, especially Nelder (1974).

The FORTRAN version of GLIM is approximately six times as large as LAMDA, in terms of executable card images.

SAS

The *Statistical Analysis System* is another large package, one procedure of which, FUNCAT, deals with discrete data analysis. The class of models which can be fitted is broader than those presented here, except that the covariates must be discrete. Estimation and test statistics are based on the δ-method (Proposition 8). Basically, one can specify models of the form $\mathbf{f}(p) = \mathbf{X}\boldsymbol{\beta}$, where $\mathbf{X}$ is a design matrix, $\boldsymbol{\beta}$ is the vector parameter, and $\mathbf{f}$ is a vector-valued function defined as a composition of log, exp, and matrix multipliers.

The primary reference is the *SAS User's Guide* (distributed by the SAS Institute, Raleigh, North Carolina). The FUNCAT procedure is described rather tersely, and one should read the papers of Grizzle, Starmer, and Koch (1969) and Forthofer and Koch (1973) to get an idea of the potential for application. There are three well-chosen examples in the manual, each with input and output.

The design matrix can be specified as a combination of "columns" and discrete classifications. Crossed and nested factorial classifications are easily specified for the design matrix, but are rather problematical when defining **f**. Models with simple response structure are much easier to specify than those with complex structure; logistic regression is particularly easy.

CHAPTER VI

Further Topics

This chapter covers a collection of topics which are important but do not appear to fit well in earlier chapters. For the most part, it is assumed here that the earlier material has been absorbed by the reader, so that lengthy explanation is not needed.

In Section A we discuss the algorithm which LAMDA uses to obtain the ML and MDI estimates. Section A.2 concerns measures which one might take when the algorithm fails, including the extended form of the `LXM` command, which is `LXX`.

The usual sampling assumption, that the experimental units are independent, is weakened in Section B. Here Markov models are shown to be formally identical to covariate models, so that in many cases the analysis of Markov-chain data can be carried out using the methods of Chapter V.

We saw in Chapter V, Section A.3 that by conditioning one transforms a full model to a conditional model, and that the latter is valid for inference. In Section C we will establish conditions under which one would draw the same conclusions, regardless of whether one conditions or not. We will also see the very important result that for certain parameters, inference does not depend on which set of factors is chosen to be conditional; this is the topic of reversed inference.

In Section D we will discuss the issue of homogeneity tests, concerned with whether some model or some set of parameters pertains to each of a collection of contingency tables. This includes the collapsibility problem.

Finally, in Section E we will see how one can attempt to analyze nonlinear discrete models by a sequence of linear approximations.

A. COMPUTATIONAL ALGORITHM

1. The Newton–Raphson Procedure

In general the maximum-likelihood (and minimum-discrimination-information) parameter estimates cannot be obtained by algebraically solving a system of equations, although in particular cases this is possible. The iterative technique used by LAMDA is a slight modification of the well-known Newton–Raphson procedure, whose convergence properties are generally superior to other methods.

The idea behind the method is that to find the value $\hat{\boldsymbol{\theta}}$ which solves the equation

$$\frac{\partial}{\partial \boldsymbol{\theta}} L(\boldsymbol{\theta}) = \mathbf{0}$$

one begins by guessing $\boldsymbol{\theta}_0$ and then using the Taylor approximation

$$\frac{\partial}{\partial \boldsymbol{\theta}} L(\boldsymbol{\theta}) = \frac{\partial}{\partial \boldsymbol{\theta}} L(\boldsymbol{\theta}_0) + \left[\frac{\partial^2}{\partial \boldsymbol{\theta}\, \partial \boldsymbol{\theta}'} L(\boldsymbol{\theta}_0) \right] (\boldsymbol{\theta} - \boldsymbol{\theta}_0)$$

to solve the linear approximation of the equation above,

$$\frac{\partial}{\partial \boldsymbol{\theta}} L(\boldsymbol{\theta}_0) + \left[\frac{\partial^2}{\partial \boldsymbol{\theta}\, \partial \boldsymbol{\theta}'} L(\boldsymbol{\theta}_0) \right] (\boldsymbol{\theta} - \boldsymbol{\theta}_0) = \mathbf{0}$$

which gives

$$\boldsymbol{\theta}_1 = \boldsymbol{\theta}_0 - \left[\frac{\partial^2}{\partial \boldsymbol{\theta}\, \partial \boldsymbol{\theta}'} L(\boldsymbol{\theta}_0) \right]^{-1} \frac{\partial}{\partial \boldsymbol{\theta}} L(\boldsymbol{\theta}_0)$$

Of course, $\boldsymbol{\theta}_1$ does not usually solve the original equation, but it tends to come closer than $\boldsymbol{\theta}_0$ does. Thus, we iterate the procedure by letting $\boldsymbol{\theta}_1$ play the role of $\boldsymbol{\theta}_0$, obtaining $\boldsymbol{\theta}_2$ from $\boldsymbol{\theta}_1$ as $\boldsymbol{\theta}_1$ was obtained from $\boldsymbol{\theta}_0$. Continuing this gives a sequence $\boldsymbol{\theta}_k$ which nearly always converges to $\hat{\boldsymbol{\theta}}$.

There is, however, no guarantee that $\boldsymbol{\theta}_k$ will converge to anything. One potential problem is that the matrix which needs to be inverted in the Newton–Raphson iteration may be singular or nearly singular. We will see below how exact singularity can be avoided, but in order to deal with the problem of trying to invert a near-singular matrix, we can change the Newton–Raphson step to

$$\boldsymbol{\theta}_{k+1} = \boldsymbol{\theta}_k - \left[\frac{\partial^2}{\partial \boldsymbol{\theta}\, \partial \boldsymbol{\theta}'} L(\boldsymbol{\theta}_k) + \varepsilon_k \mathbf{I} \right]^{-1} \frac{\partial}{\partial \boldsymbol{\theta}} L(\boldsymbol{\theta}_k)$$

where $\{\varepsilon_k\}$ is a sequence of decreasing positive numbers that tends to zero. This modification is used by LAMDA.

In the case of a model without covariates, we have the ln likelihood function

$$L(\boldsymbol{\theta}) = \sum_{i=1}^{n} \boldsymbol{\theta}'\mathbf{u}(y_i) - \delta(\boldsymbol{\theta}) + \ln q(y_i)$$

where $\{y_i : 1 \leqq i \leqq n\}$ is the sample. Compute

$$\frac{\partial}{\partial \boldsymbol{\theta}} L(\boldsymbol{\theta}) = \sum_{i=1}^{n} \mathbf{u}(y_i) - M\left[\mathbf{u} : q^{\boldsymbol{\theta}}\right]$$

so that $(\partial/\partial\boldsymbol{\theta})L(\boldsymbol{\theta}) = \mathbf{0}$ is equivalent to $M[\mathbf{u} : p] = M[\mathbf{u} : q^{\boldsymbol{\theta}}]$. In the covariate case we have

$$L(\boldsymbol{\beta}) = \sum_{i=1}^{n} \boldsymbol{\beta}'\mathbf{u}(y_i) \otimes \mathbf{t}(x_i) - \delta(x_i, \boldsymbol{\beta}) + \ln q(y_i)$$

so that

$$\frac{\partial}{\partial \boldsymbol{\beta}} L(\boldsymbol{\beta}) = \sum_{i=1}^{n} \mathbf{u}(y_i) \otimes \mathbf{t}(x_i) - M\left[\mathbf{u} \otimes \mathbf{t}(x_i) : q_{x_i}^{\boldsymbol{\beta}}\right]$$

and setting the right side to zero gives the equation of Assertion 1 in Chapter V, Section A.2. Thus the Newton–Raphson method is appropriate for both covariate and covariate-free exponential models. In fact the covariate case $t(x_i) = 1$ is equivalent to the covariate-free model, so that LAMDA always assumes the covariate model in its actual computations.

Compute that

$$\frac{\partial^2}{\partial \boldsymbol{\theta}\, \partial \boldsymbol{\theta}'} L(\boldsymbol{\theta}) = -C\left[\mathbf{u} : q^{\boldsymbol{\theta}}\right]$$

$$\frac{\partial^2}{\partial \boldsymbol{\beta}\, \partial \boldsymbol{\beta}'} L(\boldsymbol{\beta}) = -\sum_{i=1}^{n} C\left[\mathbf{u} \otimes \mathbf{t}(x_i) : q_{x_i}^{\boldsymbol{\beta}}\right]$$

This has three consequences. First, the second derivative matrices can be computed as covariance matrices with respect to discrete distributions. Second, the inverted matrix computed in the Newton–Raphson step is the negative of the estimated covariance matrix of the parameter estimates, according to Proposition 23 in Chapter III, Section A.3 and Assertion 3 of Chapter V, Section A.2. Thus the variances and covariances of the estimates are obtained as a by-product of the Newton–Raphson computations.

Finally, the form of the second-derivative matrices shows what condition must be observed to keep them nonsingular. For the covariate-free case, the function $\mathbf{u}$ must have affinely linearly independent components. In the covariate case the situation is more complex, but it can be shown that if the parameter $\boldsymbol{\beta}$ is identifiable, then the sum of the covariance matrices in the expression for the second derivative of L is nonsingular. Thus in all cases identifiability of parameters avoids the exact singularity problem.

2. Convergence Problems

Despite one's best efforts, the Newton–Raphson procedure may not yield the desired solution. There are several ways of detecting this from the output:

1. **DELTA(*U*)** is not near 0.
2. **DELTA(*C*)** is not near 0.
3. **ITERATIONS = 10.**

Recalling that the purpose is to find the value $\hat{\boldsymbol{\beta}}$ which maximizes the ln likelihood L (equivalently, minimizes the unlikelihood $U = -2L$), we define **DELTA(*U*)** to be the amount which U changed between the last and second-to-last Newton–Raphson iteration. If this is not small, then the reported value of $\hat{\boldsymbol{\beta}}$ may still be at some distance from the true $\hat{\boldsymbol{\beta}}$.

The maximum amount by which a coordinate of $\boldsymbol{\beta}$ changed in the last iteration is given as **DELTA(*C*)**. Again, a value not near zero indicates that convergence of the iterates may not have been achieved.

The maximum number of iterations permitted is 10, so that **ITERATIONS = 10** is suggestive that the iterations stopped because the maximum was achieved. Of course, it is possible that convergence just happened to occur at the 10th and last iteration.

When the Newton–Raphson procedure fails, it is usually because the MLE does not exist. Recall from Chapter III, Section A.6 that certain patterns of empty cells cause the likelihood function to increase as the parameter values drift out to infinity, so that no finite $\hat{\boldsymbol{\beta}}$ exists. A similar result holds for the covariate case. Thus, although one may have constructed a model whose parameter would be identifiable if there were no empty cells, when there are empty cells the parameter may not be identifiable with respect to the cells which actually had occupants. The easiest way to try to solve this problem is to delete various parameters from the model until the Newton–Raphson procedure converges. This approach may be tedious, but it is reasonable in the sense that on logical grounds one should not try to fit a model with unidentifiable parameters.

Experience with the Newton–Raphson algorithm suggests that when $\hat{\boldsymbol{\beta}}$ does not exist, the successive iterates $\boldsymbol{\beta}_k$ cause the distributions $p^{\boldsymbol{\beta}_k}$ (or $p_x^{\boldsymbol{\beta}_k}$) to converge to a distribution which could be considered the ML distribution in an extended sense. More precisely, it appears to be possible to write $\boldsymbol{\beta}_k = \boldsymbol{\lambda}_k + \tau_k \boldsymbol{\alpha}_k$, in which $\boldsymbol{\lambda}_k$ and $\boldsymbol{\alpha}_k$ converge to $\boldsymbol{\lambda}$ and $\boldsymbol{\alpha}$, $\|\boldsymbol{\alpha}_k\| = \|\boldsymbol{\alpha}\| = 1$, and $\tau_k \to \infty$. Moreover, it can be shown that the distributions $p^{\boldsymbol{\beta}_k}$ (or $p_x^{\boldsymbol{\beta}_k}$) tend to the element $p^{\boldsymbol{\lambda}}$ (or $p_x^{\boldsymbol{\lambda}}$) of $\mathcal{E}[q:\mathbf{u}]$ (or $\mathcal{E}[q:\mathbf{u}|\mathbf{t}]$), provided one interprets these latter models to be defined over the plateau

$$\mathcal{Y}_0 = \left\{ y \in \mathcal{Y} : \boldsymbol{\alpha}'\mathbf{u}(y) = \max_{z \in \mathcal{Y}} \boldsymbol{\alpha}'\mathbf{u}(z) \right\}$$

in the covariate-free case, or

$$\mathcal{Y}_x = \left\{ y \in \mathcal{Y} : \boldsymbol{\alpha}'\mathbf{u}(y) \otimes \mathbf{t}(x) = \max_{z \in \mathcal{Y}} \boldsymbol{\alpha}'\mathbf{u}(z) \otimes \mathbf{t}(x) \right\}$$

in the covariate case. It is possible to try to obtain the decomposition $\boldsymbol{\beta}_k = \boldsymbol{\lambda} + \tau_k \boldsymbol{\alpha}$ of the iterate into the *convergent part* $\boldsymbol{\lambda}$ and the *divergent part* $\boldsymbol{\alpha}$.

We now indicate some extensions to the **LXM** command which can be useful when convergence is a problem. The extended form is

*xx*LXX *aa*, *bb*, *cc*, *dd*, *ee*, *ff*

$b_1, b_2, \ldots, b_r$

Here *xx* is, as usual, the number of the variables in the preceding **SEL** command that define the *u*-functions. *aa* is the tolerance for **DELTA(*C*)**; that is, the Newton–Raphson iterations will stop and the results be printed when **DELTA(*C*)** is no greater than *aa*. Likewise, *bb* is the tolerance for **DELTA(*U*)**. *cc* is the value that is added to the diagonal elements of the second-derivative matrix (the ε_k) before it is inverted. With each iteration *cc* is divided by 100, and it is set to 0 when it drops below .001. *dd* is the maximum number of iterations permitted. If *ee* is set to 1 the divergent and convergent components of the MLE will be reported, along with **DELTA** = maximum change in a component of $\boldsymbol{\lambda}$ on the last iteration, and **NORM** = τ_k on the last iteration. If *ff* is set to 1, then each Newton–Raphson step is printed out. The values $b_1, b_2, \ldots, b_r$ are the initial guess required to start the Newton–Raphson process.

The **LXM** command is equivalent to

```
xxLXX .001, .001, 100, 10, 0, 0
0,0,...,0
```

It may only be desired to change the initial guess at $\boldsymbol{\beta}$ in order to try to achieve convergence, but even so, all the additional specifications (aa, bb,...) must be given.

There are situations in which the ML estimate exists, but **LXM** fails to produce convergence. Changing the initial value of $\boldsymbol{\beta}$, as suggested above, may work in such a case, but it may also be rather tedious to search for an initial value which is successful. Another approach which requires less effort is to modify the generating distribution. The idea is that if all the models one wants to fit contain $\mathcal{E}[q:\mathbf{u}]$, then one might as well use $\hat{q}$, the ML distribution in $\mathcal{E}[q:\mathbf{u}]$, as the generating distribution. This follows from Proposition 30. The advantage is that $\hat{q}$ should be closer to the desired fitted distribution than q is. An example appears in Run IV.B.8, pertaining to Chapter IV, Section B.6. Here the models between $[ASC][ASV][ACV][SCV]$ and $[AS][AC][AV][SC][SV][CV]$ were to be fitted, and so at line 45 the generating distribution was taken to be the fitted distribution in the latter model, which had been obtained on an earlier run.

B. MARKOV MODELS

It is often unrealistic to assume that various observations are independent of each other, and then one might turn to Markov models, since they permit some weak interdependence. This is particularly useful in cases in which each experimental unit makes several responses, and although different units may be considered to be independent, different responses by the same unit cannot. The phrase "repeated measures" is sometimes used to describe this kind of data.

There are many sorts of Markov models. We will consider a simple class in which there is a fixed state space S with elements s. At each time period an individual can be in any state, and so a temporal record of the states through which the individual passes is a vector, $\mathbf{s}' = (s_1, s_2, \ldots, s_k)$, assuming that there are k time periods. The set of all such vectors forms $\mathcal{Y}$. A parametric Markov model for the probability distribution on $\mathcal{Y}$ is constructed as follows. Let $\mathbf{u}_1$ be a function on S, and for $i = 2, \ldots, k$ let $\mathbf{u}_i$ be a function of pairs (s', s) of elements from S. Define

$$\delta_1(\boldsymbol{\theta}_1) = \ln \sum_{s \in S} e^{\boldsymbol{\theta}_1' \mathbf{u}_1(s)}$$

$$\delta_i(s', \boldsymbol{\theta}_i) = \ln \sum_{s \in S} e^{\boldsymbol{\theta}_i' \mathbf{u}_i(s', s)} \qquad (i = 2, \ldots, k)$$

Finally, we define

$$q^{\boldsymbol{\theta}}(\mathbf{s}) = \exp\left[\boldsymbol{\theta}_1'\mathbf{u}_1(s_1) - \delta_1(\boldsymbol{\theta}_1) + \sum_{i=2}^{k} \boldsymbol{\theta}_i'\mathbf{u}_i(s_{i-1}, s_i) - \delta_i(s_{i-1}, \boldsymbol{\theta}_i)\right]$$

From these definitions, we can compute that for any $m < k$ the probability distribution governing the first m time periods is given by the expression above, with the last summation extending only to m instead of k. Thus an initial segment of a Markov model is also a Markov model. It is then easy to compute that the conditional probability of the state s_m occupied at time m given the past states $s_1, \ldots, s_{m-1}$ is

$$q^{\boldsymbol{\theta}}(s_m | s_1, \ldots, s_{m-1}) = \exp\left[\boldsymbol{\theta}_m'\mathbf{u}_m(s_{m-1}, s_m) - \delta_m(s_{m-1}, \boldsymbol{\theta}_m)\right]$$

This is the *simple Markov property*, that the conditional probability of a "future" state s_m given the "past" history $s_1, \ldots, s_{m-1}$ depends only on the "present" s_{m-1}.

From the elementary relation, which holds for any distribution p,

$$p(s_1, \ldots, s_k) = p(s_1)p(s_2|s_1)p(s_3|s_2, s_1) \cdots p(s_k|s_1, \ldots, s_{k-1})$$

and the simple Markov property, we have

$$q^{\boldsymbol{\theta}}(s_1, \ldots, s_k) = q^{\boldsymbol{\theta}}(s_1)q^{\boldsymbol{\theta}}(s_2|s_1)q^{\boldsymbol{\theta}}(s_3|s_2) \cdots q^{\boldsymbol{\theta}}(s_k|s_{k-1})$$

The unlikelihood of a single observation is then

$$U(\mathbf{s}) = 2\left[\delta_1(\boldsymbol{\theta}_1) - \boldsymbol{\theta}_1'\mathbf{u}_1(s_1) + \sum_{i=2}^{k} \delta_i(s_{i-1}, \boldsymbol{\theta}_i) - \boldsymbol{\theta}_i'\mathbf{u}_i(s_{i-1}, s_i)\right]$$

If n independent individuals were observed to follow this model, then we would sum the above expression over all the observed vectors $\mathbf{s}_i$ to obtain the sample unlikelihood

$$U = \sum_{i=1}^{n} U(\mathbf{s}_i)$$

and then use ML inference to estimate the parameters.

The Pure Evolutionary Case. If the $\boldsymbol{\theta}_i$'s are free (there are no interrelationships among them), then to do inference on $\boldsymbol{\theta}_m$ it is enough to use the partial unlikelihood

$$2\sum_{i=1}^{n}\delta_m(s_{i,m-1},\boldsymbol{\theta}_m)-\boldsymbol{\theta}_m'\mathbf{u}_m(s_{i,m-1},s_{i,m})$$

where s_{ij} is the jth state occupied by individual i. This is equivalent to viewing the $s_{i,m-1}$ $(1 \leqq i \leqq n)$ as being fixed (that is, covariates) and the $s_{i,m}$ $(1 \leqq i \leqq n)$ as random responses. Thus it would be appropriate to look just at the table of transitions from time $m-1$ to time m, which would be a square table with two factors (state at time $m-1$ and state at time m), each having levels equal to the elements of S. Any function $\mathbf{u}_m$ on the table would be permissible, although we might be particularly interested in the case where we could write $\mathbf{u}_m(s', s) = \mathbf{v}(s) \otimes \mathbf{t}(s')$, because then the model for the transitions would be a linear exponential model. Inference on $\boldsymbol{\theta}_1$ would be carried out with the covariate-free model

$$q^{\boldsymbol{\theta}_1}(s)=\exp[\boldsymbol{\theta}_1'\mathbf{u}_1(s)-\delta_1(\boldsymbol{\theta}_1)]$$

Thus in the pure evolutionary case the entire analysis is broken down into separate independent analyses involving the initial distribution and the transition distributions.

The Pure Stationary Case. At the other extreme, suppose that all the $\boldsymbol{\theta}_i$'s and $\mathbf{u}_i$'s $(i \geqq 2)$ are equal. We then say that the Markov model has *stationary transition probabilities*, and write $\boldsymbol{\theta}_i = \boldsymbol{\sigma}$, $\mathbf{u}_i = \mathbf{v}$. The initial distribution would be analyzed as before, and the analysis of $\boldsymbol{\sigma}$ would be based on the partial unlikelihood

$$2\sum_{i=1}^{n}\sum_{j=2}^{k}\delta_j(s_{i,j-1},\boldsymbol{\sigma})-\boldsymbol{\sigma}'\mathbf{v}(s_{i,j-1},s_{i,j})$$

It is interesting to observe that the above unlikelihood makes it appear that we are treating $n(k-1)$ independent transitions, so that for the computational setup of LAMDA we may use each transition as an "independent" experimental unit. This is an extremely convenient property of Markov models with stationary transition probabilities. It is sometimes said that the transitions themselves are independent, but this is not true, because except for the first and last states, each other state appears once as a covariate and once as a response in the unlikelihood.

Mixed Cases. A class of Markov models which lies between the extremes discussed so far is of the form

$$q^{\boldsymbol{\theta},\boldsymbol{\sigma}}(\mathbf{s}) = \exp\Bigg[\boldsymbol{\theta}_1'\mathbf{u}_1(s_1) - \delta_1(\boldsymbol{\theta}_1) + \sum_{i=2}^{k} \boldsymbol{\sigma}'\mathbf{v}(s_{i-1}, s_i)$$

$$+\boldsymbol{\theta}_i'\mathbf{u}_i(s_{i-1}, s_i) - \delta_i(s_{i-1}, \boldsymbol{\theta}_i, \boldsymbol{\sigma})\Bigg]$$

Here $\boldsymbol{\sigma}$ may be thought of as the parameter of the *stationary part* of the transition probabilities, and the $\boldsymbol{\theta}_i$'s as the parameters of the *evolutionary part*. Note the special case in which all the $\mathbf{u}_i$'s are the same.

Example. As an illustration of these ideas, consider an experiment in which individuals' opinions are traced over k time periods. Assume that opinion can be coded ordinally as $-1, 0, 1$ (for example, -1 = Favors Democrats, 0 = Undecided, 1 = Favors Republicans). Assuming the initial distribution of opinion to be unimportant, we have only to consider a model for transitions. Let $y = (s', s)$ denote a typical cell of the transition table, and define the indicators

$$u_{I1}(s', s) = 1 \quad \text{if} \quad s' = -1 \qquad (0 \text{ otherwise})$$

$$u_{I2}(s', s) = 1 \quad \text{if} \quad s' = 1 \qquad (0 \text{ otherwise})$$

and

$$v(s', s) = s$$

Taking q to be the uniform distribution, $\mathcal{E}[q : \mathbf{u}_I, v, v\mathbf{u}_I]$ is a model appropriate for a singly ordered table, and upon conditioning on $\mathbf{u}_I$ we obtain $\mathcal{E}[q : v|1, \mathbf{u}_I]$ for a model of the mth transition, that is,

$$q^{\boldsymbol{\theta}}(s|s') = \exp\{[\theta_{0m} + \boldsymbol{\theta}_{Im}'\mathbf{u}_I(s', s)]v(s', s) - \delta(s', \boldsymbol{\theta})\}q(s)$$

We now have several interesting hypotheses to test:

1. That the above purely evolutionary Markov model holds.
2. $\theta_{0m} = \theta_0$ (all m); since θ_{0m} measures the shift in opinion among those with $s_{m-1} = 0$, this model says that the shift is stationary.
3. $\boldsymbol{\theta}_{Im} = \boldsymbol{\theta}_I$ (all m), which says that the shift in opinion among the $s_{m-1} = -1$ and $s_{m-1} = 1$, relative to the $s_{m-1} = 0$ group, is stationary.

4. Both hypotheses 2 and 3, so the transitions are stationary.
5. Hypothesis 4 and $\theta_0 = 0$.
6. Hypothesis 4 and $\boldsymbol{\theta}_I = \mathbf{0}$.
7. Hypotheses 5 and 6.

It would be appropriate to replace q by the observed marginal distribution of s_m, so that all the above models contained the model under which s_{m-1} and s_m were independent. It is clear that more elaborate models could be constructed by including a t-function of covariates, if the sample were sufficiently extensive to permit such an analysis.

Extended Markov Models. A more general concept of a Markov model stipulates a collection of simple functions $\{g_i\}$ such that for each m, $q^{\boldsymbol{\theta}}(s_m | s_1, \ldots, s_{m-1})$ depends only on s_m and $g_{m-1}(s_1, \ldots, s_{m-1})$. The models discussed above are the special case in which $g_{m-1}(s_1, \ldots, s_{m-1}) = s_{m-1}$. Whatever form one chooses for the transition model, one can always analyze the data as though it consisted of a sample of initial states (s_1) and a sample of transitions (s_{m-1} to s_m), and one can pretend for the purposes of computation that these samples are independent. In the transition from s_{m-1} to s_m one treats $g_{m-1}(s_1, \ldots, s_{m-1})$ as a covariate and s_m as the random response. Consequently, so long as one stays within the framework of linear exponential models for the transitions, there is no difficulty in using LAMDA to analyze data with Markov models.

As a computational example of this more general kind of model, consider the data in Table VI.1. Each laboratory animal (A) was permitted twenty attempts to complete a task on each of a number of trials (T). Trials were continued until the animal achieved a perfect score. The numbers in the table are the numbers of successes out of twenty permitted attempts. We take $S = \{0, 1, \ldots, 20\}$. For the response part of the transition model we take

$$u_1(s) = \ln[s + 0.5], \qquad u_2(s) = \ln[20.5 - s]$$

and for the covariates we use indicators of E, B, and EB, and the number of successes on the preceding trial. Thus we have a simple Markov model with stationary transitions and covariates. The program appears in Run VI.B.1, and although they were not used for this run, we include the transition number (T), the total number of successes prior to the current trial, and the number of the animal, as labeled in Table VI.1. Notice that the response part of the model is the discrete analogue of the beta distribution. The output from this run appears in Table VI.2; note that the parameters

Table VI.1

A Repeated-Measures Animal Experiment

	*E*0								*E*1							
	*B*1				*B*0				*B*1				*B*0			
	A 1	*A* 2	*A* 3	*A* 4	*A* 5	*A* 6	*A* 7	*A* 8	*A* 9	*A* 10	*A* 11	*A* 12	*A* 13	*A* 14	*A* 15	*A* 16
*T*0	6	1	2	1	1	0	2	1	2	0	0	3	0	0	2	0
*T*1	8	6	7	6	0	0	2	1	0	0	0	4	0	0	0	3
*T*2	3	16	5	15	2	0	9	0	0	10	0	4	4	7	3	2
*T*3	6	17	13	19	3	0	4	0	7	17	10	6	4	5	0	0
*T*4	6	17	19	17	16	0	1	4	18	17	15	8	15	3	0	0
*T*5	5	8	19	19	12	0	9	4	19	19	18	15	10	4	7	0
*T*6	18	18	18	17	17	0	15	5	19	19	19	17	11	6	14	3
*T*7	18	17	19	19	18	0	16	7	18	19	15	15	13	2	11	2
*T*8	17	18	17	19	15	14	17	13	18	19	19	14	19	11	15	8
*T*9	19	18	20	20	16	15	17	17	20	19	19	16	20	12	18	12
*T*10	19	19	—	—	17	18	19	15	—	20	18	19	—	6	18	15
*T*11	18	20	—	—	20	16	19	15	—	—	19	18	—	13	17	18
*T*12	20	—	—	—	—	18	20	19	—	—	20	19	—	14	20	17
*T*13	—	—	—	—	—	18	—	18	—	—	—	20	—	19	—	18
*T*14	—	—	—	—	—	17	—	20	—	—	—	—	—	18	—	18
*T*15	—	—	—	—	—	17	—	—	—	—	—	—	—	17	—	20
*T*16	—	—	—	—	—	19	—	—	—	—	—	—	—	18	—	—
*T*17	—	—	—	—	—	19	—	—	—	—	—	—	—	20	—	—
*T*18	—	—	—	—	—	19	—	—	—	—	—	—	—	—	—	—
*T*19	—	—	—	—	—	20	—	—	—	—	—	—	—	—	—	—

E = Condition
*E*0 = Not Trained
*E*1 = Pre-trained
B = Stimulus
*B*0 = Light
*B*1 = Bell
A = Animal Number
T = Trial Number

Table VI.2

LAMDA *Output for the Animal Experiment*

```
        L A M D A  - - -  VERSION 4.0


*** RUN VI.B.1 ***
--- VAR2 = BELL
--- VAR3 = EXPERIMENTAL
--- VAR4 = TRANSITION NUMBER
--- VAR5 = TOTAL SUCCESSES
--- VAR6 = PREVIOUS SUCCESSES
--- VAR7 = LOG(CURRENT SUCCESSES + .5)
--- VAR8 = LOG(20.5 - CURRENT SUCCESSES)
 MODEL 0 2 3 2*3 6
=== LXM ===
U(P,QP)    =      844.364
DELTA(C) =  0.02181584
DELTA(U) =  0.00585937
 NO OF CASES =   194.000
10 ITERATIONS

- - - A N A L Y S I S - - -
D(P,QP)    =     -198.625
D(QP,Q)    =      336.907
D(P,Q)     =      138.282

- - - COEFFICIENTS - - -

     -0.9146       2.5834
      0.6342      -0.4389
     -0.0479       0.0394
     -0.4303       0.1092
      0.3989      -0.1378

- - - T-VALUES - - -

     -3.8341       4.5926
      1.1102      -1.4025
     -0.1503       0.1229
     -0.6238       0.2437
      7.0995      -4.6482
```

relating to E, B, and EB are all nonsignificant, although one ought to do another run with them omitted to make a proper test.

One could easily use such covariates as the average number of successes per trial up to the preceding trial in order to construct a more complex model. One could also use the variable T in various ways to construct models with evolutionary transition probabilities. Interaction terms involving E, B, and a linear function of T might serve to bring out more subtle aspects of the data involving the learning rate.

C. CONDITIONAL AND REVERSED INFERENCE

Throughout this book an attempt has been made to analyze data using one or more models which are proper for the type of sampling plan which was used. The purpose of this section is to show that under certain circumstances one can use improper models and still get valid inferences. The importance of this fact will emerge in the course of the discussion.

For this section we will assume an A, B-table with typical cell $AiBj$. We will always associate the letter i with factor A and j with factor B. Thus, if $p(ij)$ represents a distribution on the table, we may abuse the notation by writing

$$p_j(i) = p(Ai|Bj) = \frac{p(ij)}{p(+j)}$$

$$p_i(j) = p(Bj|Ai) = \frac{p(ij)}{p(i+)}$$

Everything in this section applies immediately to the case in which each of A and B refers to a collection of factors, so that in a sense we have separated the factors of interest into the "A-type" and "B-type." In this case i and j will be vector subscripts.

Let $q_A(i)$ be a strictly positive distribution on the A-marginal table, and likewise for $q_B(j)$ on the B-margin. Let $\mathbf{u}_A(i)$, $\mathbf{u}_B(j)$, and $\mathbf{u}_{AB}(ij)$ be vector functions of the indicated subscripts. The *Joint Model* on the table is

$$q^{\boldsymbol{\theta}}(ij) = \exp\left[\boldsymbol{\theta}_A'\mathbf{u}_A(i) + \boldsymbol{\theta}_B'\mathbf{u}_B(j) + \boldsymbol{\theta}_{AB}'\mathbf{u}_{AB}(ij) - \delta\right] q_A(i) q_B(j)$$

where $\boldsymbol{\theta}' = (\boldsymbol{\theta}_A', \boldsymbol{\theta}_B', \boldsymbol{\theta}_{AB}')$. As was shown in Chapter V, Section A.3, the Joint Model implies the *Forward Model* for $A|B$,

$$q_j^{\boldsymbol{\theta}}(i) = \exp\left[\boldsymbol{\theta}_A'\mathbf{u}_A(i) + \boldsymbol{\theta}_{AB}'\mathbf{u}_{AB}(ij) - \delta(j, \boldsymbol{\theta})\right] q_A(i)$$

which depends only on $\boldsymbol{\theta}_A$ and $\boldsymbol{\theta}_{AB}$, and the *Backward Model* for $B|A$,

$$q_i^{\boldsymbol{\theta}}(j) = \exp\left[\boldsymbol{\theta}_B'\mathbf{u}_B(j) + \boldsymbol{\theta}_{AB}'\mathbf{u}_{AB}(ij) - \delta(i, \boldsymbol{\theta})\right] q_B(j)$$

which depends only on $\boldsymbol{\theta}_B$ and $\boldsymbol{\theta}_{AB}$.

We assume an empirically derived distribution $\hat{p}$ on the table, but for the moment we will not specify the sampling plan under which $\hat{p}$ was obtained. We can now formulate two questions:

1. *Conditional inference*: since $\boldsymbol{\theta}_A$ and $\boldsymbol{\theta}_{AB}$ appear in both the Forward and Joint Models, when will inferences about them be the same, regardless of the model used?
2. *Reversed inference*: since $\boldsymbol{\theta}_{AB}$ appears in both the Forward and Backward Models, when will inferences about it be the same, regardless of the model used?

The conditional-inference question has practical importance in the analysis of large tables where the Joint Model is proper, since the Forward Model may have many fewer parameters than the Joint Model, and so be easier to deal with from the standpoint of computation and interpretation.

The reversed-inference question is of interest in cases where the sampling plan satisfies the Forward Model (but not the Joint Model) and the B factor is much simpler than the A-factor, since the Backward Model will then have fewer parameters and thus be easier to deal with.

An important instance of reversed inference occurs in the epidemiological case-control study (sometimes called "retrospective"). Here B represents disease status and A represents a number of risk factors thought to be associated with the disease. The designation "case-control" indicates that random samples have been drawn from the various subpopulations defined by disease status, so that B is a covariate and A the random response. $\boldsymbol{\theta}_{AB}$ is the parameter which measures the interrelationship between disease and risk factors, and the Forward Model is appropriate. It is natural, however, to think in terms of the Backward Model which specifies probabilities of disease given the risk factors, and so a $B|A$ analysis is more interpretable than an $A|B$ analysis. Thus we would like to know when the $B|A$ analysis is justified despite the fact that the $A|B$ analysis is proper under the sampling plan. This will clearly happen when both analyses give the same results.

Proposition 32. Suppose that $\mathfrak{S}[q_B : \mathbf{u}_B]$ is the saturated model for the B-marginal table. Then ML inferences about $\boldsymbol{\theta}_A$ and $\boldsymbol{\theta}_{AB}$ coincide in the Forward and Joint Models.

Proof. The likelihood equations for the Joint Model are

$$\sum_{ij} \mathbf{u}_A(i)\hat{p}(ij) = \sum_{ij} \mathbf{u}_A(i)\hat{q}(ij)$$

$$\sum_{ij} \mathbf{u}_B(j)\hat{p}(ij) = \sum_{ij} \mathbf{u}_B(j)\hat{q}(ij)$$

$$\sum_{ij} \mathbf{u}_{AB}(ij)\hat{p}(ij) = \sum_{ij} \mathbf{u}_{AB}(ij)\hat{q}(ij)$$

while the likelihood equations for the Forward Model are

$$\sum_{ij} \mathbf{u}_A(i)\hat{p}(ij) = \sum_{ij} \mathbf{u}_A(i)\hat{q}_j(i)\hat{p}(+j)$$

$$\sum_{ij} \mathbf{u}_{AB}(ij)\hat{p}(ij) = \sum_{ij} \mathbf{u}_{AB}(ij)\hat{q}_j(i)\hat{p}(+j)$$

To see that the estimates of $\boldsymbol{\theta}_A$ and $\boldsymbol{\theta}_{AB}$ coincide it is sufficient to check that if $\hat{q}(ij)$ is the ML distribution under the Joint Model, then $\hat{q}_j(i)$ is the ML distribution under the Forward Model, and conversely, if $\hat{q}_j(i)$ is the ML distribution under the Forward Model, then $\hat{q}_j(i)\hat{p}(+j)$ is the ML distribution under the Joint Model. These both follow from the assumption of the proposition, which requires that $\hat{p}(+j) = \hat{q}(+j)$.

The discrepancy between the empirical distribution and the ML distribution in the Joint Model is

$$
\begin{aligned}
n\mathbb{D}_1(\hat{p}, \hat{q}) &= -2n\sum_{ij}\hat{p}(ij)\ln\frac{\hat{p}(ij)}{\hat{q}(ij)} \\
&= -2n\sum_{ij}\hat{p}_j(i)\hat{p}(+j)\ln\frac{\hat{p}_j(i)\hat{p}(+j)}{\hat{q}_j(i)\hat{q}(+j)} \\
&= -2n\sum_{j}\hat{p}(+j)\sum_{i}\hat{p}_j(i)\ln\frac{\hat{p}_j(i)}{\hat{q}_j(i)} \\
&= n\sum_{j}\hat{p}(+j)\mathbb{D}_1(\hat{p}_j, \hat{q}_j)
\end{aligned}
$$

and this latter expression is the unlikelihood of the ML distribution for the Forward Model. For any particular distribution q on the A, B-table, if $q(+j) = \hat{p}(+j)$, then by precisely the same reasoning $n\mathbb{D}_1(\hat{q}, q) = \sum_j n\hat{p}(+j)\mathbb{D}(\hat{q}_j, q_j)$, and so testing any simple hypothesis about $\boldsymbol{\theta}_A$ and $\boldsymbol{\theta}_{AB}$, given that the model holds, yields the same numerical results regardless whether the Joint or the Forward Model is used. □

Proposition 33. If $\mathfrak{S}[q_B : \mathbf{u}_B]$ is the saturated model on the B-margin, and $\mathfrak{S}[q_A : \mathbf{u}_A]$ is the saturated model on the A-margin, then inference about $\boldsymbol{\theta}_{AB}$ coincides in the Forward and the Backward Model.

Proof. The preceding proposition implies that inference about $\boldsymbol{\theta}_{AB}$ is the same in the Forward and Joint Models, and the situation is symmetric with regard to A and B. □

As an example of the usefulness of reversed inference, consider the data of Table VI.3. The sampling method for the data was of the case-control variety, so that an $E, T, X, A|C$ analysis would be appropriate. The response part of this model has $4 \times 2 \times 2 \times 2 = 32$ cells, and if we want to carry out an analysis of any complexity we are clearly going to need several

Table VI.3

A Case-Control Study of Skin Cancer

		$E0$		$E1$		$E2$		$E3$	
		$T0$	$T1$	$T0$	$T1$	$T0$	$T1$	$T0$	$T1$
					$C0$				
$A0$	$X0$	65	11	2	0	2	1	3	1
	$X1$	23	24	4	3	0	0	1	0
$A1$	$X0$	55	9	6	0	7	0	1	1
	$X1$	23	39	6	3	0	1	3	0
					$C1$				
$A0$	$X0$	1	1	1	0	0	0	1	0
	$X1$	0	1	0	2	0	0	2	0
$A1$	$X0$	9	3	1	0	3	1	0	1
	$X1$	1	11	2	3	2	2	2	8
					$C2$				
$A0$	$X0$	26	9	4	2	5	1	4	0
	$X1$	19	40	2	8	1	5	0	3
$A1$	$X0$	44	16	11	1	7	1	10	1
	$X1$	33	76	6	12	6	1	6	6

C = Skin Cancer
$C0$ = None
$C1$ = Squamous
$C2$ = Basal
E = Exposure to Sun
$0(1)3T$ = Tannability
$T0$ = Tans
$T1$ = Burns
X = Complexion
$X0$ = Dark
$X1$ = Pale
A = Age
$A0$ = Less than 59
$A1$ = More than 60

parameters and a large number of models. Moreover, it is not clear what one should make of a parameter which measures, say, the interaction between levels of A and X conditional on E and T, taking C into account as a covariate. Since there is no interest in the E, T, X, A-marginal table, it seems reasonable to fit distributions which agree with the observations on this margin, which is the condition of Proposition 32. Since C has only three levels, it is easy to saturate the C-margin, which is the condition of Proposition 33. Consequently, it is legitimate for the purpose of investigating the association between C and E, X, T, A to use a linear exponential model like $C|E, X, T, A$, provided that the u-functions for C consist of the indicators of $C1$ and $C2$ (or any equivalent set of functions). Some interesting models would be

Model	Number of Parameters
$C\|E\,T\,X\,A$	14
$C\|[ET]XA$	20
$C\|[ETX]A$	34
$C\|E'T\,X\,A$	10
$C\|[E'T]XA$	12
$C\|[E'TX]A$	18

where E' stands for a single factor linear in the levels of E. In each case if the roles of responses and covariates were reversed, then the number of parameters would increase. Moreover, all the parameters in the above models can be interpreted as effects of the risk factors and their interactions on the probabilities of contracting one of the two types of skin cancer.

Prediction. Note in this example that we have no basis for predicting skin cancer in terms of risk factors. For instance, we can compute the ln odds favoring Ci over $C0$, obtained from $C|E\,X\,T\,A$:

$$\ln \frac{\hat{q}(Ci|E, X, T, A)}{\hat{q}(C0|E, X, T, A)} = \hat{\beta}_{Ci} + \hat{\beta}_{CiE}t_E + \hat{\beta}_{CiT}t_T + \hat{\beta}_{CiX}t_X + \hat{\beta}_{CiA}t_A$$

but only $\hat{\beta}_{CiE}$ through $\hat{\beta}_{CiA}$ are properly estimated. $\hat{\beta}_{Ci}$ simply reflects the overall frequencies of Ci and $C0$ in the sample, and since the C-marginal sample was not random, $\hat{\beta}_{Ci}$ need have no relationship to the actual occurrence of Ci in the population. However, if we knew the true probabilities $\pi(C0)$ and $\pi(Ci)$, we could write

$$\frac{\hat{q}(Ci|EkTjXhAm)}{\hat{q}(C0|EkTjXhAm)} = \frac{\hat{q}(CiEkTjXhAm)}{\hat{q}(C0EkTjXhAm)}$$

$$= \frac{\hat{q}(EkTjXhAm|Ci)\hat{p}(Ci)}{\hat{q}(EkTjXhAm|C0)\hat{p}(C0)}$$

$$= \frac{\hat{q}(EkTjXhAm|Ci)\pi(Ci)}{\hat{q}(EkTjXhAm|C0)\pi(C0)} \cdot \frac{\hat{p}(Ci)\pi(C0)}{\pi(Ci)\hat{p}(C0)}$$

This shows that

$$\ln\left(\frac{\hat{q}(Ci|EkTjXhAm)}{\hat{q}(C0|EkTjXhAm)} \cdot \frac{\pi(Ci)\hat{p}(C0)}{\hat{p}(Ci)\pi(C0)}\right)$$

is a proper estimate of the ln odds for a predictive system. Thus by changing $\hat{\beta}_{Ci}$ to $\hat{\beta}_{Ci} + \ln[\pi(Ci)\hat{p}(C0)/\hat{p}(Ci)\pi(C0)]$, the probabilities $\hat{q}(Ci|E, T, X, A)$ could be used for prediction.

D. HOMOGENEITY TESTS

In its most general form, a homogeneity hypothesis states that some aspect of a distribution p on a table $\mathcal{Y}$ is the same from two different points of view. In this section we let A and B be factors, and associate the letter i with levels of A, j with levels of B, so that we may continue the notational abuses of Section C. As before, either of A or B could itself consist of a number of subfactors, so we have effectively split the factors into "A-type" and "B-type." We assume that each experimental unit is independent of the others.

1. Full Homogeneity

Under these assumptions, we may divide the sample according to the levels of B, thus obtaining independent subsamples of size $N(+j)$ for $1 \leqq j \leqq J$. We consider the conditional distributions p_j within levels of B, and define the models

$$\mathcal{H} = \{p_j : p_j = p_{j'} \text{ for all } j, j'\}$$

$$\mathcal{E}[q:\mathbf{u}|B] = \{p_j : \text{each } p_j \in \mathcal{E}[q:\mathbf{u}]\} \quad .$$

$$\mathcal{H}[q:\mathbf{u}] = \mathcal{H} \cap \mathcal{E}[q:\mathbf{u}|B]$$

$\mathcal{H}$ is the model which asserts identity of the conditional distributions, and $\mathcal{E}[q:\mathbf{u}|B]$ asserts that all the conditional distributions lie in the same exponential model, $\mathcal{E}[q:\mathbf{u}]$, where q and $\mathbf{u}$ are defined in terms of A alone. Let $\hat{q}_j$ denote the ML distribution in $\mathcal{E}[q:\mathbf{u}]$ based on the conditional empirical distribution $\hat{p}_j$ within level Bj. Let $w_j = N(+j)/N(++)$ be the sample fraction in Bj, and observe that $\hat{p}(i+) = \Sigma_j w_j \hat{p}_j(i)$ is the ML distribution in $\mathcal{H}$, since under $\mathcal{H}$ we have one homogeneous sample of size n on the A-marginal table. Let $\hat{q}$ be the ML distribution in $\mathcal{H}[q:\mathbf{u}]$; $\hat{q}$ is the ML distribution in $\mathcal{E}[q:\mathbf{u}]$ on the A-margin obtained from the data $\hat{p}(i+)$. When we use $\mathbb{D}_1$ it is implicit that it pertains to the A-marginal table, $\mathcal{Y}$.

Proposition 34.

1. If $\pi \in \mathcal{H}$, then

$$N(+j)\mathbb{D}_1(\hat{p}_j, \hat{p}) \to_d (1-\omega_j)\chi^2(\#\mathcal{Y} - 1)$$

where $\omega_j = \lim w_j$.

2. If $\pi \in \mathcal{E}[q:\mathbf{u}|B]$, then

$$N(+j)\mathbb{D}_1(\hat{p}_j, \hat{q}_j) \to_d \chi^2(\#\mathcal{Y} - 1 - \#\mathbf{u})$$

3. If $\pi \in \mathcal{H}[q:\mathbf{u}]$, then

(3a) $N(+j)\mathbb{D}_1(\hat{q}_j, \hat{q}) \to_d (1-\omega_j)\chi^2(\#\mathbf{u})$

(3b) $N(+j)\mathbb{D}_1(\hat{p}_j, \hat{q}) \to_d \chi^2(\#\mathcal{Y} - 1 - \#\mathbf{u}) + (1-\omega_j)\chi^2(\#\mathbf{u})$

where the chi-squares on the right are independent.

Proof. Note that part 2 is just a restatement of Proposition 27 in the present setting. Note also that part 1 follows from part 3, by taking $\mathbf{u}$ to be maximal. Thus we only need to establish part 3.

We first establish (3b). Assume that $(\mathbf{u}, \mathbf{v})$ is maximal π-standard, and let $(\hat{\mathbf{u}}, \hat{\mathbf{v}})$ be its $\hat{q}$-standardization. It will suffice to consider the case $i = 1$ and to replace $\mathbb{D}_1$ by $\mathbb{D}_2$. Then

$$N(+1)\mathbb{D}_2(\hat{p}_1, \hat{q}) = N(+1)\|M[\hat{\mathbf{u}}:\hat{p}_1]\|^2 + N(+1)\|M[\hat{\mathbf{v}}:\hat{p}_1]\|^2$$

The standardization argument of Proposition 27 can be applied directly to prove

$$N(+1)\|M[\hat{\mathbf{v}}:\hat{p}_1]\|^2 \to_d \chi^2(\#\mathcal{Y} - 1 - \#\mathbf{u})$$

We now use the standardization argument on the other part of the above decomposition. First, for any k

$$\sqrt{V[\hat{u}_k : \hat{q}]} \cdot \hat{u}_k = u_k - M[u_k : \hat{q}] - \sum_{j=1}^{k-1} M\left[u_k \hat{u}_j : \hat{q}\right] \hat{u}_j$$

so that

$$\sqrt{N(+1)}\{M[\hat{u}_k : \hat{p}_1] - M[u_k : \hat{p}_1] + M[u_k : \hat{q}]\} \to_p 0$$

Since $M[u_k : \hat{q}] = M[u_k : \hat{p}] = \Sigma_j w_j M[u_k : \hat{p}_j]$, we have

$$\sqrt{N(+1)}\left\{ M[\hat{u}_k : \hat{p}_1] - M[u_k : \hat{p}_1] + \sum_j w_j M\left[u_k : \hat{p}_j\right]\right\} \to_p 0$$

Now set $\mathbf{Z}_j = \sqrt{N(+j)}\, M[\mathbf{u} : \hat{p}_j]$ $(1 \leqq j \leqq J)$, so that

$$(\mathbf{Z}'_1, \ldots, \mathbf{Z}'_J) \to_d N(\mathbf{0}, \mathbf{I})$$

and we can write

$$\sqrt{N(+1)}\, M[\hat{\mathbf{u}} : \hat{p}_1] - \mathbf{Z}_1 + \sum_j \sqrt{w_1 w_j}\, \mathbf{Z}_j \to_p \mathbf{0}$$

Consequently,

$$N(+1)\|M[\hat{\mathbf{u}} : \hat{p}_1]\|^2 - \left\|\mathbf{Z}_1 - \sqrt{w_1} \sum_j \sqrt{w_j}\, \mathbf{Z}_j\right\|^2 \to_p 0$$

and since

$$\mathbf{Z}_1 - \sqrt{w_1} \sum_j \sqrt{w_j}\, \mathbf{Z}_j \to_d N\left(\mathbf{0}, \left[(1 - \omega_1)^2 + \omega_1 \sum_{j>1} \omega_j\right]\mathbf{I}\right)$$

we have shown

$$N(+1)\|M[\hat{\mathbf{u}} : \hat{p}_1]\|^2 \to_d (1 - \omega_1)\chi^2(\#\mathbf{u})$$

Finally, since $\sqrt{N(+1)}\, M[\hat{\mathbf{u}} : \hat{p}_1]$ depends asymptotically only on terms $M[\mathbf{u} : \hat{p}_j]$, while from Proposition 27, $\sqrt{N(+1)}\, M[\hat{\mathbf{v}} : \hat{p}_1]$ depends asymptotically only on $M[\mathbf{v} : \hat{p}_1]$, we have the independence assertion. This completes the proof of (3b).

For (3a) we only need to make a few observations. First, by the rapid convergence of $\hat{q}_j$, we have $\sqrt{N(+j)}\, M[\mathbf{v} : \hat{q}_j] \to_p \mathbf{0}$, and so by the argument of Proposition 27, $\sqrt{N(+j)}\, M[\hat{\mathbf{v}} : \hat{q}_j] \to_p \mathbf{0}$. Thus $N(+j)\mathbb{D}_2(\hat{q}_j, \hat{q})$ is asymptotically equivalent to $N(+j)\|M[\mathbf{u} : \hat{q}_j]\|^2$. Second, from $M[\hat{\mathbf{u}} : \hat{q}_j] = M[\hat{\mathbf{u}} : \hat{p}_j]$ the proof of (3b) can be applied. □

P-values for the statistics in (3b) can be determined as follows. Let F denote the cumulative distribution function of $\chi^2(m_0) + (1 - w)\chi^2(m)$, where the chi-squares are independent. Define

$$c_j = \begin{cases} (1-w)^{m/2} & \text{if} \quad j = 0 \\ (1-w)^{m/2} w^j \displaystyle\prod_{i=1}^{j} \frac{m/2 + i - 1}{i} & \text{if} \quad j > 0 \end{cases}$$

Then $F(x) = \Sigma_{j=0}^{\infty} c_j F_{m_0+m+2j}(x/(1-w))$, where F_ν is the cumulative distribution function of $\chi^2(\nu)$. If the sum on the right is truncated to $0 \leqq j \leqq k$, then the error in the approximation does not exceed $1 - \Sigma_{j=0}^{k} c_j$. Since $c_j \geqq 0$ and $\Sigma_{j=0}^{\infty} c_j = 1$, one can choose k to achieve a prescribed accuracy.

We now establish the appropriate Chi-squares for testing homogeneity.

Proposition 35.

1. If $\pi \in \mathcal{H}$, then

$$n \sum_j w_j \mathbb{D}_1(\hat{p}_j, \hat{p}) \to_d \chi^2((J-1)(\#\mathcal{Y} - 1))$$

2. If $\pi \in \mathcal{E}[q : \mathbf{u} | B]$, then

$$n \sum_j w_j \mathbb{D}_1(\hat{p}_j, \hat{q}_j) \to_d \chi^2(J(\#\mathcal{Y} - 1 - \#\mathbf{u}))$$

3. If $\pi \in \mathcal{H}[q : \mathbf{u}]$, then $\mathcal{E}[q : \mathbf{u} | B] \to \mathcal{H}[q : \mathbf{u}]$ is tested with

(3a) $n \Sigma_j w_j \mathbb{D}_1(\hat{q}_j, \hat{q}) \to_d \chi^2((J-1)\#\mathbf{u})$

and $\mathcal{H} \to \mathcal{H}[q : \mathbf{u}]$ is tested using

(3b) $n\mathbb{D}_1(\hat{p}, \hat{q}) \to_d \chi^2(\#\mathcal{Y} - 1 - \#\mathbf{u})$

Proof. Note that part 2 is immediate from Proposition 27 and the independence of the B-subsamples. A similar remark applies to (3b). Now note that part 1 follows from (3a), by choosing $\mathbf{u}$ maximal. Thus it suffices to prove (3a), and we give a sketch of this proof.

Set $\mathbf{Z}_j = \sqrt{N(+j)}\,M[\mathbf{u}: \hat{q}_j]$, $\hat{\mathbf{Z}}_j = \sqrt{N(+j)}\,M[\hat{\mathbf{u}}: \hat{q}_j]$, $\mathbf{Z} = [\mathbf{Z}_1\ \mathbf{Z}_2\ \cdots\ \mathbf{Z}_k]$, $\hat{\mathbf{Z}} = [\hat{\mathbf{Z}}_1\ \hat{\mathbf{Z}}_2\ \cdots\ \hat{\mathbf{Z}}_k]$. We have from the preceding proof that

$$\hat{\mathbf{Z}} - \mathbf{Z}[\mathbf{I} - \sqrt{\mathbf{w}}\sqrt{\mathbf{w}}'] \to_p \mathbf{0}$$

where $\sqrt{\mathbf{w}}$ has components $\sqrt{w_j}$. Letting ζ_i' denote the ith row of $\mathbf{Z}$, we have

$$\sum_{ij} \hat{Z}_{ij}^2 - \sum_i \left\| \zeta_i'[I - \sqrt{\mathbf{w}}\sqrt{\mathbf{w}}'] \right\|^2 \to_p 0$$

and since $\mathbf{I} - \sqrt{\mathbf{w}}\sqrt{\mathbf{w}}'$ is idempotent of rank $J - 1$, and the terms in the right-hand sum are asymptotically independent, we have

$$n \sum_j w_j \mathbb{D}_2(\hat{q}_j, \hat{q}) = \sum_{ij} \hat{Z}_{ij}^2 \to_d \chi^2((J-1)\#\mathbf{u})$$ □

It is worthwhile to note that as a consequence of the proof, we have

$$n \sum_j w_j \mathbb{D}_1(\hat{p}_j, \hat{q}) \to_d \chi^2(J(\#\mathcal{Y} - 1) - \#\mathbf{u})$$

when $\pi \in \mathcal{H}[q:\mathbf{u}]$.

We are now able to carry out the four homogeneity tests (1) saturated $\to \mathcal{H}$, (2) saturated $\to \mathcal{E}[q:\mathbf{u}|B]$, (3) $\mathcal{H} \to \mathcal{H}[q:\mathbf{u}]$, and (4) $\mathcal{E}[q:\mathbf{u}|B] \to \mathcal{H}[q:\mathbf{u}]$, and to produce the corresponding ANALYSIS table. Moreover, Proposition 34 gives the individual distributions of the components of the chi-square statistics of Proposition 35; these can be used as generalized residuals to ascertain whether the poor fit of one of the homogeneity models is due to a small collection of the B-subsamples.

2. Partial Homogeneity

It is frequently of interest to ask whether some aspect of the p_j's is homogeneous across B-subsamples, rather than asking whether the distributions agree exactly.

One very general approach to this problem is given in Problem II.21. We illustrate this approach by writing, in the notation of Problem II.21, $\mu_j = M[w: \pi_j]$ for some function w on the A-table. Then $X_j = M[w: \hat{p}_j]$,

$\hat{\sigma}_j^2 = V[w : \hat{p}_j]/N(+j)$, the homogeneity test is based on

$$\sum_j \frac{(X_j - X^*)^2}{\hat{\sigma}_j^2} \to_d \chi^2(J-1)$$

and upon confirmation of homogeneity one may test a specific common value μ of μ_j by using

$$\sum_j \frac{1}{\hat{\sigma}_j^2}(X^* - \mu)^2 \to_d \chi^2(1)$$

Note that here $X^* = (\Sigma M[w : \hat{p}_j]/\hat{\sigma}_j^2)/(\Sigma 1/\hat{\sigma}_j^2)$, which is not in general equal to $M[w : \hat{p}]$.

If one were interested in several aspects of the p_j's, tests of the above sort could be combined by using the Bonferroni procedure.

Another kind of partial homogeneity question concerns collapsibility. In the general case, the collapsibility issue revolves around quantities of the form $\Sigma_{ij} C(ij) \ln p(ij)$ for $p \in \mathcal{P}$. In Chapter IV, Section A.3 we saw that classical parametrization of log-linear models involves terms of this form. In this case the $C(ij)$ coefficients will be related to, but not identical to, the Möbius-function terms. Now the corresponding quantity defined on the A-margin is $\Sigma_i C(i+) \ln p(i+)$, and the distribution p is *collapsible along B* with respect to this quantity if and only if

$$\sum_{ij} C(ij) \ln p(ij) = \sum_i C(i+) \ln p(i+)$$

The difference between the left and right sides of this equation is in general $\Sigma_{ij} C(ij) \ln p_i(j)$, and so collapsibility is equivalent to

$$\sum_{ij} C(ij) \ln p_i(j) = 0$$

A natural test of this hypothesis is based on the conditional $B|A$ distributions, $\hat{p}_i$ $(1 \leqq i \leqq I)$. From the asymptotic results on linear combinations of log ratios in Chapter II, Section B.2, for each i we may treat

$$\sum_j C(ij) \ln \hat{p}_i(j)$$

as

$$N\left(\sum_j C(ij) \ln \pi_i(j), \left(\sum_{ij} \frac{C(ij)^2}{\pi_i(j)} - C(i+)^2\right)\frac{1}{N(i+)}\right)$$

and as a consequence $\Sigma_{ij}C(ij)\ln \hat{p}_i(j)$ can be treated as Normal with mean 0 and variance

$$\sum_{ij} \frac{C(ij)^2}{\pi_i(j)N(i+)} - \sum_i \frac{C(i+)^2}{N(i+)}$$

Replacing $\pi_i(j)N(i+)$ by $N(ij)$ we have the final form of the estimated variance,

$$\sum_{ij} \frac{C(ij)^2}{N(ij)} - \sum_i \frac{C(i+)^2}{N(i+)}$$

The collapsibility question for a vector of quantities $\Sigma_{ij}\mathbf{C}(ij)\ln p(ij)$ would be tested by treating $\Sigma_{ij}\mathbf{C}(ij)\ln \hat{p}_i(j)$ as Normal with mean $\mathbf{0}$ and estimated covariance

$$\sum_{ij} \frac{\mathbf{C}(ij)\mathbf{C}(ij)'}{N(ij)} - \sum_i \frac{\mathbf{C}(i+)\mathbf{C}(i+)'}{N(i+)}$$

and obtaining a P-value from the corresponding Chi-square statistic.

3. Example of Collapsibility

Suppose that A represents a 2^2 table, and $i = (i_1 i_2)$, with i_1 and i_2 in $\{0, 1\}$. Let $C(ij) = (-1)^{i_1+i_2}$, so that $C(i+) = J(-1)^{i_1+i_2}$ and collapsibility means

$$\frac{1}{J}\sum_j \sum_i (-1)^{i_1+i_2} \ln p(i_1 i_2 j) = \sum_i (-1)^{i_1+i_2} \ln p(i_1 i_2 +)$$

so that the averages of the ln odds ratios in B-subsamples coincides with the ln odds ratio in the A-margin. The test is based on

$$\sum_{ij} (-1)^{i_1+i_2} \ln \hat{p}_i(j) \to_d N\left(0, \sum_{ij} \frac{1}{N(ij)} - \sum_i \frac{J^2}{N(i+)}\right)$$

A variant of this procedure is based on

$$C(ij) = \frac{(-1)^{i_1+i_2}}{v_j}$$

where

$$v_j = \frac{1}{N(00j)} + \frac{1}{N(01j)} + \frac{1}{N(10j)} + \frac{1}{N(11j)}$$

is the estimate of the variance of the estimated ln odds ratios in the Bj-subsample. Then

$$C(i+) = (-1)^{i_1+i_2} \sum_j \frac{1}{v_j}$$

and collapsibility means

$$\frac{\sum_{ij} (-1)^{i_1+i_2} \frac{1}{v_j} \ln p(ij)}{\sum_j \frac{1}{v_j}} = \sum_i (-1)^{i_1+i_2} \ln p(i+)$$

which is tested with

$$\sum_{ij} (-1)^{i_1+i_2} \frac{1}{v_j} \ln \hat{p}_i(j) \to_d N\left(0, \sum_{ij} \frac{1}{v_j^2 N(ij)} - \sum_i \frac{1}{N(i+)} \left[\sum_j \frac{1}{v_j}\right]^2\right)$$

4. Homogeneity Data Example

In Chapter IV, Section B.6 we collapsed the data of Table IV.20 in order to analyze the A, S, C, V-margin. By a process of model fitting we arrived at the model $[ASC][AV][SV]$, but one cannot rest with this as an analysis without checking whether the results might have been different if the U-factor had been taken into account. One way to phrase this question is to ask whether the model fits homogeneously over levels of U.

We first assess the fit of $[ASC][AV][SV]$ within levels of U, using the Chi-squares of Proposition 34, part 2, and their sum, which is the Chi-square of Proposition 35, part 2. This is done in the first part of Table VI.4, and we would certainly conclude that the model fits. We then test homogeneity of the fitted model across levels of U with the Chi-squares of (3a) in Proposition 34 and their sums in (3a) in Proposition 35. Again, the large P-values indicate a good fit. We conclude that the collapsing of Table IV.20 for the purpose of analysis was justified.

Table VI.4

Tests of Homogeneity Across Levels of U

j	$N(+j)$	$1-w_j$	df	$N(+j)\mathbb{D}_1(\hat{p}_j, \hat{q}_j)$	$\langle P\rangle^a$	df	$N(+j)\mathbb{D}_1(\hat{q}_j, \hat{q})$	$\langle P\rangle^a$
$U0$	605	.7845	8	9.47	⟨.304⟩	15	9.40	⟨.681⟩
$U1$	504	.8204	8	11.87	⟨.156⟩	15	2.13	⟨1.000⟩
$U2$	513	.8172	8	5.39	⟨.717⟩	15	10.21	⟨.642⟩
$U3$	420	.8504	8	5.40	⟨.716⟩	15	8.56	⟨.816⟩
$U4$	764	.7278	8	5.33	⟨.724⟩	15	8.43	⟨.711⟩
	2807		40	37.46	⟨.585⟩	60[a]	38.73	⟨.985⟩

[a]Note that 60 is not the sum of the individual dfs, and that the P-values are based on the "effective sample sizes," $N(+j)/(1-w_j)$.

We could have addressed the homogeneity question before collapsing the table. In this case, we would have used the Chi-squares of Proposition 34, part 1 and Proposition 35, part 1, and upon confirmation of homogeneity gone on to fit the marginal table.

5. Collapsibility Data Example

Consider the problem of collapsibility with regard to the ln odds ratio between C and V in the data of Table IV.20. Denoting a typical cell of the table by $AiSjCkVmUr$, we define

$$C(ijkmr) = \frac{(-1)^{k+m}}{\dfrac{1}{N(ij00r)} + \dfrac{1}{N(ij01r)} + \dfrac{1}{N(ij10r)} + \dfrac{1}{N(ij11r)}}$$

so that we weight the ln odds ratio between C and V by the inverse of its estimated variance. Now define the quantities

$$C^*(ASU) = \sum_{ijkmr} C(ijkmr)\ln N(ijkmr)$$

$$C^*(AS) = \sum_{ijkm} C(ijkm+)\ln N(ijkm+)$$

$$C^*(AU) = \sum_{ikmr} C(i+kmr)\ln N(i+kmr)$$

$$C^*(SU) = \sum_{jkmr} C(+jkmr)\ln N(+jkmr)$$

$$C^*(A) = \sum_{ikm} C(i+km+)\ln N(i+km+)$$

$$C^*(S) = \sum_{jkm} C(+jkm+)\ln N(+jkm+)$$

$$C^*(U) = \sum_{kmr} C(++kmr)\ln N(++kmr)$$

$$C^*(\varnothing) = \sum_{km} C(++km+)\ln N(++km+)$$

Also define the terms

$$V^*(ASU) = \sum_{ijkmr} \frac{C(ijkmr)^2}{N(ijkmr)}$$

$$V^*(AS) = \sum_{ijkm} \frac{C(ijkm+)^2}{N(ijkm+)}$$

and so on, using the same notational convention for the summation over subscripts as for C^*. We now perform the tests for collapsibility in the following form:

To collapse across	treat	as Normal with variance
A	$C^*(ASU) - C^*(SU)$	$V^*(ASU) - V^*(SU)$
AS	$C^*(SU) - C^*(U)$	$V^*(SU) - V^*(U)$
ASU	$C^*(U) - C^*(\varnothing)$	$V^*(U) - V^*(\varnothing)$

The test for AS assumes that A has been confirmed, and likewise the test for ASU assumes that AS has been confirmed. Thus there are six possible orders in which the factors can be tested.

In Table VI.5 we find the values of C^* and V^*; here each value in Table IV.20 was increased by 0.5 before the computations. We begin by testing each factor separately for collapsibility:

Collapsed Factor	$\frac{C^*\text{-difference}}{\sqrt{V^*\text{-difference}}}$	$\langle P \rangle$
A	4.50	$\langle .000 \rangle$
S	−0.42	$\langle .673 \rangle$
U	−0.26	$\langle .798 \rangle$

Table VI.5

Terms for Use in the Collapsibility Test

i	$C^*(i)$	$V^*(i)$
ASU	10.8818	46.1349
AS	11.6480	37.1750
AU	12.7512	26.4992
SU	−8.7561	27.0819
A	11.3640	22.7300
S	−7.8945	23.1750
U	−6.7290	14.3532
$\varnothing$	−6.9060	13.5000

We cannot collapse across A, but we can test

Collapsed Factor	$\dfrac{C^*\text{-difference}}{\sqrt{V^*\text{-difference}}}$	$\langle P \rangle$
S after U	.075	$\langle .940 \rangle$
U after S	.715	$\langle .475 \rangle$

Thus, we conclude that it is permissible to collapse over S and U simultaneously.

E. LOCAL APPROXIMATION BY EXPONENTIAL MODELS

Although exponential models are of considerable generality, there are certainly many nonexponential models which are appropriate in various situations. We will indicate in this section how a smooth parametric model can be approximated locally by an exponential model, so that in principle LAMDA can be used to analyze models of great generality.

Suppose that $\{p^{\boldsymbol{\alpha}}\}$ is a parametric model on $\mathcal{Y}$, and that $\boldsymbol{\alpha}$ ranges in some open subset of $\mathbb{R}^k$. Assume that each $p^{\boldsymbol{\alpha}}$ is in $\mathcal{P}$, and that the model is smooth in the sense that

$$\mathbf{u}(y, \boldsymbol{\alpha}) = \frac{\partial}{\partial \boldsymbol{\alpha}} \ln p^{\boldsymbol{\alpha}}(y)$$

exists and has a continuous derivative with respect to $\boldsymbol{\alpha}$. One expects a

$\mathbb{D}_1$-MDE to satisfy

$$\mathbf{0} = \frac{\partial}{\partial \boldsymbol{\alpha}} \mathbb{D}_1(\hat{p}, p^{\boldsymbol{\alpha}}) = -2\frac{\partial}{\partial \boldsymbol{\alpha}} \sum_y \hat{p}(y) \ln p^{\boldsymbol{\alpha}}(y)$$

$$= -2\sum_y \mathbf{u}(y, \boldsymbol{\alpha})\hat{p}(y)$$

Note that

$$\sum_y \mathbf{u}(y, \boldsymbol{\alpha}) p^{\boldsymbol{\alpha}}(y) = \sum_y \frac{\partial}{\partial \boldsymbol{\alpha}} p^{\boldsymbol{\alpha}}(y) = \frac{\partial}{\partial \boldsymbol{\alpha}} \sum_y p^{\boldsymbol{\alpha}}(y) = \mathbf{0}$$

so that the equation which the MDE $\hat{\boldsymbol{\alpha}}$ should satisfy can formally be written

$$\sum_y \mathbf{u}(y, \boldsymbol{\alpha})\hat{p}(y) = \sum_y \mathbf{u}(y, \boldsymbol{\alpha}) p^{\boldsymbol{\alpha}}(y)$$

by analogy with the likelihood equation for an exponential model.

The local approximation to $\{p^{\boldsymbol{\alpha}}\}$ at the fixed value $\boldsymbol{\alpha}_0$ is defined to be the exponential model

$$q^{\boldsymbol{\theta}}(y) - \exp[\boldsymbol{\theta}'\mathbf{u}(y, \boldsymbol{\alpha}_0) - \delta(\boldsymbol{\theta})]\, p^{\boldsymbol{\alpha}_0}(y)$$

If we write $\boldsymbol{\theta} = \boldsymbol{\alpha} - \boldsymbol{\alpha}_0$, then

$$\ln q^{\boldsymbol{\theta}}(y) = \ln p^{\boldsymbol{\alpha}_0}(y) + (\boldsymbol{\alpha} - \boldsymbol{\alpha}_0)'\mathbf{u}(y, \boldsymbol{\alpha}_0) - \delta(\boldsymbol{\theta})$$

and the first two terms on the right form the Taylor approximation to $\ln p^{\boldsymbol{\alpha}}(y)$ near $\boldsymbol{\alpha}_0$. Note that $q^{\mathbf{0}} = p^{\boldsymbol{\alpha}_0}$.

Now let $\boldsymbol{\theta}_1$ be the MLE in the exponential model $\{q^{\boldsymbol{\theta}}\}$. Define $\boldsymbol{\alpha}_1 = \boldsymbol{\theta}_1 + \boldsymbol{\alpha}_0$, and shift attention to the local approximation at $\boldsymbol{\alpha}_1$,

$$q^{\boldsymbol{\theta}}(y) = \exp[\boldsymbol{\theta}'\mathbf{u}(y, \boldsymbol{\alpha}_1) - \delta(\boldsymbol{\theta})]\, p^{\boldsymbol{\alpha}_1}(y)$$

Let $\boldsymbol{\theta}_2$ be the MLE in this exponential model, and set $\boldsymbol{\alpha}_2 = \boldsymbol{\theta}_2 + \boldsymbol{\alpha}_1$. Iterate this procedure, and assume that $\boldsymbol{\alpha}_k \to \hat{\boldsymbol{\alpha}}$ and $\boldsymbol{\theta}_k \to \mathbf{0}$ as $k \to \infty$. Since by the ML equation

$$\sum_y \mathbf{u}(y, \boldsymbol{\alpha}_{k-1})\hat{p}(y) = \sum_y \mathbf{u}(y, \boldsymbol{\alpha}_{k-1}) p^{\boldsymbol{\alpha}_k}(y)$$

when we let $k \to \infty$ we obtain $\lim q^{\boldsymbol{\theta}_k} = \lim p^{\boldsymbol{\alpha}_k} = p^{\hat{\boldsymbol{\alpha}}}$ and so

$$\sum_y \mathbf{u}(y, \hat{\boldsymbol{\alpha}}) \hat{p}(y) = \sum_y \mathbf{u}(y, \hat{\boldsymbol{\alpha}}) p^{\hat{\boldsymbol{\alpha}}}(y)$$

so that $\hat{\boldsymbol{\alpha}}$ satisfies the equation which we expect the MDE to satisfy. This does not prove that $\hat{\boldsymbol{\alpha}}$ is the MDE, but in many cases it is possible to show that a solution of the above equation is automatically the MDE, and so we will not pursue the problem. The estimated covariance matrix of $\hat{\boldsymbol{\alpha}}$ can be computed as

$$\frac{1}{n}\left[\sum_y \mathbf{u}(y, \hat{\boldsymbol{\alpha}}) \mathbf{u}(y, \hat{\boldsymbol{\alpha}})' p^{\hat{\boldsymbol{\alpha}}}(y)\right]^{-1}$$

which would be the reported covariance matrix on the last iteration.

It is worth remembering that this procedure may be sensitive to the choice of initial $\boldsymbol{\alpha}_0$, and that for some choices of $\boldsymbol{\alpha}_0$ one may even obtain an exponential model with unidentifiable parameter.

As an example of this procedure, consider a rectangular table with factors A and B, and let $\mathbf{u}_A$ and $\mathbf{u}_B$ stand for the customary factorial indicators. A model which is more general than independence, but is not saturated, is

$$p(ij) = \exp\left[\boldsymbol{\theta}_A' \mathbf{u}_A(i) + \boldsymbol{\theta}_B' \mathbf{u}_B(j) + \mathbf{u}_A(i)' \boldsymbol{\alpha} \boldsymbol{\beta}' \mathbf{u}_B(j) - \delta\right]$$

subject to the identifiability condition $\|\boldsymbol{\alpha}\| = 1$. It is easy to check that

$$\ln \frac{p(ij) p(00)}{p(i0) p(0j)} = \alpha_i \beta_j$$

so that some structure is being imposed on the ln odds ratios. Only the $\boldsymbol{\alpha}$- and $\boldsymbol{\beta}$-parameters involve nonlinearity. The local approximation at $\boldsymbol{\alpha}_0$ and $\boldsymbol{\beta}_0$, expressed in terms of $\boldsymbol{\theta}_1$ and $\boldsymbol{\theta}_2$, is

$$q(ij) = \exp\left[(\boldsymbol{\theta}_A' + \boldsymbol{\theta}_1'\{\mathbf{I} - \boldsymbol{\alpha}_0 \boldsymbol{\alpha}_0'\} \boldsymbol{\beta}_0' \mathbf{u}_B(j)) \mathbf{u}_A(i) + \{\boldsymbol{\theta}_B' + \boldsymbol{\alpha}_0' \mathbf{u}_A(i) \boldsymbol{\theta}_2'\} \mathbf{u}_B(j) - \delta\right] p(ij)$$

where $p(ij)$ corresponds to the values $\boldsymbol{\theta}_A = \mathbf{0}$, $\boldsymbol{\theta}_B = \mathbf{0}$, $\boldsymbol{\alpha} = \boldsymbol{\alpha}_0$, $\boldsymbol{\beta} = \boldsymbol{\beta}_0$.

Since LAMDA only solves the ML problem for an exponential model, each run will form one step in the iterative procedure, and this could become tedious because new values of $\mathbf{u}(y, \boldsymbol{\alpha}_k)$ and $p^{\boldsymbol{\alpha}_k}$ need to be entered for each iteration. Nonetheless, this method may well be faster than writing a computer program specifically tailored for the model under consideration.

PROBLEMS

VI.1. Let $\mathcal{Y} = \{0, 1, \ldots, r\}$ $(r > 2)$, and consider the covariate model

$$q_x^{\boldsymbol{\alpha}, \beta}(y) = \exp[\boldsymbol{\alpha}'\mathbf{u}(y) + \beta x v(y) - \delta]$$

where u_i is the indicator of i for $1 \leqq i \leqq r$, and $v = \Sigma_i u_i$.

1. Show that this is not a linear exponential model, so that it cannot be fitted directly by LAMDA.
2. Define

$$p^{\boldsymbol{\theta}}(y) = \exp\left[\sum_{i=2}^{r} \theta_i u_i(y) - \delta\right] \qquad \text{on} \quad \{y \in \mathcal{Y} : y \geqq 1\}$$

Show that

$$M\left[v : q_x^{\boldsymbol{\alpha}, \beta}\right] = \frac{e^{\beta x}(\Sigma e^{\alpha_i})}{1 + e^{\beta x}(\Sigma e^{\alpha_i})}$$

and

$$M\left[\mathbf{u} : q_x^{\boldsymbol{\alpha}, \beta}\right] = M\left[v : q_x^{\boldsymbol{\alpha}, \beta}\right] M\left[\mathbf{u} : p^{\boldsymbol{\theta}}\right]$$

where $\theta_i = \alpha_i - \alpha_1$ $(2 \leqq i \leqq r)$.

3. Let $\hat{\tau}$, $\hat{\beta}$ be the MLEs in the logistic regression of v on x, so that

$$\frac{e^{\hat{\tau} + \hat{\beta} x}}{1 + e^{\hat{\tau} + \hat{\beta} x}}$$

is the fitted probability of $v = 1$ given covariate x. Let $\hat{\boldsymbol{\theta}}$ be the MLE in the model $\{p^{\boldsymbol{\theta}}\}$ with respect to the data for which $y \geqq 1$. Define $\hat{\alpha}_1 = \hat{\tau} - \ln(1 + \Sigma_{i \geqq 2} e^{\hat{\theta}_i})$ and $\hat{\alpha}_i = \hat{\theta}_i + \hat{\alpha}_1$ for $2 \leqq i \leqq r$. Show that $\hat{\beta}$ and $\hat{\boldsymbol{\alpha}}$ are the MLEs in the original model.

VI.2. Consider a Markov model for $S = \{-1, 0, 1\}$. We would like the conditional transition distributions to be of the following form:

		State at Time m		
		-1	0	1
State	-1	$\exp[-\delta_{-1}]$	$\exp[\alpha - \delta_{-1}]$	0
at Time	0	$\exp[\beta - \delta_0]$	$\exp[-\delta_0]$	$\exp[\alpha - \delta_0]$
$m - 1$	1	0	$\exp[\beta - \delta_1]$	$\exp[-\delta_1]$

where

$$\delta_0 = \ln[1 + e^{\alpha} + e^{\beta}]$$

$$\delta_{-1} = \ln[1 + e^{\alpha}]$$

$$\delta_1 = \ln[1 + e^{\beta}]$$

Explain how this model might be fitted to some data using LAMDA.

VI.3. Let $\mathcal{Y} = \{0, 1, \ldots, m\}$, and let $\mathbf{u}$ consist of the indicators of $1, 2, \ldots, m$. For each $y \in \mathcal{Y}$ let $f_y(x)$ be a probability density. Suppose that

1. for each $y \in \mathcal{Y}$ there is a random sample of size $N(y)$ from a population of x having distribution f_y;
2. $\ln[f_y(x)/f_0(x)] = \boldsymbol{\beta}_y'\mathbf{t}(x)$ for known $\mathbf{t}$ and unknown $\boldsymbol{\beta}_y$;
3. the population distribution π on $\mathcal{Y}$ is known.

Then explain how the $\boldsymbol{\beta}_y$ could be estimated using a linear exponential model. Show that assumption 2 holds for the normal case

$$f_y(\mathbf{x}) = \frac{\exp\left[-\frac{1}{2}(\mathbf{x} - \boldsymbol{\mu})'\mathbf{A}^{-1}(\mathbf{x} - \boldsymbol{\mu})\right]}{(2\pi)^{\#\mathbf{x}/2}|\det \mathbf{A}|^{1/2}}$$

and the multinomial case

$$f_y(\mathbf{x}) = \frac{r!}{x_1!x_2! \cdots x_r!} \prod_{i=1}^{k} p_i^{x_i}$$

If an observation were made from $\sum_y \pi(y) f_y(x)$, how should one decide from which population f_y the observation came?

VI.4. Suppose that $\mathcal{Y} = \mathcal{Y}_1 \times \mathcal{Y}_2$, so that a typical cell is $y = (y_1, y_2)$. Consider the model

$$q^{\boldsymbol{\theta}_1, \boldsymbol{\alpha}}(y) = \exp\left[\boldsymbol{\theta}_1'\mathbf{u}_1(y_1) + \boldsymbol{\alpha}'\mathbf{v}(y_1, y_2) - \delta(\boldsymbol{\theta}_1, \boldsymbol{\alpha})\right] q_1(y_1) q_2(y_2)$$

Suppose that y_2 cannot be observed, so that inference must be carried out in terms of the marginal model

$$q^{\boldsymbol{\theta}_1, \boldsymbol{\alpha}}(y_1) = \sum_{y_2} q^{\boldsymbol{\theta}_1, \boldsymbol{\alpha}}(y_1, y_2)$$

Discuss the identifiability of $\boldsymbol{\alpha}$, and show that the local approximation can be written

$$q^{\boldsymbol{\theta}}(y_1) = \exp[\boldsymbol{\theta}_1'\mathbf{u}_1(y_1) + \boldsymbol{\theta}_2'\mathbf{v}(y_1, \boldsymbol{\alpha}_0) - \delta(\boldsymbol{\theta})]q^{\mathbf{0}, \boldsymbol{\alpha}_0}(y_1)$$

where

$$\mathbf{v}(y_1, \boldsymbol{\alpha}) = M\left[\mathbf{v} : q_{y_1}^{\mathbf{0}, \boldsymbol{\alpha}}\right]$$

and $q_{y_1}^{\mathbf{0}, \boldsymbol{\alpha}}$ refers to the conditional distribution of y_2 given y_1.

VI.5. Fix $q \in \mathcal{P}$ and $\mathbf{u}$, and consider the model

$$\{p \in \mathcal{P} : M[\mathbf{u} : p] = M[\mathbf{u} : q]\}$$

Explain an iterative procedure by which the ML distribution in this model could be computed. Using an appropriate parametrization of the model, explain how the MLE of the parameter and its covariance matrix might be obtained.

VI.6. Reconsider Problem III.14. Determine the model of the type $W, B|A$ which results from conditioning the model given there, and fit this model to the data. Compare your results with those you obtained earlier.

VI.7. Reanalyze the data of Chapter V, Section B.7, using a model of the type $D|X, A, S$. Explain what restrictions, if any, must be placed on the functions of (X, A, S) in order to have a valid analysis. (Remember that the sampling plan here was retrospective.)

VI.8. Consider an A, B, C-cross-classification. Let $C(ij)$ be such that $C(+j) = 0$ for each j. Show that if p is in $AC|B$, then p is collapsible across C with respect to $\Sigma_{ijk} C(ij)\ln p(ijk)$.

VI.9. Consider the data of Table I.9.

1. Fit a Markov model, using saturated models for the (S, D, A) initial distribution and transition tables, and test that the model holds.
2. Repeat the procedure, using the independence model for the transition table, and thus test that childbirth and adoption patterns are independent from one child to the next.
3. Let the covariate x denote the current size of the family when the transition took place. Fit a Markov model containing terms in x, and test that family size has no effect.

VI.10. For $i = 1, 2$, let q_i be Binomial with k_i trials and success probability $\frac{1}{2}$. Consider the model

$$q^{\boldsymbol{\theta}}(ij) = \exp[\theta_1 i + \theta_2(i + j) - \delta]\, q_1(i) q_2(j)$$

Determine the form of the marginal models $q^{\boldsymbol{\theta}}(i+)$ and $q^{\boldsymbol{\theta}}(+j)$. Assuming that θ_2 is a nuisance parameter, form the appropriate conditional model, and suggest how inference about θ_1 might be carried out.

VI.11. Some preparation is required for this problem. If $z_1, \ldots, z_n$ are independent random variables whose probability distributions depend on a parameter $\boldsymbol{\theta}$,

$$P[z_i \in A] = \begin{cases} \sum\{f_i(\zeta; \boldsymbol{\theta}) : \zeta \in A\} & \text{(discrete case)} \\ \int_A f_i(\zeta; \boldsymbol{\theta})\, d\zeta & \text{(continuous case)} \end{cases}$$

then the log likelihood of the sample is

$$L(\boldsymbol{\theta}) = \sum_i \ln f_i(z_i; \theta)$$

Under regularity conditions, the ML estimate $\hat{\boldsymbol{\theta}}$ is found as the unique solution of the likelihood equation

$$\frac{\partial}{\partial \boldsymbol{\theta}} L(\boldsymbol{\theta}) = \mathbf{0}$$

and then an appropriate estimate of the covariance matrix of $\hat{\boldsymbol{\theta}}$ is given by

$$-\left[\frac{\partial^2}{\partial \boldsymbol{\theta}\, \partial \boldsymbol{\theta}'} L(\hat{\boldsymbol{\theta}})\right]^{-1}$$

This is called general maximum likelihood.

A standard result from matrix theory is that if we partition a square symmetric matrix as

$$\begin{bmatrix} \mathbf{A} & \mathbf{B} \\ \mathbf{B}' & \mathbf{D} \end{bmatrix}$$

then that part of the inverse matrix corresponding to $\mathbf{D}$ is $(\mathbf{D} -$

$\mathbf{B}'\mathbf{A}^{-1}\mathbf{B})^{-1}$. This permits one to compute the estimated covariance matrix of a subvector of $\hat{\boldsymbol{\theta}}$.

1. Let $N(y)$ be independent Poisson random variables for $y = 1, 2, \ldots, r$, with parameters $b(y)\exp[\alpha + \boldsymbol{\theta}'\mathbf{u}(y)]$, where $b(y)$ and $\mathbf{u}(y)$ are known and α and $\boldsymbol{\theta}$ are parameters. Suppose we make the identifications $\hat{p}(y) = N(y)/N(+)$, $q(y) = b(y)/b(+)$, and determine $\hat{\boldsymbol{\theta}}$ as the MLE in $\mathcal{E}[q:\mathbf{u}]$ and $\hat{\mathbf{C}}$ as its covariance matrix (as reported by LAMDA). Show that $\hat{\boldsymbol{\theta}}$ and $\hat{\mathbf{C}}$ are numerically the same as would have been obtained by applying general maximum likelihood to the original Poisson variables.
2. Let $T(yi)$ $[1 \leqq i \leqq N(y)]$ be independent exponential random variables with parameters $\exp[\alpha + \boldsymbol{\theta}'\mathbf{u}(y)]$, where $\mathbf{u}(y)$ is known and α and $\boldsymbol{\theta}$ are parameters. (T is exponential with parameter λ if

$$P[T \leqq t] = \int_0^t \lambda e^{-\lambda v}\, dv = 1 - e^{-\lambda t}$$

and so the log likelihood of a single observation T is $\ln \lambda - \lambda T$.) Suppose we make the identification $\hat{p}(y) = N(y)/N(+)$, $q(y) = T(y+)/T(++)$, and let $\hat{\boldsymbol{\theta}}$ and $\hat{\mathbf{C}}$ be the MLE and its covariance matrix in $\mathcal{E}[q:\mathbf{u}]$. Show that again $\hat{\boldsymbol{\theta}}$ and $\hat{\mathbf{C}}$ are numerically equal to the results of applying general maximum likelihood to the original exponential variables.
3. Suppose that in part 2 we replace $T(yi)$ by $S(yi) = \min\{T(yi), \tau(yi)\}$, where $\tau(yi)$ are fixed, known values [the $S(yi)$ are called censored versions of the $T(yi)$]. Show that if we change $N(y)$ to mean the number of uncensored observations [that is, the number for which $S(yi) < \tau(yi)$], and use the approximate likelihood

$$L(\lambda) = \begin{cases} \ln \lambda - \lambda S & \text{(if uncensored)} \\ -\lambda S & \text{(if censored)} \end{cases}$$

then the conclusion of part 2 remains true.

REMARKS

Some alternatives to the Newton–Raphson algorithm are called iterative proportional fitting (IPF) or scaling; see Bishop et al. (1975), Gokhale

(1971), and Darroch and Ratcliff (1972). These methods generally fit the frequencies $n\hat{q}$ directly, and then solve for the parameters, whereas the Newton–Raphson method determines the parameter estimates, from which $n\hat{q}$ can be computed. IPF can be better than the Newton–Raphson method for large problems, since it does not need to invert matrices, but IPF is restricted to hierarchical log-linear models on cross-classifications, and has no capacity for fitting covariate models.

An additional note on the use of **LXM** is that one may expect better results if all of the u- and t-functions are in approximately the same range. Thus for a linear term t_A taking values between -10 and 10, it would be better to use $t_A^2/10$ and $t_A^3/100$ in place of t_A^2 and t_A^3. If the parameter estimates are needed, then one would divide those reported by **LXM** by 10 and 100, respectively. When parameter estimates are not required, it is not a bad idea to use standardized sets of functions.

For more information on Markov models, see Billingsley (1961), Anderson and Goodman (1957) and Kullback, Kupperman and Ku (1962).

The conditional-inference result (Proposition 32) was more or less established in Anderson (1972). Other important references are Breslow (1976) and Breslow and Powers (1978).

Some of the material on homogeneity is classical, and some has been patterned after Murthy and Gafarian (1970). The method for obtaining P-values is from Robbins and Pitman (1949). See also the collapsibility references in the Remarks at the end of Chapter IV.

It may be worthwhile to point out that the collapsibility problem really has at least two aspects. On the one hand, given ample data, one wants to know whether it is legitimate to collapse the table; that is, would it be misleading to show only the collapsed table? The approach taken in the literature is to specify models which imply collapsibility, and presumably one would fit the model, and if it fitted well, collapse the table. This would be more or less equivalent to performing a collection of the tests presented here (except that our definition of collapsibility is more general than the classical one). The other aspect is that one may have too few data to have confidence in any test based on asymptotic theory, but there has been no formulation of how much information we should require before going ahead with such a test.

APPENDIX A

Using LAMDA

INSTRUCTIONS

Many of the computations which need to be performed on discrete data are tedious and error-prone if they are done by hand or by pocket calculator. Moreover, the analysis of data by the maximum-likelihood and minimum-discrimination-information principles, which was introduced in Chapter III, Section A.3, involves numerical work which is totally impractical for a machine limited by the speed at which a human can punch keys. Thus a computer program tailored for discrete data analysis is a necessity as well as a convenience, and LAMDA (Linear And Manipulative Data Analysis) is such a program. Instructions for acquiring a copy of LAMDA may be found in the preface. Other programs that are directed towards discrete data are mentioned in the Remarks of Chapter V.

Typically an analysis using LAMDA will consist of four stages. The first is to think through the kind of analysis that is desired, which means considering what models are to be fitted to the data. The second step is to work out the LAMDA coding which will fit the models. Since the data will often not be in correct form, you may need to do some preparatory transformations or other manipulations. The third step is to run the program, and frequently it will be necessary to return to step two to reconsider the LAMDA coding, or perhaps to step one to rethink what the purpose of the overall analysis is. After one or more successful runs, at step four you may want to produce clean computer output displaying the results, which can be reproduced in a report.

1. Introduction to LAMDA

On some computer systems it will be possible for you to use LAMDA interactively, that is, each command will be executed as you type it in, while on other systems only batch processing is allowed. In a batch run, each command will be punched on a card or stored in a file in card-image format, and then LAMDA will be called in to execute all the instructions at once. For runs of any complexity batch processing is recommended. However, for the purpose of describing and illustrating LAMDA we will speak as if the processing were interactive.

Assuming now that you have learned from the computer consultants at your installation how to call up LAMDA, the next goal is to understand in general terms what it can accomplish. First of all, you have to imagine a file that contains the data you want to analyze. This is called the current file, or *C-file* for short, and the next section will cover methods for getting data into it.

The *C*-file is a rectangle having a certain number of rows and columns. Usually, but not always, each row will correspond to one unit in your data, that is, one subject, one case, one cell in a table, or one subsample, and the values stored in different columns along the row will be values of different variables associated with the unit. Thus, the value at the intersection of the *i*th row and *j*th column is the value of variable *j* associated with case *i*, as shown in Figure A.1.

One purpose of LAMDA is to make it fairly easy to move the data in the *C*-file around, reorganize them, transform them, compute some simple statistics, and plot interrelationships among the variables. It will be possible to take the reorganized data and output them to another file, which you may want to save, or perhaps feed as input to another program. You will need to learn the vocabulary of LAMDA (see Appendix D), and be aware of the important distinction between active and passive commands.

For convenience, all of the commands in LAMDA are of identical form, which is

*xx***IOP***yyyyy*

where *xx* is a two-digit number (both can be blanks), **IOP** stands for the name of an operator, and *yyyyy* stands for a string of information whose

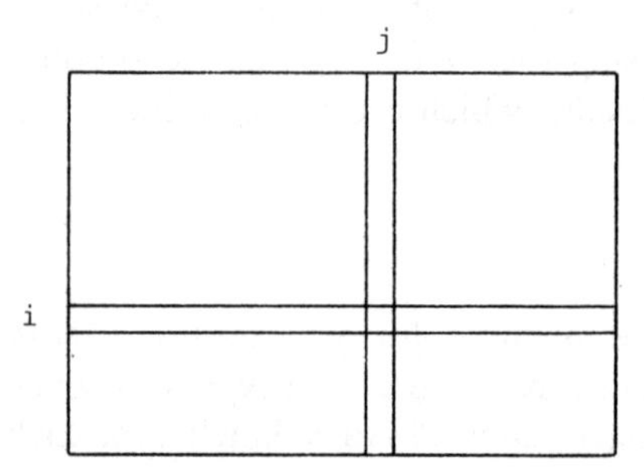

Fig. A.1 Conceptualization of the *i*th *row and* *j*th *column of the C-file.*

meaning depends on the nature of the operator. It is important to remember the *xx* part of the command. For example, to get out of LAMDA and back to the operating system, you must enter

END

and although it does not show, there are two blanks in front of **END** (that is, *xx* = two blanks, **IOP** = **END**, *yyyyy* = blanks).

2. Entering Data

The purpose of this section is to explain how the *C*-file may be filled with data. There are several ways to do this, and we will begin with the easiest.

Suppose you have entered LAMDA and you now have a sheet of paper in front of you with the data you would like to transfer to the *C*-file. The *C*-file is, at the beginning, a shapeless mass, because the numbers of rows and columns are not yet specified. The command for specifying them is

DIM *yy zz*

Note again that there are two blanks before **DIM**, in conformity with the *xx*__IOP__*yyyyy* format. In **DIM** *yy zz* the *yy* part of the information field specifies the number of rows, and the number of columns is given by *zz*. For example,

DIM 115 3

sets 115 rows and 3 columns. At this point the *C*-file has shape but is empty. In order to begin entering data do

ENT

This causes LAMDA to consider subsequent lines to be data for storage in the *C*-file. The data are taken from the lines you enter and written one row at a time into the *C*-file, just as the letters are written across this page. It is not important to put each case on a different line; you may even put several rows on each line you enter, or break off in the middle of a case and resume it at the beginning of the next line. What is crucial is that the data be entered in the correct order, along rows. Since you have already told LAMDA how many rows and columns there are to be in the *C*-file, it knows how many values to expect. When it has read enough in, it will then expect further commands. An error message at this point probably means that you have miscounted the number of entries required to fill the *C*-file.

Example

```
    DIM 5 3
    ENT
1 5 7 2 3 7 3 4 1 4 5
8 5 2 3
```

The *C*-file then looks like

```
1 5 7
2 3 7
3 4 1
4 5 8
5 2 3
```

The second way of entering data is essential if the data set is large, since you do not want to spend several hours entering several thousand values each time you do an analysis. The second approach therefore assumes that you have your data already on a system file, catalogued in your computer operating system. LAMDA refers to this file as the non-rewindable file, or *N-file*, because it is never rewound at any point by LAMDA. The command which causes your data to be read into the *C*-file is

*xx*N*C*yyyyy*

in which *xx* is the number of variables (that is, columns) for each case, and *yyyyy* is the floating-point FORTRAN format under which each case is to be read. If you do not know what a FORTRAN *F*-type format specification is, then read about it in an early section of a FORTRAN manual or obtain help from computer consultants. On some systems it will be possible to use some special symbols for *yyyyy* to indicate a free-field format.

Example

```
12N*C(4X,3F2.0,2X,3F2.1,6F1.0)
```

In this example, there are 12 variables per case. The data in the *N*-file occur in the form of four blanks, three two-digit numbers, two blanks, three two-digit numbers with one digit after the decimal, and six one-digit numbers.

You will have noticed that there is no specification of the number of cases (rows) which should be read from the *N*-file. If the computer system you use has been competently constructed, your file will have an end-of-file mark at the end, and when LAMDA hits this mark it will stop reading cases. Another way to accomplish this is to put at the end of your data

```
  EOF
```

(with two leading blanks), and LAMDA will stop reading when it comes to this card image.

If you enter your data from the *N*-file, you may not know how many cases were on the file you read from. To find out at any time what the dimensions of the *C*-file are, just enter

```
STS
```

and LAMDA will respond with the number of rows and columns.

The one detail necessary to complete the picture is how you get your system file identified with LAMDA's *N*-file. This varies from one system to another, and so again you will need to talk to a consultant to find out how your local system works. What you will need to know is that your file must be associated with logical unit 8. This means that when LAMDA reads data from the *N*-file it does so with the statement "READ(8,1000)."

An additional convenience feature of LAMDA permits you to merge files which were created under different formats. For example, in

```
6N*C(6F2.0)
6N+C(6F3.0)
6N+C(2F3.0,2X,4F1.0)
```

the first line commences reading the *N*-file under the indicated format until the first EOF or system end-of-file mark. The second line continues reading the *N*-file under the new format until the second EOF or system end-of-file, and finally the last line continues reading under the third format.

3. Modifying Data

Assuming that you have entered your data into the *C*-file, in this section we will see how it can be modified in certain ways. First, the question arises whether or not you really got your data in correctly. In order to find out, do

```
PRT
```

which will cause the contents of the *C*-file to be printed out. It is good practice to obtain a complete copy of the *C*-file no matter how large or small it is; otherwise, when you find an apparent anomaly in the analysis, you will have no way to go back and assure yourself that the source of the problem was not a data-entry error.

If you have already screened your data file and are convinced that it has been purged of error, you may still want to see the first few lines, which you can do with

PRS *yy*

where *yy* is the number of rows you want to see.

If upon **PRT**ing the *C*-file you discover an error in one specific value, you may correct it by determining the row *yy* and column *zz* in which it lies, and doing

CHG *yy zz ww*

where *ww* is the correct value. If a particular column, *yy*, contains so many errors that you would like to reenter the entire column, then do

COL *yy*

and on the subsequent lines enter the numbers to replace column *yy*. Be sure to enter the values according to their rows. Similarly, if row *yy* needs to be replaced entirely, do

ROW *yy*

and on subsequent lines enter the new values, in the order of their columns.

The **ROW** and **COL** commands are dissimilar in a certain respect. If you enter **ROW** *yy* and *yy* is larger than the number of rows in the *C*-file, then the *C*-file is enlarged to contain your new row. You must be careful, since if there are 20 rows and you enter **ROW 25**, then at the end any garbage left in the computer's memory will occupy lines 21 through 24 of your *C*-file. If you enter **COL** *yy* and *yy* is larger than the number of columns in the *C*-file, you will have a mess. (Columns can be appended by use of the **SEL** command, to be discussed shortly.)

Example

```
DIM 4 2
ENT
```

```
1 2 3 5 5 6 7 8
  PRT
1.0000  2.0000  3.0000  5.0000
5.0000  6.0000  7.0000  8.0000
  CHG 1 4 4
  PRS 1
1.0000  2.0000  3.0000  4.0000
  COL 3
2 6
  PRT
1.0000  2.0000  2.0000  4.0000
5.0000  6.0000  6.0000  8.0000
  ROW 3
9 10 11 12
  PRT
1.0000  2.0000  2.0000  4.0000
5.0000  6.0000  6.0000  8.0000
9.0000 10.0000 11.0000 12.0000
```

Up to this point the commands presented have been for the purpose of changing values or adding new values directly by inputting them from your terminal. There are other ways of modifying the data discussed in Sections 7 and 8 below. The remainder of this section is devoted to a special class of commands called transformations.

Transformations are used to modify an entire column of the *C*-file at once. For example, suppose that you want to make column *xx* equal to the sum of columns *yy* and *zz*. The proper command is

*xx*ADD *yy zz*

The numbers *xx*, *yy*, and *zz* are usually columns already in the *C*-file. The complete set of transformations may be found in the dictionary (Appendix D). The general rule is that *xx* refers to the column where the result of the computation is to be stored, while *yy*, *zz* refer to columns which will be used in the calculation. These need not be distinct. For example

*xx***MPY** *xx yy*

has the effect of replacing variable *xx* with its product by variable *yy*, and

*xx***MPY** *xx xx*

replaces variable *xx* with its square. In the case of transformations it is possible to reference columns not in the *C*-file. These act as temporary variables that are not retained. For instance, assume that there are three columns in the *C*-file, and do

```
4ADD 1 2
5SUB 4 3
3MPY 4 5
```

The variables 4 and 5 are only temporary, and at the end of the computation the *C*-file will still have only three columns. Note that the special column number **0** can be used on the right side of a transformation (see the **SEL** command below).

The class of transformations has a feature that has not been discussed yet. When you do a **ROW** command the modification takes place immediately, but when you do **ADD** nothing actually happens. In fact, the transformations are merely stored up for execution at some future time. After you have entered a list of transformations, you can cause them to be executed by doing

```
C*C
```

As one might expect, this command causes the *C*-file to be written onto itself, but the important thing is that as this is done all pending transformations are carried out. For example, if there are four variables in the *C*-file and variable 4 is to be replaced by the sum of variables 1 through 4, then do

```
4ADD 4 3
4ADD 4 2
4ADD 4 1
 C*C
```

The first line adds variables 3 and 4 and stores the result in variable 4. The second line then adds the current variable 4 (which is now equal to the sum of the old variables 3 and 4) to variable 2 and stores the result in variable 4, and similarly for the third line. None of these additions is actually carried out until the **C*C** command is given. The commands **N*C** and **N+C** also cause all pending transformations to be done.

It is important to remember that the transformations are carried out in the order they are given, and that this order is usually not arbitrary.

Another type of command which is extremely useful is the select command, which is written

*xx*SEL*yyyyy*

Here *yyyyy* stands for a string of column numbers, and *xx* specifies how many numbers there are in this string. For example, in a *C*-file with four variables

```
4SEL 4 3 2 1
 C*C
```

causes the order of the variables to be reversed. Note that **C*C** is required for **SEL** to be carried out. The **SEL** command may either decrease, increase, or leave unchanged the number of columns in the *C*-file. For example,

```
3SEL 1 2 3
```

would have dropped the fourth column in the above example, while

```
6SEL 1 2 3 4 4 2
```

would have increased the number of columns to six. Here the new variable 4 and new variable 5 will both equal the old variable 4, and new variable 6 will equal old variable 2. Note that **SEL** can be used to add a column to the *C*-file, since after the **SEL** command below

```
5SEL 1 2 3 4 4
 C*C
 COL5
```

there is a column 5 for **COL** to change. It is implicit in any occurrence of **C*C**, **N*C**, or **N+C** that if no **SEL** command has been given then LAMDA leaves the columns alone.

There are two special column numbers which can be used on the right side of **SEL**. One is **0**, which designates a column of ones. The other is **−1**, which specifies a sequential numbering of the rows of the *C*-file. The special variable **0** can also be used with transformations.

4. Outputting Data

For several reasons it is useful to be able to save the *C*-file either permanently or temporarily. For this purpose there is another LAMDA file, called the rewindable file, or *R-file*. As its name implies, the *R*-file is rewound by LAMDA every time it is read or written on. In order to write the *C*-file to the *R*-file, do

```
C*R---
```

This causes the data in the *C*-file to be written onto file 9. If you want to save the results for later, you will have to associate one of your permanent files with file 9, just as your input file had to be associated with file 8.

The output format is assumed to be `(8F10.4)`. If you want another floating-point format, do

C*R*yyyyy*

where *yyyyy* is your format. After you have written on the *R*-file, you can read it back to the *C*-file with

*xx*R*C ---

where *xx* is the number of variables per case. Again the `(8F10.4)` format is assumed unless you do

*xx*R*C*yyyyy*

with your format *yyyyy*. Remember that the *R*-file is always rewound before reading or writing.

There is another feature of the `C*R` command. Under ordinary circumstances all input and output formats for LAMDA are *F*-type. The only exception is the `C*R` command with an integer specification. To accomplish this, do

INT*yyyyy*
C*R*zzzzz*

Here *yyyyy* is a string of integers specifying how many places to move the decimal in each of the columns before writing them out to the *R*-file, and *zzzzz* is an *I*-type format. For example,

```
INT 0 0 1 -2
C*R(4I5)
```

assumes four variables in the *C*-file, and when these are written out under the *I*-format the first two will be left unchanged, the third will be multiplied by 10 (that is, its decimal point will be shifted one digit to the right) and the fourth will be divided by 100 (that is, its decimal point will be shifted two digits to the left).

5. Active and Passive Commands

All but a few of the LAMDA commands can be classified as either active or passive, and this distinction is quite important in practice. The two ways in which they differ are:

1. An active command causes something to be done as soon as it is given, while passive commands have no immediate effect.
2. An active command erases all pending passive commands after carrying them out.

Active commands are like verbs (they specify an action), while passive commands are like adverbs (they modify or add meaning to the verbs). Transformations form a special subclass of the passive commands. To illustrate this, in

```
3MPY 2 3
 C*C
2SEL 2 3
 C*C
```

the first occurrence of `C*C` (active) erases the `MPY` (passive) after carrying out the multiplication. Thus, when the second `C*C` is executed it sees only the `SEL` command. If this were not the case, the second `C*C` would cause `MPY` to be executed with a nonexistent third variable.

If you have entered a string of passive commands, but not executed them, and you want to wipe them out, do

```
NOP
```

which means "no operation" and is an active command whose sole effect is to erase the passive commands pending. Inserting `NOP` in a batch run is useful because it marks a point in the program at which no passive commands are pending.

Conversely, suppose that you have a long sequence of transformations you want implemented as the *N*-file is written onto the *C*-file, and there are some more data on the *N*-file in a new format, and you would like to apply the same sequence of transformations to the second part of the data, too. To avoid having to type the transformations again, do

```
(sequence of transformations)
3N*C(3F1.0)
 RPT
3N+C(3F2.0)
```

The RPT command repeats the passive commands in effect just before the active command which precedes it. RPT must immediately follow the command it is to modify—otherwise all passive commands will be lost.

Transformations are distinguished from other passive commands in that they are cumulative. As was mentioned before, when a transformation is carried out, it is assumed that all preceding transformations have been executed. This does not happen with the other passive commands; in fact, the passive nontransformations are arranged so that at most one of each type will be relevant at any one time, and as a consequence whenever a passive nontransformation of a certain type is given, it overwrites any preceding command of that type. Thus in

```
3SEL 1 2 3
3SEL 3 2 1
 C*C
```

it is the second SEL which has effect.

6. Screening Data

A very useful command is DIS, which chooses which variables will be displayed. For example

```
3DIS 2 4 1
 PRT
```

would cause columns 2, 4, and 1 (in that order) to be printed out from the *C*-file.

The command

*xx*SUM*yyyyy*

produces a list of (1) the *xx* variable numbers given in *yyyyy*, (2) the means and standard deviations of these variables, and (3) their covariance and inverse covariance matrices. (See Section 8 and Example 2 below, and Chapter II, Section A.1, for details.)

A feature of LAMDA which is convenient for labeling the output is that if *xx*IOP are left blank, then *yyyyy* is printed out. The printing field begins, thus, in the sixth column. One needs to exercise some care, because on most systems a 1 in the first printing position causes the printer to move to the top of the next form, while a 0 causes double spacing, and a blank, single spacing. For example

```
3N*C(3F4.0)
     1ECONOMIC DATA
     0
 PRT
```

would read in the data, move to the top of the next page and print the heading ECONOMIC DATA, double-space, and then print the data. This command (in which *xx*IOP is blank) is neither active nor passive; it does not erase pending passive commands.

The final command which is useful for screening permits several variables to be plotted against a distinguished variable, which can be useful for visualizing relationships. It is possible to construct a plot by just doing

PLT

This will cause the printing of a plot of all data cases with variable 1 plotted along the horizontal axis and the variables 2 through 10 plotted along the vertical axis. Variable 2 will be plotted with the symbol A, variable 3 with B, and so on down to variable 10 with I.

The default size of the plot will just fit on standard $8\frac{1}{2}$-by-11-inch paper, with horizontal and vertical scales determined by the data. You may want to have a plot of a different size, with different vertical scales and plotted using different symbols. All of these may be specified by the appropriate commands. The command

HOR *xx yy zz*

specifies that the horizontal scale will have *xx* spaces with a scale running from *yy* to *zz*. Likewise,

VER *xx yy zz*

specifies *xx* lines with a scale from *yy* to *zz*. In order to choose which variables to plot, use

*xx*DIS*yyyyy*

where *xx* is the number (≤ 10) of variables appearing in the string *yyyyy*. The first variable mentioned will be plotted on the horizontal axis, while the remainder will be plotted on the vertical axis. To obtain a different set of plotting symbols, do

xx`SYM`*yyyyy*

where *yyyyy* is a string containing the *xx* symbols to be used for plotting the vertical-axis variables. The string *yyyyy* must follow `SYM` immediately and with no spaces, since a blank is itself a valid printing character.

Example

```
 VER 20 -10  10
 HOR 40 0 80
4SYM+OX-
5DIS 10 1 2 3 4
 PLT
```

Note that there will be one less plotting symbol than the number of variables in `DIS`. Notice also that there are no spaces after `SYM`. If some of the plotting points lie outside the range specified by `VER` and `HOR`, the plot will be enlarged to accommodate them. If this feature is to be suppressed, so that points out of range are ignored, then do

`1PLT`

instead of `PLT`. If a particular plotting position is occupied by more than one point, the symbol nearest the end of the `SYM` list will appear.

7. Creating, Sorting, and Ordering Data

In the preceding sections it was assumed that the data were entered either by hand or from a preexisting file. It is also possible to create data out of thin air with LAMDA. One way to do this is with the "create" command

xx`CRT`

which specifies that *xx* variables are to be created. It is assumed that the *C*-file is empty, since `CRT` begins writing at the top of the *C*-file. Exactly *xx* lines must follow `CRT`, and on each line must appear the lowest value for the created variable, the step size, and the highest value, in that order. For example,

xx yy zz

would specify a variable whose lowest value was *xx* and which proceeded in steps of *yy* until the first time *zz* was passed. Thus

```
2CRT
0 1 10
1 5 25
```

would produce two variables, the first running from 0 to 10 in steps of 1 (0, 1,..., 10) and the second from 1 to 26 in steps of 5 (1, 6, 11,..., 21, 26). The first line following `CRT` specifies variable 1, the second variable 2, and so on. The `CRT` command is especially useful in producing the coding for cells in a contingency table.

So far all operations on the *C*-file have been applied to columns. We now come to the only two which apply to rows. The first of these is called `PAK` and can be used to pack the data or sort them into categories. To see how this works, imagine that you have a raw data file with one row for each of 1000 subjects. On each row are three variables that specify the coding for a three-way contingency table. We suppose that there are twenty possible codes, that is, twenty cells in the table. When you read the data into the *C*-file it will have dimensions 1000 by 3, but there is considerable redundancy here if you only need to know the frequencies in the 20-cell table. The first step is to create a new column variable for the purpose of counting cases. Do this with

```
4SEL 0 1 2 3
 C*C
```

which makes the first column a column of 1's. Then upon doing

```
PAK
```

the *C*-file will be reduced to 20 by 4, and the first variable will contain the frequency in the raw data of the code given by variables 2, 3, and 4. If some cells are empty, there will be fewer than 20 rows.

`PAK` and several other commands (`SUM, NRM, STD, FIT, LXM`) treat the first variable differently than the other variables. In the above example, `PAK` used variable 1 to count cases, and so we call variable 1 the *weight variable*. You can specify another variable as the weight variable by doing

xx`WGT`

where *xx* is the column number of the new weight variable. Remember that any active command erases the passive `WGT` command, and thus resets the weight variable to variable 1.

In detail, the **PAK** command has the following effect: Each case is compared with all preceding cases. If a match is found, then both cases are fused into one by adding their **WGT** variables. By combining the **CRT** and **PAK** commands you can create contingency tables from raw data in such a way that empty cells will not be excluded. For example, suppose there are four factors represented in an unsorted *N*-file. Do

```
 5CRT
0 0 0
0 1 2
1 1 3
1 1 2
1 1 7
 5SEL 0 1 2 3 4
 4N+C(4F3.0)
  PAK
```

The **CRT** begins writing the *C*-file by creating the classes into which the data is going to be sorted. The first variable is to be identically 0, since the table is initially empty. As the *N*-file is read in, variable 1 is made equal to 1 for each case, essentially counting the cases. Apparently the first variable in the *N*-file assumes the values 0, 1, and 2, while the second can be 1, 2, and 3, and so on. After the data are added to the *C*-file, the **PAK** command sorts the data which were in the *N*-file into the contingency table created by **CRT**.

The second of the two commands which operate on rows is one that permits reordering of the data. It is of the form

*xx***LEX***yyyyy*

where *xx* is the number of column numbers in the string *yyyyy*. It causes the rows to be reordered lexicographically according to the columns given in *yyyyy*. (If we took $1 = A$, $2 = B$, and so on, then lexicographical ordering is just alphabetization.) For example

```
3LEX 2 1 3
```

would order the rows so that the values of variables 1, 2, and 3 were in order of increasing magnitude, with the values of variable 3 changing the most rapidly and those of variable 2 changing the least rapidly as one proceeded through the file.

8. Standardization and Fitting of Variables

The summary statistics computed by the **SUM** command are calculated as follows. Let $A_i (1 \leq i \leq n)$ stand for the values of one of the variables in the string *yyyyy*. Let W_i be the corresponding values of the **WGT** variable, by default equal to variable 1. Then the weighted mean is

$$\bar{A} = \frac{\sum W_i A_i}{\sum W_i}$$

and the weighted standard deviation is

$$S_A = \sqrt{\frac{\sum W_i (A_i - \bar{A})^2}{\sum W_i}}$$

and these are the values reported by **SUM**. Note that if you want the "unweighted" mean and standard deviation, then you must make sure that the **WGT** variable is the constant 1.

A variable is normed when $\bar{A} = 0$ and $S_A = 1$. The data A_i can be normed by replacing each A_i with the value $(A_i - \bar{A})/S_A$. This is accomplished by doing

*xx***NRM***yyyyy*

in which *yyyyy* is the string of *xx* variables which are to be normed. The weighted means, standard deviations, and covariances are printed out, just as with **SUM**.

Suppose that A_i and B_i are the values of two variables. If A_i and B_i are both normed, and if $\Sigma W_i A_i B_i = 0$, then they form a *standard pair*. A collection of variables is a *standard set* provided all its pairs are standard pairs. The command

*xx***STD***yyyyy*

produces a standard set from the string of *xx* variables *yyyyy*. Each variable in the process is replaced by its corresponding standardized variable. The method by which this is carried out is explained in Chapter II, Section A.2.

When **STD** is used some statistics are printed out. First, all of the quantities produced by the commands **SUM** and **NRM** are printed out. Next a lower triangular matrix of coefficients is given. The procedure which produces the standardization first normalizes the entire set of variables, and then successively finds for each variable the linear combination of the preceding variables which fits the given variable best in the sense of weighted least squares. The reported coefficients in the matrix are the coefficients of the best-fitting linear combination. (These are the regression coefficients of the last standardized variable on the standardized variables which precede it; see Problem II.1).

Another command which accomplishes something similar is

xx`FIT`*yyyyy*

where as usual there are *xx* variables in the string *yyyyy*. First, `FIT` standardizes all but the last named variable in the *yyyyy* string. It then replaces this last variable with the best-fitting linear combination of the preceding standardized variables. (This is the weighted least-squares fit.) The last line of the output from `FIT` contains the coefficients used to produce the fitted variable.

Error Messages. There are only two LAMDA-generated error messages in Version 4.0. The first reports that the `LXM` or `LXX` command was given when the *C*-file exceeded 1000 rows, which is the limit for these commands. The commands are not executed in this situation. The second message is

```
SYNTAX ERROR - LAST INSTRUCTION IGNORED
```

which means that the `IOP` part of *xx*`IOP`*yyyyy* could not be identified. In this case no command is executed.

All other error messages will have been produced by FORTRAN or by the operating system. An error on input unit 10 means that there is a mistake in the *yyyyy* part of an instruction. Rerunning the program with printout statements (six blanks followed by a message) will help to trap the erroneous instruction.

Continuations. Most LAMDA commands must be given completely on one line. The exceptions are: (1) data given after `ENT`, `ROW`, or `COL`, (2) data definition lines given after `CRT`, and (3) the variable numbers given on a `SEL` command may continue onto one additional line.

EXAMPLES

In this section we see three examples which illustrate some of the features of LAMDA. In each case the rows on the left are input and those on the right are output. The line numbers on the input rows are merely for reference, and would not be included in an actual run.

1. A Contingency Table

Create a 4^2 contingency table coding, and print it out:

```
1:          *** RUN I.C.1 ***
2:    2CRT
3:  0 1 3
4:  0 1 3
5:     PRT
```

```
L A M D A  - - -  VERSION 4.0


*** RUN I.C.1 ***
   0.0         0.0
   0.0         1.0000
   0.0         2.0000
   0.0         3.0000
   1.0000      0.0
   1.0000      1.0000
   1.0000      2.0000
   1.0000      3.0000
   2.0000      0.0
   2.0000      1.0000
   2.0000      2.0000
   2.0000      3.0000
   3.0000      0.0
   3.0000      1.0000
   3.0000      2.0000
   3.0000      3.0000
```

Add a first variable, and then put the frequencies from Table I.5 into variable 1:

```
 6:   3SEL 0 1 2
 7:    C*C
 8:    COL 1
 9:  11 18 6 5
10:  19 62 33 40
11:  1 8 9 9
12:  10 28 31 66
```

Read the *C*-file out to the *R*-file. Delete the column variable (variable 3), and pack in order to get the row frequencies; print them out:

```
13:    C*R---
14:   2SEL 1 2
15:    C*C
16:    PAK
17:    PRT
```

```
 40.0000     0.0
154.0000     1.0000
 27.0000     2.0000
135.0000     3.0000
```

Read the original data back from the *R*-file to the *C*-file. Create a new "mobility" variable measuring the change between grandson's and grandfather's status.

```
18:   3R*C---
19:   4SEL 1 2 3 0
20:   4SUB 3 2
21:    C*C
```

Compute and print out the marginal frequencies for the change variable:

```
22:   2SEL 1 4
23:    C*C
24:    PAK
25:    PRT
```

```
148.0000     0.0
 60.0000     1.0000
 46.0000     2.0000
  5.0000     3.0000
 58.0000    -1.0000
 29.0000    -2.0000
 10.0000    -3.0000
```

Compute the mean and variance of the change variable with respect to the empirical distribution in variable 1:

```
26:   1SUM 2
```

```
---VAR-----MEAN-------STDEV--
   2       0.0590      1.2449
-----------------------------
```

Read the original data back in again, save only the frequencies, re-dimension the *C*-file to be a 4^2 table, and print it out:

```
27:   3R*C---
28:   1SEL 1
29:    C*C
30:    DIM 4 4
31:    PRT
```

```
11.0000   18.0000    6.0000    5.0000
19.0000   62.0000   33.0000   40.0000
 1.0000    8.0000    9.0000    9.0000
10.0000   28.0000   31.0000   66.0000
```

Change each column to a probability distribution, and print out again:

```
32:   1RFQ
33:   2RFQ
34:   3RFQ
35:   4RFQ
36:    PRT
37:    END
```

```
0.2683    0.1552    0.0759    0.0417
0.4634    0.5345    0.4177    0.3333
0.0244    0.0690    0.1139    0.0750
0.2439    0.2414    0.3924    0.5500
```

2. Orthogonal Polynomials

Create the functions 1, x, x^2, and x^3 on $x = -4(1)4$:

```
 1:         *** RUN I.C.2 ***
 2:      DIM 9 1
 3:     1SEL -1
 4:     1LIN1 1 -5
 5:      C*C
 6:     4SEL 0 1 1 1
 7:     3MPY 2 3
 8:     4MPY 4 3
 9:      C*C
10:      PRT
```

```
              L A M D A  - - -  VERSION 4.0

          *** RUN I.C.2 ***
             1.0000    -4.0000   16.0000  -64.0000
             1.0000    -3.0000    9.0000  -27.0000
             1.0000    -2.0000    4.0000   -8.0000
             1.0000    -1.0000    1.0000   -1.0000
             1.0000     0.0       0.0       0.0
             1.0000     1.0000    1.0000    1.0000
             1.0000     2.0000    4.0000    8.0000
             1.0000     3.0000    9.0000   27.0000
             1.0000     4.0000   16.0000   64.0000
```

Plot them against x (the numbers on the plot have been connected by hand):

```
11:    5DIS 2 1 2 3 4
12:    4SYM0123
13:     VER 30 0 0
14:     HOR 40 0 0
15:     PLT
```

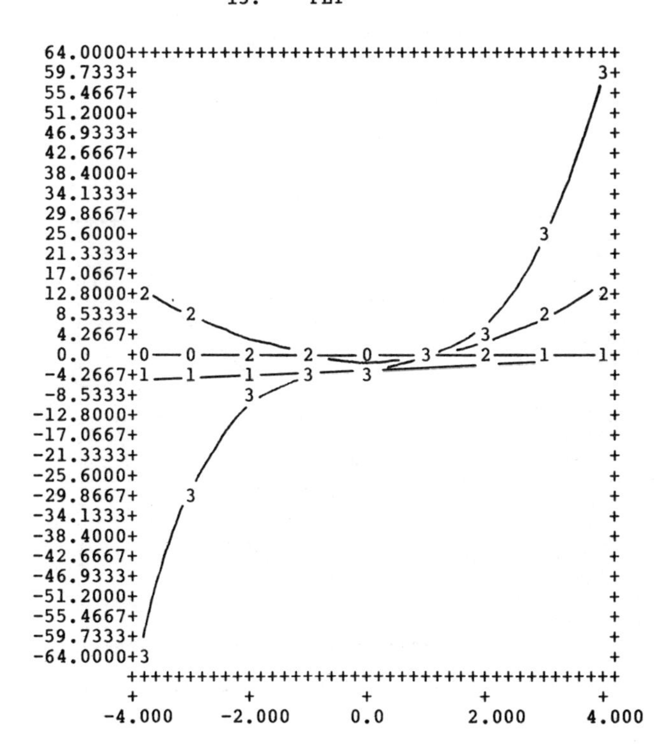

Standardize x, x^2, and x^3, using the constant weights in variable 1. These are the orthogonal polynomials. Plot them against x:

```
16:     5SEL 1 2 2 3 4
17:       C*C
18:     3STD 3 4 5
19:     5DIS 2 1 3 4 5
20:     4SYM0123
21:       VER 20 0 0
22:       HOR 40 0 0
23:       PLT
```

```
---VAR-----MEAN-------STDEV--
   3       0.0         2.5820
   4       6.6667      5.8500
   5       0.0        32.9646

---COVARIANCE---
        3           4           5
  1.0000
  0.0         1.0000
  0.9242      0.0000      1.0000

---INVERSE COVARIANCE---
  6.8603
  0.0000      1.0000
 -6.3406     -0.0000      6.8603

---STANDARDIZATION-----
        3           4           5
  0.0000
  0.9242      0.0000

------------------------------
```

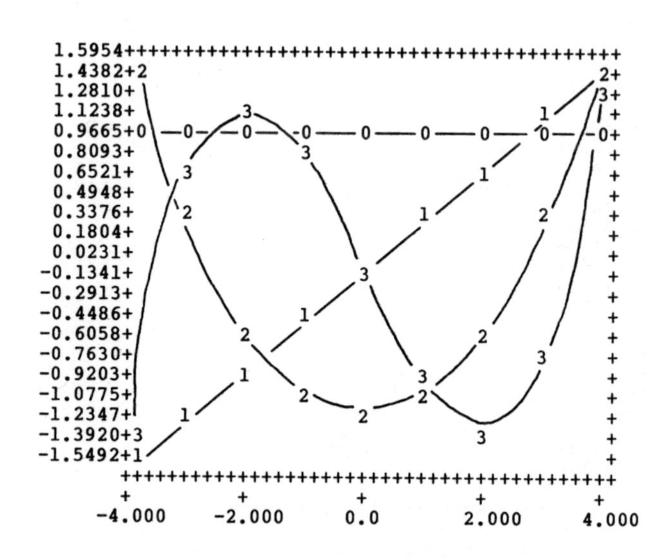

First the mean and standard deviation of each variable are printed out. Then the covariance and inverse covariance matrices of the variables are given (see Chapter II, Section A.1 for the definition of the covariance matrix.). Second, each variable is normed. Third, the variables are standardized (see Chapter II, Section A.2 for the procedure by which this is done), and in the process the weighted regression coefficients of each variable on its predecessors are printed out under **STANDARDIZATION**. A comparison of the plot above with the plot that preceded it should show why the standardized functions are more reasonable for fitting data. Now put a new function in variable 6:

```
24:    6SEL 1 2 3 4 5 0
25:     C*C
26:     COL 6
27:    2 1 0 0 1 2 3 2 2
```

Fit the new function with the orthogonal polynomials, while saving the new function in variable 7:

```
28:    7SEL 1 2 3 4 5 6 6
29:     C*C
30:    4FIT 3 4 5 6
```

```
---VAR-----MEAN-------STDEV--
   3     -0.0000      1.0000
   4     -0.0000      1.0000
   5      0.0000      1.0000

---COVARIANCE---
       3         4         5
  1.0000
  0.0       1.0000
  0.0000   -0.0000    1.0000

---INVERSE COVARIANCE---
  1.0000
 -0.0000    1.0000
 -0.0000    0.0000    1.0000

---STANDARDIZATION-----
       3         4         5
  0.0000
  0.0000   -0.0000

------------------------------

FIT VARIABLE     6
MEAN =  1.4444  ST DEV =  0.9558
COEFFICIENTS

  0.4734    0.3482   -0.6780

------------------------------
```

Note that the **FIT** command first standardizes all the variables but the last (variable 6), but this happens to be redundant in this case, because they were already standardized. The bottom row contains the coefficients of the fitted function with respect to the orthogonal polynomials. Now plot the original function (**O**) and the fitted function (**F**):

```
31:    3DIS 2 6 7
32:    2SYMFO
33:      VER 30 0 0
34:      HOR 40 0 0
35:      PLT
36:      END
```

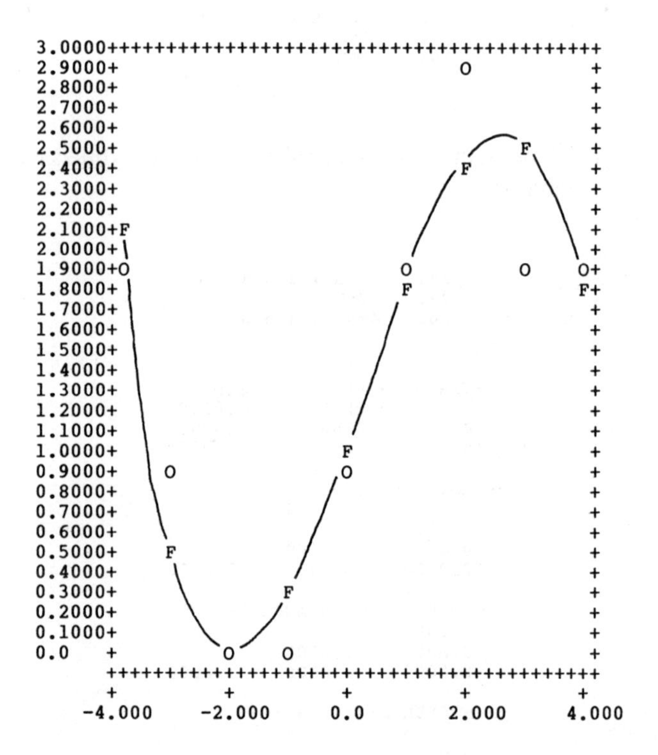

3. Binomial Probabilities

We compute and plot the Binomial probabilities

$$p(y) = \binom{8}{y} \rho^y (1 - \rho)^{8-y} \qquad (y = 0, 1, \ldots, 8)$$

for $\rho = .3(.1).7$. For convenience we use the computational formula

$$p(y) = \binom{8}{y}\left(\frac{\rho}{1-\rho}\right)^y (1-\rho)^8$$

and note that $(1-\rho)^8$ is just the factor required to make the probabilities sum to 1.

Generate the y's:

```
1:         *** RUN I.C.3 ***
2:     DIM 9 1
3:    1SEL -1
4:    1LIN1 1 -1
5:     C*C
```

Put the $\binom{8}{y}$'s in variable 2:

```
6:    2SEL 1 0
7:     C*C
8:     COL 2
9:   1 8 28 56 70 56 28 8 1
```

Set up the variables for the distributions, and use variable 8 as a dummy to help the computation; use the computation formula given above:

```
10:     7SEL 1 2 0 0 0 0 0
11:     3LIN 3 .3 0
12:     4LIN 4 .4 0
13:     5LIN 5 .5 0
14:     6LIN 6 .6 0
15:     7LIN 7 .7 0
16:     8SUB 0 3
17:     3DIV 3 8
18:     8SUB 0 4
19:     4DIV 4 8
20:     8SUB 0 5
21:     5DIV 5 8
22:     8SUB 0 6
23:     6DIV 6 8
24:     8SUB 0 7
25:     7DIV 7 8
26:     3LOG 3
27:     4LOG 4
28:     5LOG 5
29:     6LOG 6
30:     7LOG 7
31:     3MPY 3 1
32:     4MPY 4 1
33:     5MPY 5 1
34:     6MPY 6 1
35:     7MPY 7 1
36:     3EXP 3
37:     4EXP 4
38:     5EXP 5
39:     6EXP 6
40:     7EXP 7
41:     3MPY 3 2
42:     4MPY 4 2
43:     5MPY 5 2
44:     6MPY 6 2
45:     7MPY 7 2
46:      C*C
47:     3RFQ
48:     4RFQ
49:     5RFQ
50:     6RFQ
51:     7RFQ
```

Note that the reason this part of the program appears so long is that the same sequence of steps must be done for each of the distributions to be generated; there is no "looping" capability in LAMDA. Now display the distributions and plot them. The use of **.499999** at line 54 (instead of **.5000**) will have the effect that the plot will round border values down; ordinarily it rounds them up to the next higher line.

```
52:    6DIS 1 3 4 5 6 7          62:     RPT
53:     PRT                      63:    2DIS 1 5
54:     VER 20 0 .499999         64:    1PLT
55:     HOR 44 -1 10             65:     RPT
56:    2DIS 1 3                  66:    2DIS 1 6
57:    1SYM_                     67:    1PLT
58:    1PLT                      68:     RPT
59:     RPT                      69:    2DIS 1 7
60:    2DIS 1 4                  70:    1PLT
61:    1PLT                      71:     END
```

```
   L A M D A  - - -   VERSION 4.0

*** RUN I.C.3 ***
  0.0        0.0576     0.0168     0.0039     0.0007     0.0001
  1.0000     0.1977     0.0896     0.0313     0.0079     0.0012
  2.0000     0.2965     0.2090     0.1094     0.0413     0.0100
  3.0000     0.2541     0.2787     0.2188     0.1239     0.0467
  4.0000     0.1361     0.2322     0.2734     0.2322     0.1361
  5.0000     0.0467     0.1239     0.2188     0.2787     0.2541
  6.0000     0.0100     0.0413     0.1094     0.2090     0.2965
  7.0000     0.0012     0.0079     0.0313     0.0896     0.1977
  8.0000     0.0001     0.0007     0.0039     0.0168     0.0576
```

```
0.5000+++++++++++++++++++++++++++++++++++++++++++++
0.4750+                                           +
0.4500+                                           +
0.4250+                                           +
0.4000+                                           +
0.3750+                                           +
0.3500+                                           +
0.3250+                                           +
0.3000+                                           +
0.2750+          _                                +
0.2500+              _                            +
0.2250+                                           +
0.2000+                                           +
0.1750+      _                                    +
0.1500+                                           +
0.1250+                  _                        +
0.1000+                                           +
0.0750+                                           +
0.0500+  _                                        +
0.0250+                      _                    +
0.0    +                         _   _   _        +
       +++++++++++++++++++++++++++++++++++++++++++++
       +         +         +         +         +
    -1.000     1.500     4.000     6.500     9.000
```

PROBLEMS

1. Using **CRT**, construct a three-way contingency table, and put the frequencies of Table I.6 in it.

1. Using **PAK**, display all two-way and one-way marginal tables and marginal empirical distributions.
2. Within each level of Age, the variables Breathlessness and Wheeze determine a 2^2 table. Using the data entered above, compute the ln odds ratio between these variables for each of the 2^2 tables, and plot the ln odds ratios against Age.

2. Write a LAMDA program for computing and displaying the generalized binomial distributions

$$p(y) = \binom{10}{y}\left(\frac{\rho}{1-\rho}\right)^{y}\left(\frac{\tau}{1-\tau}\right)^{y^2} K \qquad (y = 0, 1, \ldots, 10)$$

where ρ and τ are the parameters, both between 0 and 1, and K is the constant required for the probabilities to sum to one. Do runs with different values of the parameters, and try to determine by experimentation what aspects of the distributions are controlled by ρ and τ.

3. Consider the data in Table I.7. For a particular trial with a particular animal, if we let S be the number of correct responses, then $\ln[(S + 0.5)/(20.5 - S)]$ is a version of the ln odds favoring a correct response.

1. Using LAMDA, produce a plot of the ln odds versus trial number, with all four animals on the same plot, using different letters for different animals.
2. For each animal separately, determine the least-squares-fitted quadratic; that is, try to write

$$\ln\left(\frac{S + 0.5}{20.5 - S}\right) = \alpha + \beta x + \gamma x^2$$

where x is the trial number. Plot the observed and fitted values.

APPENDIX B

Example **LXM** Output

① U stands for "unlikelihood," which is defined in Chapter V, Section A.2. In general, the higher the value of U, the worse the model fits, and differences between values of U for different (nested) models have chi-square distributions.

② When these values are near zero, they indicate that the model was successfully fitted. See Chapter VI, Section A.2 for the definitions and for remedial measures to be taken when the algorithm fails to fit the model.

③ The number of cases, as indicated by the **WGT** variable.

④ The number of Newton–Raphson iterations required to fit the model (see Chapter VI, Section A.1).

⑤ The simple ANALYSIS table. In the notation of Table III.4, $n\mathbb{D}_1(\hat{p}, \hat{q}) = D(P, QP)$, $n\mathbb{D}_1(\hat{q}, q) = D(QP, Q)$, and $n\mathbb{D}_1(\hat{p}, q) = D(P, Q)$. For covariate models (such as this example) it is best to ignore the ANALYSIS table and use differences of $U(P, QP)$ or $D(QP, Q)$ for inference.

⑥ Parameter estimates when an $\mathcal{E}$-model is being fitted. For covariate-free models, there will be a single row ($\hat{\boldsymbol{\theta}}'$ in the notation of Chapter III). For logistic regression, there will be a single column (see Chapter V, Section A.1). In all other cases, the matrix $\hat{\mathbf{B}}$ appears, as defined in Chapter V, Section A.1. The columns correspond to the u-functions and the rows to the t-functions, *in the order in which they were given in the* **SEL** *command*. (See Chapter V, Section B.5).

⑦ Parameter estimates divided by their estimated standard deviations. These are referred to the normal distribution for significant departure from 0.

⑧ **POS** stands for cells in $\mathcal{Y}$, *in the order in which they appeared in the C-file* (**POS** means *position*).
GEN = frequencies under the generating distribution;

```
L A M D A  - - -  VERSION 4.0


*** RUN V.B.10 ***
 M,R,MR/G,F
=== LXM ===
U(P,QP)    =     1699.301     <---- (1)
DELTA(C) =  0.00002584        <---- (2)
DELTA(U) =  0.00097656        <---- (2)
 NO OF CASES =    772.000     <---- (3)
 5 ITERATIONS                 <---- (4)

- - - A N A L Y S I S - - -
D(P,QP)    =       -71.652
D(QP,Q)    =       441.137    <---- (5)
D(P,Q)     =       369.484

- - - COEFFICIENTS - - -

     -0.3894      -0.1751       1.5572
      0.3391       1.0819      -1.0049    <---- (6)
      1.3830      -0.0317      -0.0007

- - - T-VALUES - - -

     -1.8514      -0.8583       5.7909
      1.1342       3.7670      -2.7719    <---- (7)
      3.9618      -0.0869      -0.0016
                                                               (8)
POS     GEN       OBS        FIT        RES        F-RES  G-RES   RES-F  RES-G

 1    193.000    90.000     90.000      0.000      0.000-11.551  0.000 -8.561
 2    193.000   108.000    108.001     -0.001     -0.000 -8.819 -0.000 -7.065
 3    193.000   132.000    132.000      0.000      0.000 -5.831  0.000 -5.070
 4    193.000   442.000    441.998      0.002      0.000 18.115  0.000 20.696

- - - COVARIANCE MATRIX - - -

   0.04424
  -0.03475     0.08937                              <---- (9)
  -0.02288    -0.01993     0.12185
   0.01935    -0.01708    -0.00946     0.04163
  -0.01679     0.04799    -0.01109    -0.03791     0.08249
  -0.01025    -0.00943     0.08266    -0.01507    -0.01775     0.13312
  -0.04388     0.03401     0.02283    -0.04163     0.03791     0.01505     0.07231
   0.03378    -0.08738     0.02008     0.03792    -0.08248     0.01768    -0.05921     0.13142
   0.02354     0.01857    -0.12195     0.01506     0.01774    -0.13300    -0.03160    -0.02885     0.18178
```

OBS = frequencies under the observed distribution;
FIT = frequencies under the fitted distribution;
RES = OBS − FIT. The remaining standardized residuals are described in Chapter IV, Section A.4.

⑨ The estimated covariance matrix of the parameter estimates. The ordering here is determined by the columns of the matrix $\hat{\mathbf{B}}$. For example, here -0.0317 is the sixth parameter (three down the first column, then three more down the second column). Thus the sixth row of the covariance matrix corresponds to this parameter estimate, and so its variance is 0.13312.

APPENDIX C

LAMDA Runs

```
 1:          *** RUN I.C.1 ***
 2:     2CRT
 3:   0 1 3
 4:   0 1 3
 5:     PRT
 6:    3SEL 0 1 2
 7:     C*C
 8:     COL 1
 9:   11 18 6 5
10:   19 62 33 40
11:   1 8 9 9
12:   10 28 31 66
13:     C*R---
14:    2SEL 1 2
15:     C*C
16:     PAK
17:     PRT
18:    3R*C---
19:    4SEL 1 2 3 0
20:    4SUB 3 2
21:     C*C
22:    2SEL 1 4
23:     C*C
24:     PAK
25:     PRT
26:    1SUM 2
27:    3R*C---
28:    1SEL 1
29:     C*C
30:     DIM 4 4
31:     PRT
32:    1RFQ
33:    2RFQ
34:    3RFQ
35:    4RFQ
36:     PRT
37:     END
```

```
 1:          *** RUN I.C.2 ***
 2:     DIM 9 1
 3:    1SEL -1
 4:    1LIN1 1 -5
 5:     C*C
 6:    4SEL 0 1 1 1
 7:    3MPY 2 3
 8:    4MPY 4 3
 9:     C*C
10:     PRT
11:    5DIS 2 1 2 3 4
12:    4SYM0123
13:     VER 30 0 0
14:     HOR 40 0 0
15:     PLT
16:    5SEL 1 2 2 3 4
17:     C*C
18:    3STD 3 4 5
19:    5DIS 2 1 3 4 5
20:    4SYM0123
21:     VER 20 0 0
22:     HOR 40 0 0
23:     PLT
24:    6SEL 1 2 3 4 5 0
25:     C*C
26:     COL 6
27:    2 1 0 0 1 2 3 2 2
28:    7SEL 1 2 3 4 5 6 6
29:     C*C
30:    4FIT 3 4 5 6
31:    3DIS 2 6 7
32:    2SYMFO
33:     VER 30 0 0
34:     HOR 40 0 0
35:     PLT
36:     END
```

```
 1:        *** RUN I.C.3 ***
 2:    DIM 9 1
 3:   1SEL -1
 4:   1LIN1 1 -1
 5:    C*C
 6:   2SEL 1 0
 7:    C*C
 8:    COL 2
 9:  1 8 28 56 70 56 28 8 1
10:   7SEL 1 2 0 0 0 0 0
11:   3LIN 3 .3 0
12:   4LIN 4 .4 0
13:   5LIN 5 .5 0
14:   6LIN 6 .6 0
15:   7LIN 7 .7 0
16:   8SUB 0 3
17:   3DIV 3 8
18:   8SUB 0 4
19:   4DIV 4 8
20:   8SUB 0 5
21:   5DIV 5 8
22:   8SUB 0 6
23:   6DIV 6 8
24:   8SUB 0 7
25:   7DIV 7 8
26:   3LOG 3
27:   4LOG 4
28:   5LOG 5
29:   6LOG 6
30:   7LOG 7
31:   3MPY 3 1
32:   4MPY 4 1
33:   5MPY 5 1
34:   6MPY 6 1
35:   7MPY 7 1
36:   3EXP 3
37:   4EXP 4
38:   5EXP 5
39:   6EXP 6
40:   7EXP 7
41:   3MPY 3 2
42:   4MPY 4 2
43:   5MPY 5 2
44:   6MPY 6 2
45:   7MPY 7 2
46:    C*C
47:   3RFQ
48:   4RFQ
49:   5RFQ
50:   6RFQ
51:   7RFQ
52:   6DIS 1 3 4 5 6 7
53:    PRT
54:    VER 20 0 .499999
55:    HOR 44 -1 10
56:   2DIS 1 3
57:   1SYM_
58:   1PLT
59:    RPT
60:   2DIS 1 4
61:   1PLT
62:    RPT
63:   2DIS 1 5
64:   1PLT
65:    RPT
66:   2DIS 1 6
67:   1PLT
68:    RPT
69:   2DIS 1 7
70:   1PLT
71:    END
```

```
 1:       1*** RUN II.B.1 ***
 2:     NOP --- 4 ROWS BY 4 COLUMNS
 3:     DIM 4 4
 4:     NOP --- ENTER DATA
 5:     ENT
 6:    1  1  1  70
 7:    1 -1 -1  93
 8:   -1  1 -1  63
 9:   -1 -1  1  74
10:     NOP --- DECLARE VAR4 AS WEIGHT VARIABLE
11:    4WGT
12:       0EMPIRICAL MEANS AND VARIANCES
13:    3SUM 1 2 3
14:     END
```

```
 1:       1*** RUN II.B.2 ***
 2:     NOP --- 4 ROWS BY 5 COLUMNS
 3:     DIM 4 5
 4:     ENT
 5:   .0370  1  1  1  18
 6:   .1047  1 -1 -1  50
 7:   .2242 -1  1 -1  78
 8:   .6341 -1 -1  1 373
 9:     NOP --- VIEW THE DATA
10:     PRT
11:       1STANDARDIZE WITH RESPECT TO Q
12:     NOP --- STANDARIZE VARS2,3,4
13:    3STD 2 3 4
14:         --- VIEW THE STANDARDIZED DATA
15:     PRT
16:       0COMPUTE P-HAT MEANS
17:     NOP --- COMPUTE MEANS OF VARS2,3,4 WRT VAR5
18:    5WGT
19:    3SUM 2 3 4
20:     END
```

```
 1:       1*** RUN II.B.3 ***
 2:     NOP --- INSERT TWO BLANK COLUMNS
 3:   14SEL 0 0 1 2 3 4 5 6 7 8 9 10 11 12
 4:         --- READ 12 VARIABLES (SINES AND COSINES) FROM FILE 8
 5:   12N*C(12F8.4)
 6:     NOP --- ENTER POPULATION FREQUENCIES
 7:     COL 1
 8:   1460 1446 1635 1547 1619 1586
 9:   1630 1625 1611 1603 1529 1659
10:     NOP --- ENTER OBSERVED FREQUENCIES
11:     COL 2
12:   29 22 14 19 18 23
13:   28 36 27 39 38 30
14:         --- STANDARDIZE SEPARATELY
15:   12NRM 3 4 5 6 7 8 9 10 11 12 13 14
16:     PRT
17:     NOP --- DECLARE VAR2 AS WEIGHT
18:    2WGT
19:         --- COMPUTE EMPIRICAL MEANS
20:   12SUM 3 4 5 6 7 8 9 10 11 12 13 14
21:         --- STANDARDIZE ALL FUNCTIONS
22:   11STD 3 6 13 10 12 14 4 7 11 9 5
23:     NOP --- DECLARE VAR2 AS WEIGHT
24:    2WGT
25:         --- COMPUTE EMPIRICAL MEANS
26:   11SUM 3 6 13 10 12 14 4 7 11 9 5
27:     PRT
28:     END
```

```
 1:       1*** RUN II.B.4 ***
 2:     NOP --- REPLACE X WITH NO. OF VALUES
 3:     DIM X 1
 4:     NOP --- INSERT X POSITIVE NORMAL VALUES
 5:     NOP --- FOLLOWING ENT
 6:     ENT
 7:    2SEL 1 0
 8:    2MPY 1 1
 9:    2LIN 2 .04417 1
10:    2LIN 2 1.5976 0
11:    2MPY 2 1
12:    2EXP 2
13:    2ADD 2 0
14:    2POW 2 -1
15:    2LIN 2 2 0
16:     C*C
17:       0       Z           P
18:     PRT
19:     END
```

```
 1:       1*** RUN II.B.5 ***
 2:     NOP --- CHANGE X TO NUMBER OF VALUES INPUT
 3:     DIM X 2
 4:     ENT
 5:     NOP --- ENTER X ROWS HERE
 6:     NOP     IN THE FORM
 7:     NOP
 8:     NOP     C D
 9:     NOP
10:     NOP     WHERE C IS THE CHI-SQUARE VALUE
11:     NOP     WITH D DEGREES OF FREEDOM
12:    3SEL1 2 0
13:         --- WILSON-HILFERTY APPROXIMATION
14:    4LIN3 9 0
15:    3DIV1 2
16:    3POW 3 .3333333
17:    3SUB3 0
18:    4MPY4 2
19:    4POW 4 -1
20:    4LIN4 2 0
21:    3ADD3 4
22:    4POW 4 .5
23:    3DIV 3 4
24:    4MPY3 3
25:    4LIN4 .04417 1
26:    4MPY4 3
27:    4LIN4 1.5976 0
28:    4EXP4
29:    4ADD4 0
30:    3POW4 -1
31:     C*C
32:             CHISQ     DF        P
33:     PRT
34:     END
```

```
 1:        1*** RUN III.B.1 ***
 2:     NOP --- READ IN SINES AND COSINES ALTERNATELY
 3:    6N*C(6F8.4)
 4:     STS
 5:     NOP --- RE-DIMENSION TO GET SINES AND
 6:     NOP --- COSINES ON THE SAME ROW
 7:     DIM12 12
 8:   14SEL0 1 2 3 4 5 6 7 8 9 10 11 12 0
 9:     C*C
10:     COL1
11:   1460 1446 1635 1547 1619 1586
12:   1630 1625 1611 1603 1529 1659
13:     COL14
14:   29 22 14 19 18 23
15:   28 36 27 39 38 30
16:     PRT
17:         SINE FUNCTIONS
18:    6DIS2 3 4 5 6 7
19:     PRT
20:         COSINE FUNCTIONS
21:    6DIS8 9 10 11 12 13
22:     PRT
23:     NOP --- SET GEN DIST TO POP FREQUENCIES
24:    1G=V
25:         FIT MODEL WITH SIN1,SIN4,COS5,COS2
26:    4SEL2 5 12 9
27:   14WGT
28:    4LXM
29:     END
```

```
 1:        1*** RUN III.B.2 ***
 2:     DIM 12 3
 3:     NOP --- ENTER EXPECTED FREQUENCIES UNDER
 4:     NOP --- POPULATION DISTRIBUTION (FROM LAMDA OUTPUT)
 5:     COL1
 6:   24.885 24.647 27.868 26.368 27.596 27.033
 7:   27.783 27.698 27.459 27.323 26.062 28.277
 8:     NOP --- ENTER OBSERVED FREQUENCIES
 9:     COL2
10:   29 22 14 19 18 23
11:   28 36 27 39 38 30
12:     NOP --- ENTER FITTED FREQUENCIES (ALSO FROM LAMDA)
13:     COL3
14:   27.437 21.607 17.275 16.113 19.897 21.247
15:   27.317 38.952 30.691 35.234 37.993 29.237
16:     NOP --- CONVERT TO PROBABILITIES
17:    1RFQ
18:    2RFQ
19:    3RFQ
20:    6SEL-1 1 2 3 3 3
21:    5DIV 5 2
22:    6DIV 3 6
23:     C*C
24:     NOP --- VAR1 = ROW NUMBER
25:     NOP --- VAR2 = Q
26:     NOP --- VAR3 = P-HAT
27:     NOP --- VAR4 = Q-HAT
28:     NOP --- VAR5 = Q-HAT/Q
29:     NOP --- VAR6 = P-HAT/Q-HAT
30:        1 Y       Q      P-HAT       Q-HAT
31:    4DIS 1 2 3 4
32:     PRT
33:        1       PLOT OF Q-HAT OVER Q
34:     VER25 1 1
35:     HOR50 0 12
36:    2DIS 1 5
37:     PLT
38:     RPT
39:        1       PLOT OF P-HAT OVER Q-HAT
40:    2DIS 1 6
41:     PLT
42:     END
```

```
 1:       1*** RUN III.B.3 ***
 2:     DIM7 1
 3:     ENT
 4:  2 13 8 11 6 12 5
 5:     NOP --- CREATE INDICATORS
 6:    7SEL1 -1 -1 -1 -1 -1 -1
 7:    2IND 2 1.5 2.5
 8:    3IND 3 2.5 3.5
 9:    4IND 4 3.5 4.5
10:    5IND 5 4.5 5.5
11:    6IND 6 5.5 6.5
12:    7IND 7 6.5 7.5
13:     C*C
14:     PRT
15:    7GEN
16:  53 56 33 26 30 52 25
17:    6SEL2 3 4 5 6 7
18:    6LXM
19:     END
```

```
 1:       1*** RUN III.B.4 ***
 2:     DIM7 1
 3:     ENT
 4:  2 13 8 11 6 12 5
 5:    7SEL1 -1 -1 -1 -1 -1 -1
 6:    2IND 2 1.5 2.5
 7:    3IND 3 2.5 3.5
 8:    4IND 4 3.5 4.5
 9:    5IND 5 4.5 5.5
10:    6IND 6 5.5 6.5
11:    7IND 7 6.5 7.5
12:         --- POOL GROUPS  1,2,4,5,6
13:    2ADD 2 3
14:    2ADD 2 5
15:    2ADD 2 6
16:    2ADD 2 7
17:     C*C
18:     PRT
19:    7GEN
20:  53 56 33 26 30 52 25
21:    2SEL 2 4
22:    2LXM
23:     END
```

```
 1:       1*** RUN III.B.5 ***
 2:     DIM 9 1
 3:    6SEL  0 -1 -1 0 0 0
 4:    2LOG2
 5:    3LIN3 -1 0
 6:    6ADD 2 3
 7:    6EXP 6
 8:    4MPY3 3
 9:    5MPY4 3
10:     C*C
11:     COL1
12:  2 32 53 55 23 18 7 5 1
13:       1DIPHTHERIA DATA
14:     PRT
15:    6G=V
16:    4SEL 2 3 4 5
17:    4LXM
18:     RPT
19:    2SEL 2 3
20:    2LXM
21:     END
```

```
 1:        1*** RUN III.B.6 ***
 2:     DIM 7 5
 3:     ENT
 4:   4 0 0 -1 5
 5:   8 0 -1 0 10
 6:   18 -1 0 0 20
 7:   40 0 0 0 30
 8:   10 1 0 0 20
 9:   2 0 1 0 10
10:   3 0 0 1 5
11:        1SYMMETRY TEST
12:         DATA
13:     PRT
14:    5RFQ
15:    5LIN5 85 0
16:     C*C
17:    1G=V
18:    5WGT
19:    3SEL2 3 4
20:    3LXM
21:     END
```

```
 1:        1*** RUN III.B.7 ***
 2:     DIM 4 2
 3:     ENT
 4:   0 0 0 1 1 0 1 1
 5:    4SEL 0 0 1 2
 6:     C*C
 7:     COL 1
 8:   18 50 78 373
 9:     COL 2
10:   .046 .119 .232 .604
11:     PRT
12:    2G=V
13:    2SEL 3 4
14:    2LXM
15:     END
```

```
 1:        1*** RUN III.B.8 ***
 2:     3CRT
 3:   0 1 1
 4:   0 1 1
 5:   0 1 2
 6:    6SEL0 1 2 2 3 3
 7:    4MPY2 3
 8:    5IND5 .5 1.5
 9:    6IND6 1.5 2.5
10:     C*C
11:        1WOMEN'S EMPLOYMENT
12:     COL1
13:   4 7 2 5 8 7 2 5 4 3 13 9
14:     PRT
15:    5SEL2 3 4 5 6
16:    5LXM
17:    4SEL 2 3 5 6
18:    4LXM
19:     END
```

```
 1:       1*** RUN III.B.9 ***
 2:   3CRT
 3:  0 1 1
 4:  0 1 1
 5:  0 1 1
 6:   7SEL 0 1 2 3 1 2 3
 7:   5MPY 3 5
 8:   6MPY 4 6
 9:   7MPY 2 7
10:    C*C
11:       1GUN REGISTRATION DATA
12:        G = FAVORS GUN REG
13:        Y = YEAR
14:        F = FORM
15:    COL1
16:  126 141 152 182 319 290 463 403
17:    PRT
18:       0MODEL [GF][YF]
19:   5SEL 2 3 4 6 7
20:   5LXM
21:    END
```

```
 1:       1*** RUN III.B.10 ***
 2:   4CRT
 3:  0 1 1
 4:  0 1 1
 5:  0 1 1
 6:  0 1 1
 7:       1MARIJUANA DATA
 8:        1=FAMILY STATUS (F)
 9:        2=RELIGION (R)
10:        3=GEOGRAPHICAL AREA (G)
11:        4=MARIJUANA USE (M)
12:   5SEL0 1 2 3 4
13:    C*C
14:    COL1
15:  52  35 23  23
16:  37 130 69 109
17:   3  15 12  35
18:   9  67 17 136
19:    PRT
20:  13SEL1 2 3 4 5 0 0 0 0 0 0 0 0
21:   6MPY 2 4
22:   7MPY 2 3
23:   8MPY 3 4
24:   9MPY 2 5
25:  10MPY 4 5
26:  11MPY 3 5
27:  12MPY 2 10
28:  13MPY 2 8
29:    C*C
30:       0 MODEL [FGM][FGR][RM]
31:  12SEL2 3 4 5 6 7 8 9 10 11 12 13
32:  12LXM
33:    END
```

```
 1:       1*** RUN III.B.11 ***
 2:    2CRT
 3:  0 1 3
 4:  0 1 3
 5:    9SEL 0 1 2 1 1 1 2 2 2
 6:    4IND 4  .5 1.5
 7:    5IND 5 1.5 2.5
 8:    6IND 6 2.5 3.5
 9:    7IND 7  .5 1.5
10:    8IND 8 1.5 2.5
11:    9IND 9 2.5 3.5
12:    4SUB 4 7
13:    5SUB 5 8
14:    6SUB 6 9
15:     C*C
16:     COL 1
17:  5 5 4 1
18:  8 11 4 3
19:  .5 3 8 5
20:  .5 1 5 9
21:       1DYSPNEA DATA
22:    4DIS 1 4 5 6
23:     PRT
24:    5SEL0 1 4 5 6
25:    1LIN 1 4.5 0
26:     C*C
27:    2G=V
28:    3SEL 3 4 5
29:    3LXM
30:     END
```

```
 1:        1*** RUN III.B.12 ***
 2:    2CRT
 3:  0 1 4
 4:  0 1 4
 5:  10SEL 1 2 0 0 0 0 0 0 0 0
 6:    3IND1   .5 1.5
 7:    4IND1 1.5 2.5
 8:    5IND1 2.5 3.5
 9:    6IND1 3.5 4.5
10:    7IND2   .5 1.5
11:    8IND2 1.5 2.5
12:    9IND2 2.5 3.5
13:   10IND2 3.5 4.5
14:    3SUB 7 3
15:    4SUB 8 4
16:    5SUB 9 5
17:    6SUB 10 6
18:     C*C
19:  16SEL0 3 4 5 6 0 0 0 0 0 0 0 0 0 0
20:     C*C
21:     COL6
22:  0 1 0 0 0
23:  1 0 0 0 0
24:  0 0 0 0 0
25:  0 0 0 0 0
26:  0 0 0 0 0
27:     COL7
28:  0 0 1 0 0
29:  0 0 0 0 0
30:  1 0 0 0 0
31:  0 0 0 0 0
32:  0 0 0 0 0
33:     COL8
34:  0 0 0 1 0
35:  0 0 0 0 0
36:  0 0 0 0 0
37:  1 0 0 0 0
38:  0 0 0 0 0
39:     COL9
40:  0 0 0 0 1
41:  0 0 0 0 0
42:  0 0 0 0 0
43:  0 0 0 0 0
44:  1 0 0 0 0
45:     COL10
46:  0 0 0 0 0
47:  0 0 1 0 0
48:  0 1 0 0 0
49:  0 0 0 0 0
50:  0 0 0 0 0
51:     COL11
52:  0 0 0 0 0
53:  0 0 0 1 0
54:  0 0 0 0 0
55:  0 1 0 0 0
56:  0 0 0 0 0
57:     COL12
58:  0 0 0 0 0
59:  0 0 0 0 1
60:  0 0 0 0 0
61:  0 0 0 0 0
62:  0 1 0 0 0
63:     COL13
64:  0 0 0 0 0
65:  0 0 0 0 0
66:  0 0 0 1 0
67:  0 0 1 0 0
68:  0 0 0 0 0
69:     COL14
70:  0 0 0 0 0
71:  0 0 0 0 0
72:  0 0 0 0 1
73:  0 0 0 0 0
74:  0 0 1 0 0
75:     COL1
76:  -1 19 18 16 20
77:   1 -1   9   9 10
78:   2 11  -1 15 13
79:   4 11   5 -1   8
80:   0 10   7 12  -1
81:     BDY 1 0 100
82:     C*C
83:  13SEL2 3 4 5 6 7 8 9 10 11 12 13 14
84:  13LXM
85:     END
```

```
 1:        1*** RUN III.B.13 ***
 2:    2CRT
 3:   0 1 5
 4:   0 1 2
 5:   20SEL0 1 2 0 0 0 0 0 0 0 0 0 0 0 0 0 0 0 0 0
 6:    4MPY 3 3
 7:    5IND 2  .5 1.5
 8:    6IND 2 1.5 2.5
 9:    7IND 2 2.5 3.5
10:    8IND 2 3.5 4.5
11:    9IND 2 4.5 5.5
12:   10MPY 3 5
13:   11MPY 3 6
14:   12MPY 3 7
15:   13MPY 3 8
16:   14MPY 3 9
17:   15MPY 4 5
18:   16MPY 4 6
19:   17MPY 4 7
20:   18MPY 4 8
21:   19MPY 4 9
22:     C*C
23:   17STD3 5 6 7 8 9 10 11 12 13 14 4 15 16 17 18 19
24:     COL1
25:   4 9 20 9 20 24 3 14 34
26:   2 4 10 14 10 15 4 2 3
27:   20RFQ
28:   20LIN20 201 0
29:     C*C
30:    1G=V
31:   20WGT
32:   11SEL 10 11 12 13 14 4 15 16 17 18 19
33:   11LXM
34:     END
```

```
 1:        1*** RUN III.B.14 ***
 2:     DIM 11 1
 3:     ENT
 4:   63 23 14 13 17 14 12 11 6 5 4
 5:    4SEL 1 -1 0 0
 6:    3LIN 3 .75 0
 7:    3LOG 3
 8:    3MPY 2 3
 9:    3EXP 3
10:    4IND 2 .5 1.5
11:     C*C
12:    3RFQ
13:     PRT
14:    3G=V
15:    2SEL 2 4
16:    2LXM
17:     END
```

```
 1:       1*** RUN IV.B.1 ***
 2:         --- UNIFORM PLOT
 3:    NOP  --- CHANGE X TO NO. OF VALUES
 4:    NOP  --- AND Y TO X + 1
 5:    DIM X 1
 6:    NOP --- INSERT X STANDARDIZED RESIDUALS
 7:    NOP --- WITHOUT SIGNS
 8:    ENT
 9:   2MPY 1 1
10:   2LIN2 .04417 1
11:   2LIN2 1.5976 0
12:   1MPY1 2
13:   1EXP1
14:   1ADD 1 0
15:   2LIN2 0 1
16:   1DIV2 1
17:   1LIN1 2 0
18:    C*C
19:   1LEX 1
20:   3SEL -1 1  -1
21:   4LIN4 0 Y
22:   3DIV3 4
23:    C*C
24:   2DIS 1 2
25:    PRT
26:    VER20 0 .99999
27:    HOR Y 0 Y
28:   3DIS1 3 2
29:   2SYM*P
30:   1PLT
31:    END
```

```
 1:        1*** RUN IV.B.2 ***
 2:   3CRT
 3:  0 1 1
 4:  0 1 1
 5:  0 1 1
 6:   7SEL 0 1 2 3 1 2 3
 7:   5MPY 3 5
 8:   6MPY 4 6
 9:   7MPY 2 7
10:    C*C
11:       1 GUN REGISTRATION
12:        G = FAVORS REGISTRATION
13:        Y = YEAR
14:        F = FORM OF QUESTION
15:    COL 1
16:  126 141 152 182 319 290 463 403
17:    PRT
18:       0 (GY),(YF),(FG)
19:   6SEL 2 3 4 5 6 7
20:   6LXM
21:       0 (GY),(FY)
22:   5SEL 2 3 4 5 6
23:   5LXM
24:       0 (GY),GF)
25:   5SEL 2 3 4 5 7
26:   5LXM
27:       0 (FY),GF)
28:   5SEL 2 3 4 6 7
29:   5LXM
30:       0 F,(GY)
31:   4SEL 2 3 4 5
32:   4LXM
33:       0 G,(FY)
34:   4SEL 2 3 4 6
35:   4LXM
36:       0 Y,(GF)
37:   4SEL 2 3 4 7
38:   4LXM
39:       0 G,F,Y
40:   3SEL 2 3 4
41:   3LXM
42:    END
```

```
 1:       1*** RUN IV.B.3 ***
 2:   3CRT
 3:  0 1 3
 4:  0 1 1
 5:  0 1 2
 6:  22SEL 0 1 1 1 1 2 3 3 0 0 0 0 0 0 0 0 0 0 0 0 0 0
 7:   3IND 3 .5 1.5
 8:   4IND 4 1.5 2.5
 9:   5IND 5 2.5 3.5
10:   7IND 7 .5 1.5
11:   8IND 8 1.5 2.5
12:   9MPY 3 6
13:  10MPY 4 6
14:  11MPY 5 6
15:  12MPY 3 7
16:  13MPY 4 7
17:  14MPY 5 7
18:  15MPY 3 8
19:  16MPY 4 8
20:  17MPY 5 8
21:  18MPY 6 7
22:  19MPY 6 8
23:  20MPY 2 6
24:  21MPY 2 7
25:  22MPY 2 8
26:    C*C
27:       1CODES
28:        VAR2 = J
29:        VARS3,4,5 = I
30:        VAR6 = V
31:        VARS7,8 = P
32:        VARS9,10,11 = IV
33:        VARS12,13,14,15,16,17 = IP
34:        VARS18,19 = PV
35:        VAR20 = JV
36:        VARS21,22 = JP
37:    COL 1
38:  112 67 5 7 37 35 83 75 3 13 45 28
39:  76 67 9 8 57 35 86 67 14 11 63 71
40:       0 IV,IP,PV
41:  17SEL 3 4 5 6 7 8 9 10 11 12 13 14 15 16 17 18 19
42:  17LXM
43:       0 IV,IP
44:  15SEL 3 4 5 6 7 8 9 10 11 12 13 14 15 16 17
45:  15LXM
46:       0 IV,PV
47:  11SEL 3 4 5 6 7 8 9 10 11 18 19
48:  11LXM
49:       0 IP,PV
50:  14SEL 3 4 5 6 7 8 12 13 14 15 16 17 18 19
51:  14LXM
52:       0 I,PV
53:   8SEL 3 4 5 6 7 8 18 19
54:   8LXM
55:       0 P,IV
56:   9SEL 3 4 5 6 7 8 9 10 11
57:   9LXM
58:       0 V,IP
59:  12SEL 3 4 5 6 7 8 12 13 14 15 16 17
60:  12LXM
61:       0 I,P,V
62:   6SEL 3 4 5 6 7 8
63:   6LXM
64:       0 JV,JP,PV
65:   9SEL 2 6 7 8 18 19 20 21 22
66:   9LXM
67:       0 JV,JP
68:   7SEL 2 6 7 8 20 21 22
69:   7LXM
70:       0 JV,PV
71:   7SEL 2 6 7 8 18 19 20
72:   7LXM
73:       0 JP,PV
74:   8SEL 2 6 7 8 18 19 21 22
75:   8LXM
76:       0 J,PV
77:   6SEL 2 6 7 8 18 19
78:   6LXM
79:       0 P,JV
80:   5SEL 2 6 7 8 20
81:   5LXM
82:       0 V,JP
83:   6SEL 2 6 7 8 21 22
84:   6LXM
85:       0 J,P,V
86:   6SEL 2 6 7 8 21 22
87:   6LXM
88:    END
```

```
 1:       1*** RUN IV.B.4 ***
 2:         --- UNIFORM PLOT
 3:    NOP  --- CHANGE 24 TO NO. OF VALUES
 4:    NOP  --- AND 25 TO 24 + 1
 5:    DIM 24 1
 6:    NOP --- INSERT 24 STANDARDIZED RESIDUALS
 7:    ENT
 8:  1.252 .646 .166 1.196 1.277
 9:  1.949 1.031 1.080 1.316 .916 .536
10:  1.092 1.067 .515 .164 .439 .694
11:  1.719 .821 1.101 .726 .811 .987 1.047
12:   2MPY 1 1
13:   2LIN2 .04417 1
14:   2LIN2 1.5976 0
15:   1MPY1 2
16:   1EXP1
17:   1ADD 1 0
18:   2LIN2 0 1
19:   1DIV2 1
20:   1LIN1 2 0
21:    C*C
22:   1LEX 1
23:   3SEL -1 1  -1
24:   4LIN4 0 25
25:   3DIV3 4
26:    C*C
27:   2DIS 1 2
28:    PRT
29:    VER20 0 .99999
30:    HOR 25 0 25
31:   3DIS1 3 2
32:   2SYM*P
33:   1PLT
34:    END
```

```
 1:       1*** RUN IV.B.5 ***
 2:   3CRT
 3:  0 1 2
 4:  0 1 1
 5:  0 1 2
 6:  20SEL0 1 2 3 0 0 0 0 0 0 0 0 0 0 0 0 0 0 0 0
 7:   5MPY 4 4
 8:   6IND 2  .5 1.5
 9:   7IND 2 1.5 2.5
10:   8MPY 4 3
11:   9MPY 4 6
12:  10MPY 4 7
13:  11MPY 3 6
14:  12MPY 3 7
15:  13MPY 4 11
16:  14MPY 4 12
17:  15MPY 5 3
18:  16MPY 5 6
19:  17MPY 5 7
20:  18MPY 5 11
21:  19MPY 5 12
22:    C*C
23:  17STD3 4 6 7 9 10 8 11 12 13 14 5 15 16 17 18 19
24:    COL 1
25:  4 9 20 9 20 24 3 14 34
26:  2 4 10 14 10 15 4 2 3
27:  20RFQ
28:  20LIN20 201 0
29:    C*C
30:   1G=V
31:  20WGT
32:        (SAD)
33:   6SEL 5 15 16 17 18 19
34:   6LXM
35:    RPT
36:        (SA)(SD)(AD)
37:   8SEL 13 14 5 15 16 17 18 19
38:   8LXM
39:    RPT
40:        (SA)(SD)
41:  10SEL 11 12 13 14 5 15 16 17 18 19
42:  10LXM
43:    RPT
44:        D(SA)
45:  11SEL 8 11 12 13 14 5 15 16 17 18 19
46:  11LXM
47:    RPT
48:        (SA)(AD)
49:   9SEL 8 13 14 5 15 16 17 18 19
50:   9LXM
51:    RPT
52:        (SD)(AD)
53:  10SEL 9 10 13 14  5 15 16 17 18 19
54:  10LXM
55:    RPT
56:        S(AD)
57:  11SEL 8 9 10 13 14 5 15 16 17 18 19
58:  11LXM
59:    END
```

```
 1:        1*** RUN IV.B.6 ***
 2:    3CRT
 3:  0 1 2
 4:  0 1 1
 5:  0 1 2
 6:  20SEL0 1 2 3 0 0 0 0 0 0 0 0 0 0 0 0 0 0 0 0
 7:    5MPY 4 4
 8:    6IND 2  .5 1.5
 9:    7IND 2 1.5 2.5
10:    8MPY 4 3
11:    9MPY 4 6
12:  10MPY 4 7
13:  11MPY 3 6
14:  12MPY 3 7
15:  13MPY 4 11
16:  14MPY 4 12
17:  15MPY 5 3
18:  16MPY 5 6
19:  17MPY 5 7
20:  18MPY 5 11
21:  19MPY 5 12
22:    C*C
23:   9SUM 3 4 6 7 8 9 10 11 12
24:    COL 1
25:  2.979 12.286 21.592 12.198 17.321 22.444
26:  4.598 14.256 23.915 1.107 6.582 12.057
27:  10.311 13.460 16.609 4.130 3.095 2.061
28:   9SUM 3 4 6 7 8 9 10 11 12
29:    END
```

```
 1:        1*** RUN IV.B.7 ***
 2:     DIM6 4
 3:     COL1
 4:        1 0
 5:   1      0
 6:   0  0
 7:     COL2
 8:        0 1
 9:   0      0
10:   1  0
11:     COL3
12:        1  0
13:  -1     -1
14:   0  1
15:     COL4
16:        0  1
17:   0       1
18:  -1 -1
19:  10SEL0 1 2 3 4 0 1 2 3 4
20:    C*C
21:    COL1
22:       2  1
23:   6       6
24:   1 12
25:    COL6
26:       6  2
27:  15       9
28:  15 14
29:    DIM12 5
30:  10SEL1 2 3 4 5 0 2 3 4 5
31:    C*C
32:    COL6
33:  0 1 0 1 0 1 0 1 0 1 0 1
34:   7MPY6 7
35:   8MPY6 8
36:   9MPY6 9
37:  10MPY6 10
38:    C*C
39:         (FS),T
40:   7SEL2 3 4 5 6 7 8
41:   7LXM
42:         F,S,T
43:   5SEL2 3 4 5 6
44:   5LXM
45:         (FST)
46:   9SEL2 3 4 5 6 7 8 9 10
47:   9LXM
48:         F,S
49:   3SEL2 3 6
50:   3LXM
51:         (FS)
52:   5SEL2 3 6 7 8
53:   5LXM
54:         (FS),FT1,FT2
55:   7SEL2 3 6 7 8 9 10
56:   7LXM
57:   6SUB 0 6
58:   4MPY4 6
59:   5MPY5 6
60:   6SUB 0 6
61:    C*C
62:         (FS),(1-F)T1,(1-F)T2
63:   7SEL2 3 4 5 6 7 8
64:   7LXM
65:    END
```

```
 1:        1*** RUN IV.B.8 ***
 2:    4CRT
 3:  0 1 2
 4:  0 1 1
 5:  0 1 1
 6:  0 1 1
 7:  22SEL0 1 1 2 3 4 2 2 3 3 4 4
 8:       0 0 0 3 3 4 4 4 4 4
 9:   2IND 2 .5 1.5
10:   3IND 3 1.5 2.5
11:   7MPY 2 4
12:   8MPY 3 4
13:   9MPY 2 5
14:  10MPY 3 5
15:  11MPY 2 6
16:  12MPY 3 6
17:  13MPY 4 5
18:  14MPY 4 6
19:  15MPY 5 6
20:  16MPY 7 16
21:  17MPY 8 17
22:  18MPY 7 18
23:  19MPY 8 19
24:  20MPY 9 20
25:  21MPY 10 21
26:  22MPY 13 22
27:    C*C
28:    NOP  A = 2,3
29:    NOP  S = 4
30:    NOP  C = 5
31:    NOP  V = 6
32:    NOP  AS = 7,8
33:    NOP  AC = 9,10
34:    NOP  AV = 11,12
35:    NOP  SC = 13
36:    NOP  SV = 14
37:    NOP  CV = 15
38:    NOP  ASC = 16,17
39:    NOP  ASV = 18,19
40:    NOP  ACV = 20,21
41:    NOP  SCV = 22
42:    COL 1
43:  8 27 94 275 3 19 30 335 40 245 18 217
44:  4.5 65 11 219 54 666 15 173 7 152 2.5 127
45:  24GEN
46:  12.171 31.057 85.055 275.716
47:  1.65 12.122 36.124 337.103
48:  35.597 240.438 25.465 218.5
49:  3.838 74.627 8.6 212.435
50:  58.53 662.206 12.182 175.081
51:  4.714 153.548 3.073 127.164
52:        (ASC)(ASV)(ACV)(SCV)
53:  21SEL 2 3 4 5 6 7 8 9 10 11 12 13 14 15
54:        16 17 18 19 20 21 22
55:  21LXM
56:    RPT
57:        (ASC)(ASV)
58:  18SEL 2 3 4 5 6 7 8 9 10 11 12 13 14 15
59:        16 17 18 19
60:  18LXM
61:    RPT
62:        (ASC)(ACV)
63:  18SEL 2 3 4 5 6 7 8 9 10 11 12 13 14 15
64:        16 17 20 21
65:  18LXM
66:    RPT
67:        (ASC)(SCV)
68:  17SEL 2 3 4 5 6 7 8 9 10 11 12 13 14 15
69:        16 17 22
70:  17LXM
71:    RPT
72:        (ASV)(ACV)
73:  18SEL 2 3 4 5 6 7 8 9 10 11 12 13 14 15
74:        18 19 20 21
75:  18LXM
76:    RPT
77:        (ASV)(SCV)
78:  17SEL 2 3 4 5 6 7 8 9 10 11 12 13 14 15
79:        18 19 22
80:  17LXM
81:    RPT
82:        (ACV)(SCV)
83:  17SEL 2 3 4 5 6 7 8 9 10 11 12 13 14 15
84:        20 21 22
85:  17LXM
86:    END
```

```
 1:       1*** RUN V.B.1 ***
 2:     DIM 14 3
 3:     ENT
 4:  51 0 0
 5:  2 1 0
 6:  43 0 1
 7:  13 1 1
 8:  25 0 2
 9:  8 1 2
10:  15 0 3
11:  11 1 3
12:  24 0 4
13:  6 1 4
14:  40 0 5
15:  12 1 5
16:  20 0 6
17:  5 1 6
18:   9SEL 1 2 0 3 3 3 3 3 3
19:   4IND 4 .5 1.5
20:   5IND 5 1.5 2.5
21:   6IND 6 2.5 3.5
22:   7IND 7 3.5 4.5
23:   8IND 8 4.5 5.5
24:   9IND 9 5.5 6.5
25:    C*C
26:   8SEL 2 3 4 5 6 7 8 9
27:   1LXM
28:    END
```

```
 1:       1*** RUN V.B.2 ***
 2:     DIM 14 3
 3:     ENT
 4:  51 0 0
 5:  2 1 0
 6:  43 0 1
 7:  13 1 1
 8:  25 0 2
 9:  8 1 2
10:  15 0 3
11:  11 1 3
12:  24 0 4
13:  6 1 4
14:  40 0 5
15:  12 1 5
16:  20 0 6
17:  5 1 6
18:   9SEL 1 2 0 3 3 3 3 3 3
19:   4IND 4 .5 1.5
20:   5IND 5 1.5 2.5
21:   6IND 6 2.5 3.5
22:   7IND 7 3.5 4.5
23:   8IND 8 4.5 5.5
24:   9IND 9 5.5 6.5
25:   4ADD 4 5
26:   4ADD 4 7
27:   4ADD 4 8
28:   4ADD 4 9
29:    C*C
30:   4SEL 2 3 4 6
31:   1LXM
32:    END
```

```
 1:       1*** RUN V.B.3 ***
 2:     DIM 31 4
 3:     ENT
 4:  0 0 0 2
 5:  0 0 1 1
 6:  0 1 0 2
 7:  1 2 0 2
 8:  1 2 1 1
 9:  1 3 0 2
10:  2 1 0 2
11:  2 1 1 1
12:  3 0 1 1
13:  3 1 0 3
14:  3 2 0 3
15:  3 2 1 1
16:  3 3 0 2
17:  3 3 1 2
18:  3 4 0 2
19:  3 4 1 1
20:  4 1 0 2
21:  4 2 1 2
22:  4 3 0 2
23:  4 3 1 5
24:  5 3 0 1
25:  5 4 0 1
26:  5 4 1 3
27:  5 5 1 1
28:  6 3 1 2
29:  6 4 1 2
30:  6 5 1 2
31:  7 4 1 1
32:  7 5 1 4
33:  7 1 1 2
34:    7SEL 4 3 1 2 1 2 2
35:    3LIN 3 .25 1.625
36:    4LIN 4 .20  2.70
37:    5MPY 3 3
38:    6MPY 4 4
39:    7MPY 3 4
40:     C*C
41:    7SEL 2 0 3 4 5 6 7
42:    1LXM
43:    6SEL 2 0 3 4 5 6
44:    1LXM
45:    4SEL2 0 3 4
46:    1LXM
47:     END
```

```
 1:        1*** RUN V.B.4 ***
 2:    1CRT
 3:  1.625 .04 3.74
 4:    7SEL 0 1 0 0 0 0 0
 5:    3LIN 2 -2.9519 7.3023
 6:    4LIN 2 .69898 0.
 7:    4MPY 4 2
 8:    5LIN 2 -3.5195 0.
 9:    4ADD 4 5
10:    4LIN 4 1. 4.52787
11:    4POW 4 .5
12:    7LIN 4 2.448 0.
13:    7ADD 7 3
14:    6LIN 4 -2.448 0.
15:    6ADD 6 3
16:    5LIN 4 -1.177 0
17:    5ADD 5 3
18:    4LIN 4 1.177 0.
19:    4ADD 4 3
20:    3EXP 3
21:    3ADD 3 1
22:    3POW 3 -1
23:    4EXP 4
24:    4ADD 4 1
25:    4POW 4 -1
26:    5EXP 5
27:    5ADD 5 1
28:    5POW 5 -1
29:    6EXP 6
30:    6ADD 6 1
31:    6POW 6 -1
32:    7EXP 7
33:    7ADD 7 1
34:    7POW 7 -1
35:     C*C
36:    6DIS 2 7 4 3 5 6
37:     PRT
38:    6DIS 2 7 6 5 4 3
39:    5SYM--==*
40:     VER 30 0 1
41:     HOR 60 2 2
42:     PLT
43:     END
```

```
 1:        1*** RUN V.B.5 ***
 2:     DIM 22 3
 3:     ENT
 4:   0 1 3
 5:   0 0 2
 6:   1 1 3
 7:   1 0 2
 8:   2 1 3
 9:   3 1 1
10:   3 0 3
11:   3 1 4
12:   3 2 4
13:   3 1 3
14:   4 0 2
15:   4 2 2
16:   4 5 7
17:   5 0 1
18:   5 3 4
19:   5 1 1
20:   6 2 2
21:   6 2 2
22:   6 2 2
23:   7 1 1
24:   7 4 4
25:   7 2 2
26:     NOP VAR1 = G
27:     NOP VAR2 = S COUNT
28:     NOP VAR3 = TOTAL
29:    3SEL 2 3 1
30:    3LIN 3 .25 1.625
31:    3LIN 3 -2.9519 7.3023
32:    3EXP 3
33:    3ADD 3 0
34:    3POW 3 -1
35:    1DIV 1 2
36:    1SUB 1 3
37:    2POW 2 .5
38:    1MPY 1 2
39:    2SUB 0 3
40:    2MPY 2 3
41:    2POW 2 .5
42:    1DIV 1 2
43:     C*C
44:     NOP VAR1 = STD RESIDUAL
45:    1SEL 1
46:    1POW 1 2
47:    1POW 1 .5
48:     C*C
49:     NOP --- INSERT UNIFORM PLOT
50:     NOP --- PROGRAM HERE
```

```
 1:        1*** RUN V.B.6 ***
 2:     DIM 84 4
 3:     NOP  VAR1 = COUNT
 4:     NOP  VAR2 = S
 5:     NOP  VAR3 = D
 6:     NOP  VAR4 = P
 7:     ENT
 8:    9 0  7 400
 9:   31 1  7 400
10:   19 0  7 281
11:   21 1  7 281
12:   18 0  7 196
13:   27 1  7 196
14:   12 0  7 143
15:   26 1  7 143
16:   24 0  7 104
17:   16 1  7 104
18:   29 0  7  73
19:   16 1  7  73
20:   35 0  7  49
21:    6 1  7  49
22:    8 0 10 400
23:   35 1 10 400
24:    9 0 10 281
25:   31 1 10 281
26:    6 0 10 196
27:   34 1 10 196
28:   20 0 10 143
29:   23 1 10 143
30:   23 0 10 104
31:   22 1 10 104
32:   20 0 10  73
33:   21 1 10  73
34:   19 0 10  49
35:   19 1 10  49
36:   38 1 10  35
37:    4 1 10  35
38:    5 0 14 400
39:   37 1 14 400
40:    5 0 14 281
41:   39 1 14 281
42:    3 0 14 196
43:   41 1 14 196
44:    6 0 14 143
45:   33 1 14 143
46:   11 0 14 104
47:   34 1 14 104
48:   16 0 14  73
49:   29 1 14  73
50:   15 0 14  49
51:   23 1 14  49
52:   24 0 14  35
53:   16 1 14  35
```

```
 54:   27 0 14  26
 55:    1 1 14  26
 56:    1 0 20 196
 57:   37 1 20 196
 58:    3 0 20 143
 59:   41 1 20 143
 60:    3 0 20 104
 61:   35 1 20 104
 62:    6 0 20  73
 63:   40 1 20  73
 64:    7 0 20  49
 65:   30 1 20  49
 66:    5 0 20  35
 67:   27 1 20  35
 68:   32 0 20  26
 69:    7 1 20  26
 70:   35 0 20  20
 71:   11 1 20  20
 72:    3 0 28 104
 73:   34 1 28 104
 74:    4 0 28  73
 75:   41 1 28  73
 76:    4 0 28  49
 77:   40 1 28  49
 78:    6 0 28  35
 79:   32 1 28  35
 80:   20 0 28  26
 81:   21 1 28  26
 82:   21 0 28  20
 83:   24 1 28  20
 84:    1 0 40  35
 85:   39 1 40  35
 86:    7 0 40  26
 87:   26 1 40  26
 88:   13 0 40  20
 89:   30 1 40  20
 90:   66 0 40  10
 91:   27 1 40  10
 92:    8SEL1 2 3 4 0 3 4 4
 93:    4LIN 4 .01 0
 94:    3LIN3 .1 0
 95:    6MPY 3 3
 96:    6LIN6 .1 0
 97:    7MPY4 4
 98:    8MPY3 4
 99:     C*C
100:     NOP  VAR5 = 1
101:     NOP  VAR6 = D**2
102:     NOP  VAR7 = P**2
103:     NOP  VAR8 = PD
104:    7SEL2 5 3 4 6 7 8
105:    1LXM
106:     END
```

```
 1:       1*** RUN V.B.7 ***
 2:   2CRT
 3:  7 7 42
 4:  .1 .1 4.
 5:   8SEL 0 1 2 1 2 2 1 2
 6:   2LIN2 .1 0
 7:   4MPY2 2
 8:   4LIN4 .1 0
 9:   5MPY 3 3
10:   6MPY 2 3
11:   2LIN2 -.4608 1.3688
12:   3LIN3 -.2917 0
13:   4LIN4 .6823 0
14:   5LIN5 .2218 0
15:   6LIN6 -1.2736 0
16:   2ADD2 3
17:   2ADD2 4
18:   2ADD2 5
19:   2ADD2 6
20:   2EXP2
21:   2LIN2 1 1
22:   1DIV1 2
23:    C*C
24:   3SEL1 7 8
25:    C*C
26:   2LEX3 2
27:   1SEL1
28:    C*C
29:    DIM40 6
30:       0-----
31:   7SEL-1 1 2 3 4 5 6
32:   1LIN1 .1 0
33:    C*C
34:    ROW41
35:  0 7 14 21 28 35 42
36:    PRT
37:    END
```

```
 1:        1*** RUN V.B.8 ***
 2:         --- INSERT IN PLACE OF LINES 96-98
 3:         --- IN RUN V.B.6
 4:    8SEL1 2 5 3 4 6 7 8
 5:     C*C
 6:    3LIN4 .4608 -1.3688
 7:    5LIN5 .2917 0
 8:    6LIN6 -.6823 0
 9:    7LIN7 -.2218 0
10:    8LIN8 1.2736 0
11:    3ADD3 5
12:    3ADD3 6
13:    3ADD3 7
14:    3ADD3 8
15:    4MPY2 3
16:    3EXP3
17:    3LIN3 1 1
18:    4EXP4
19:    3DIV4 3
20:     C*C
21:    3SEL1 2 3
22:     C*C
23:     NOP  VAR1 = COUNT
24:     NOP  VAR2 = S
25:     NOP  VAR3 = FITTED PROB
26:     DIM42 6
27:     NOP --- EACH ROW NOW ONE P,D COMBINATION
28:    4SEL4 1 6 0
29:    4ADD1 2
30:    1DIV1 4
31:    1SUB 3 1
32:    4POW4 .5
33:    1MPY1 4
34:    4SUB 0 3
35:    3MPY3 4
36:    3POW3 .5
37:    1DIV1 3
38:     C*C
39:     NOP --- VAR1 = STD RES FOR S=1 CASES
40:    1SEL1
41:    2IND 1 -100 0
42:    2LIN 2 -2 1
43:    1MPY 1 2
44:     C*C
45:     NOP -- BEGIN UNIFORM PLOT
46:    2MPY 1 1
47:    2LIN2 .04417 1
48:    2LIN2 1.5976 0
49:    1MPY1 2
50:    1EXP1
51:    1ADD 1 0
52:    2LIN2 0 1
53:    1DIV2 1
54:    1LIN1 2 0
55:     C*C
56:    1LEX 1
57:    3SEL -1 1  -1
58:    4LIN4 0 43
59:    3DIV3 4
60:     C*C
61:    2DIS 1 2
62:     PRT
63:     VER20 0 .99999
64:     HOR43 0 43
65:    3DIS1 3 2
66:    2SYM*P
67:    1PLT
68:     END
```

```
 1:       1*** RUN V.B.9 ***
 2:    5CRT
 3:  0 1 1
 4:  0 1 1
 5:  0 1 1
 6:  0 1 1
 7:  0 1 1
 8:  16SEL0 1 2 3 4 5 0 0 0 0 0 0 0 0 0 0
 9:   7MPY 3 4
10:   8MPY 3 5
11:   9MPY 3 6
12:  10MPY 4 5
13:  11MPY 4 6
14:  12MPY 5 6
15:  13MPY 7 5
16:  14MPY 7 6
17:  15MPY 10 6
18:  16MPY 8 6
19:    C*C
20:    COL1
21:  2 2 10 23 2 4 2 16
22:  8 12 45 91 4 23 28 76
23:  38 55 21 22 6 12 4 2
24:  44 61 37 30 9 23 10 6
25:  16SEL 2 0 3 4 5 6 7 8 9 10 11 12 13 14 15 16
26:   1LXM
27:  12SEL2 0 3 4 5 6 7 8 9 10 11 12
28:   1LXM
29:   6SEL2 0 3 4 5 6
30:   1LXM
31:    END
```

```
 1:       1*** RUN V.B.10 ***
 2:   4CRT
 3:  0 1 1
 4:  0 1 1
 5:  0 1 1
 6:  0 1 1
 7:   5SEL 0 4 3 2 1
 8:    C*C
 9:    COL 1
10:  52 35 37 130
11:  23 23 69 109
12:  3 15 9 67
13:  12 35 17 136
14:   8SEL 1 2 3 0 0 4 5 0
15:   4MPY 2 3
16:   8MPY 6 7
17:    C*C
18:       0 M,R,MR/G,F,GF
19:   7SEL 2 3 4 5 6 7 8
20:   3LXM
21:       0 M,R/G,F,GF
22:   6SEL 2 3 5 6 7 8
23:   2LXM
24:       0 M,R,MR/G,F
25:   6SEL 2 3 4 5 6 7
26:   3LXM
27:       0 M,R,MR/G
28:   5SEL 2 3 4 5 6
29:   3LXM
30:       0 M,R,MR/F
31:   5SEL 2 3 4 5 7
32:   3LXM
33:    END
```

```
 1:        1*** RUN V.B.11 ***
 2:   3CRT
 3:  0 1 1
 4:  0 1 1
 5:  0 1 4
 6:   5SEL0 0 1 2 3
 7:    C*C
 8:    COL 1
 9:  22 52 17 2 1
10:  58 27 12 2 0
11:  53 31 7 1 1
12:  81 19 1 0 0
13:   5GEN
14:  .5825 .2913 .0971 .0247 .0049
15:         LINEAR MODEL
16:   4SEL 5 2 3 4
17:   1LXM
18:    RPT
19:   8SEL1 2 3 4 5 0 0 0
20:   6LIN5 .1 0
21:   6MPY5 6
22:   7LIN5 .1 0
23:   7MPY6 7
24:   8LIN5 .1 0
25:   8MPY7 8
26:    C*C
27:    RPT
28:         QUADRATIC MODEL
29:   5SEL5 6 2 3 4
30:   2LXM
31:    RPT
32:         THIRD ORDER MODEL
33:   6SEL5 6 7 2 3 4
34:   3LXM
35:    RPT
36:         FOURTH ORDER MODEL
37:   7SEL5 6 7 8 2 3 4
38:   4LXM
39:    END
```

```
 1:       1*** RUN V.B.12 ***
 2:    DIM5 1
 3:   1SEL-1
 4:   1LIN1 1 -1
 5:    C*C
 6:   5SEL1 1 1 1 1
 7:   6MPY2 2
 8:   6LIN6 -.4343 0
 9:   2LIN2 1.6775 0
10:   2ADD2 6
11:   6MPY3 3
12:   6LIN6 -.1423 0
13:   3LIN3 .2827 0
14:   3ADD3 6
15:   6MPY4 4
16:   6LIN6 -.1559 0
17:   4LIN4 .2775 0
18:   4ADD4 6
19:   6MPY5 5
20:   6LIN6 .1461 0
21:   5LIN5 -1.1173 0
22:   5ADD5 6
23:    C*C
24:   6SEL1 2 3 4 5 0
25:    C*C
26:    COL6
27:  .5825,.2913,.0471,.0243,.0049
28:   2EXP2
29:   3EXP3
30:   4EXP4
31:   5EXP5
32:   2MPY2 6
33:   3MPY3 6
34:   4MPY4 6
35:   5MPY5 6
36:    C*C
37:   2RFQ
38:   3RFQ
39:   4RFQ
40:   5RFQ
41:   2DIV2 6
42:   2LOG2
43:   3DIV3 6
44:   3LOG3
45:   4DIV4 6
46:   4LOG4
47:   5DIV5 6
48:   5LOG5
49:    C*C
50:   5DIS1 2 3 4 5
51:   4SYMOLSB
52:    VER 30 0 0
53:    HOR60 -1 5
54:    PLT
55:    END
```

```
 1:        1*** RUN V.B.13 ***
 2:         --- CREATE 12 3X3 TABLES
 3:   5CRT
 4:  0 1 1
 5:  0 1 2
 6:  0 1 1
 7:  0 1 2
 8:  0 1 2
 9:   4LIN4 -1 0
10:   4ADD4 5
11:    C*C
12:         --- ELIMINATE BELOW DIAGONAL CELLS
13:    BDY 4 0 10
14:    C*C
15:         --- VAR4 LOOKS LIKE
16:         ---   0  1  2
17:         ---      0  1
18:         ---         0
19:         --- AND VAR5 LOOKS LIKE
20:         ---   0  1  2
21:         ---      1  2
22:         ---         2
23:    STS
24:  13SEL5 5 4 4 2 2 3 1 0 0 0 0 0
25:   9IND1 1.5 2.5
26:   1IND1 .5 2.5
27:   2IND2 .5 1.5
28:   3LIN3 -1 1
29:   3MPY3 2
30:   4LIN4 -1 2
31:   4MPY4 9
32:   5IND5 .5 1.5
33:   6IND6 1.5 2.5
34:   9MPY5 7
35:  10MPY6 7
36:  11MPY5 8
37:  12MPY6 8
38:  13MPY7 8
39:    C*C
40:  15SEL0 1 2 3 4 0 5 6 7 8 9 10 11 12 13
41:    C*C
42:    NOP --- VAR1 = COUNT
43:    NOP --- VAR2 = S
44:    NOP --- VAR3 = Q
45:    NOP --- VAR4 = P
46:    NOP --- VAR5 = R
47:    NOP --- VAR6 = 1
48:    NOP --- VAR7 = A1
49:    NOP --- VAR8 = A2
```

```
50:     NOP --- VAR9 = D
51:     NOP --- VAR10 = X
52:     NOP --- VAR11 = A1D
53:     NOP --- VAR12 = A2D
54:     NOP --- VAR13 = A1X
55:     NOP --- VAR14 = A2X
56:     NOP --- VAR15 = DX
57:     COL1
58:    3  2  4
59:       3  9
60:         20
61:   23  9  9
62:      11 20
63:         24
64:    3  5  3
65:       1 14
66:         34
67:   15  8  2
68:       1  4
69:         10
70:   12  8 14
71:       3 10
72:         15
73:   15 14  4
74:       2  2
75:          3
76:    3  1 12
77:       0 22
78:         14
79:  135 13 42
80:      11 45
81:         16
82:   18  3 21
83:       1 28
84:         19
85:  165 15 28
86:       8 20
87:         15
88:   58  7 18
89:       3 10
90:          8
91:  126 18 11
92:       3  3
93:          2
94:  14SEL2 3 4 5 6 7 8 9 10 11 12 13 14 15
95:   4LXM
96:   9SEL2 3 4 5 6 7 8 9 10
97:   4LXM
98:    END
```

```
 1:        1*** RUN V.B.14 ***
 2:   5CRT
 3:  0 1 1
 4:  0 1 2
 5:  0 1 1
 6:  0 1 2
 7:  0 1 2
 8:   4LIN4 -1 0
 9:   4ADD4 5
10:    C*C
11:    BDY4 0 10
12:    C*C
13:  13SEL5 5 4 4 2 2 3 1 0 0 0 0 0
14:   9IND1 1.5 2.5
15:   1IND1 .5 2.5
16:   2IND2 .5 1.5
17:   3SUB 0 3
18:   3MPY3 2
19:   4LIN4 -1 2
20:   4MPY4 9
21:   5IND5 .5 1.5
22:   6IND6 1.5 2.5
23:    C*C
24:  10SEL0 1 2 3 4 5 0 6 7 8
25:    C*C
26:    COL1
27:    3  2  4
28:       3  9
29:         20
30:   23  9  9
31:      11 20
32:         24
33:    3  5  3
34:       1 14
35:         34
36:   15  8  2
37:       1  4
38:         10
39:   12  8 14
40:       3 10
41:         15
42:   15 14  4
43:       2  2
44:          3
45:    3  1 12
46:       0 22
47:         14
48:  135 13 42
49:      11 45
50:         16
51:   18  3 21
52:       1 28
53:         19
54:  165 15 28
55:       8 20
56:         15
57:   58  7 18
58:       3 10
59:          8
60:  126 18 11
```

```
 61:        3  3
 62:           2
 63:   11LIN6 -.7449 1.1129
 64:   12LIN8 -1.4983 0
 65:   13LIN9 -1.7416 0
 66:   14LIN10 -.4071 0
 67:   11ADD11 12
 68:   11ADD11 13
 69:   11ADD11 14
 70:    2MPY2 11
 71:   11LIN6 .9018 -1.5686
 72:   12LIN8 1.7737 0
 73:   13LIN9 1.5331 0
 74:   14LIN10 -1.3086 0
 75:   11ADD11 12
 76:   11ADD11 13
 77:   11ADD11 14
 78:    3MPY3 11
 79:   11LIN6 -1.0989 .3832
 80:   12LIN8 -1.6104 0
 81:   13LIN9 -.4426 0
 82:   14LIN10 .0390 0
 83:   11ADD11 12
 84:   11ADD11 13
 85:   11ADD11 14
 86:    4MPY4 11
 87:   11LIN6 .0884 .7882
 88:   12LIN8 -.5490 0
 89:   13LIN9 -.3679 0
 90:   14LIN10 -.7963 0
 91:   11ADD11 12
 92:   11ADD11 13
 93:   11ADD11 14
 94:    5MPY5 11
 95:    2ADD2 3
 96:    2ADD2 4
 97:    2ADD2 5
 98:    2EXP2
 99:     C*C
100:    2SEL1 2
101:     C*C
102:     DIM12 12
103:   12SEL1 3 5 7 9 11 2 4 6 8 10 12
104:     C*C
105:     DIM24 6
106:    7ADD1 2
107:    7ADD7 3
108:    7ADD7 4
109:    7ADD7 5
110:    7ADD7 6
111:    1DIV1 7
112:    2DIV2 7
113:    3DIV3 7
114:    4DIV4 7
115:    5DIV5 7
116:    6DIV6 7
117:     C*C
118:     PRT
119:        0-----
```

```
120:    DIM12 12
121:  12SEL1 7 2 8 3 9 4 10 5 11 6 12
122:    C*C
123:    DIM72 2
124:   3SEL0 1 2
125:    C*C
126:    COL1
127:  41 41 41 41 41 41
128:  96 96 96 96 96 96
129:  60 60 60 60 60 60
130:  40 40 40 40 40 40
131:  62 62 62 62 62 62
132:  40 40 40 40 40 40
133:  52 52 52 52 52 52
134:  262 262 262 262 262 262
135:  90 90 90 90 90 90
136:  251 251 251 251 251 251
137:  104 104 104 104 104 104
138:  163 163 163 163 163 163
139:   1POW1 .5
140:   2SUB 2 3
141:   2MPY2 1
142:   4SUB 0 3
143:   3MPY3 4
144:   3POW3 .5
145:   2DIV2 3
146:    C*C
147:   1SEL2
148:    C*C
149:    DIM12 6
150:    PRT
151:    END
```

```
 1:        1*** RUN VI.B.1 ***
 2:     DIM 194 7
 3:     ENT
 4:   1 1 1 1 2 2 0
 5:   1 1 1 2 2 0 0
 6:   1 1 1 3 2 0 7
 7:   1 1 1 4 9 7 18
 8:   1 1 1 5 27 18 19
 9:   1 1 1 6 46 19 19
10:   1 1 1 7 65 19 18
11:   1 1 1 8 83 18 18
12:   1 1 1 9 101 18 20
13:   2 1 1 1 0 0 0
14:   2 1 1 2 0 0 10
15:   2 1 1 3 10 10 17
16:   2 1 1 4 27 17 17
17:   2 1 1 5 44 17 19
18:   2 1 1 6 63 19 19
19:   2 1 1 7 82 19 19
20:   2 1 1 8 101 19 19
21:   2 1 1 9 120 19 19
22:   2 1 1 10 139 19 20
23:   3 1 1 1 0 0 0
24:   3 1 1 2 0 0 0
25:   3 1 1 3 1 1 10
26:   3 1 1 4 10 10 15
27:   3 1 1 5 25 15 18
28:   3 1 1 6 43 18 19
29:   3 1 1 7 62 19 15
30:   3 1 1 8 77 15 19
31:   3 1 1 9 96 19 19
32:   3 1 1 10 105 19 18
33:   3 1 1 11 123 18 19
34:   3 1 1 12 142 19 20
35:   4 1 1 1 3 3 4
36:   4 1 1 2 7 4 4
37:   4 1 1 3 11 4 6
38:   4 1 1 4 15 6 8
39:   4 1 1 5 21 8 15
40:   4 1 1 6 36 15 17
41:   4 1 1 7 53 17 15
42:   4 1 1 8 68 15 14
43:   4 1 1 9 82 14 16
44:   4 1 1 10 98 16 19
45:   4 1 1 11 117 19 18
46:   4 1 1 12 135 18 19
47:   4 1 1 13 154 19 20
48:   5 0 1 1 0 0 0
49:   5 0 1 2 0 0 4
50:   5 0 1 3 4 4 4
51:   5 0 1 4 8 4 15
52:   5 0 1 5 23 15 10
53:   5 0 1 6 33 10 11
54:   5 0 1 7 44 11 13
55:   5 0 1 8 57 13 19
56:   5 0 1 9 76 19 20
57:   6 0 1 1 0 0 0
58:   6 0 1 2 0 0 7
59:   6 0 1 3 7 7 5
60:   6 0 1 4 12 5 3
```

61: 6 0 1 5 15 3 4
62: 6 0 1 6 19 4 6
63: 6 0 1 7 25 6 2
64: 6 0 1 8 27 2 11
65: 6 0 1 9 38 11 12
66: 6 0 1 10 50 12 6
67: 6 0 1 11 56 6 13
68: 6 0 1 12 69 13 14
69: 6 0 1 13 83 14 19
70: 6 0 1 14 102 19 18
71: 6 0 1 15 120 18 17
72: 6 0 1 16 137 17 18
73: 6 0 1 17 155 18 20
74: 7 0 1 1 2 2 0
75: 7 0 1 2 2 0 3
76: 7 0 1 3 5 3 0
77: 7 0 1 4 5 0 0
78: 7 0 1 5 5 0 7
79: 7 0 1 6 12 7 12
80: 7 0 1 7 24 12 11
81: 7 0 1 8 35 11 15
82: 7 0 1 9 50 15 18
83: 7 0 1 10 68 18 18
84: 7 0 1 11 86 18 17
85: 7 0 1 12 103 17 20
86: 8 0 1 1 0 0 3
87: 8 0 1 2 3 3 2
88: 8 0 1 3 5 2 0
89: 8 0 1 4 5 0 0
90: 8 0 1 5 5 0 0
91: 8 0 1 6 5 0 3
92: 8 0 1 7 8 3 2
93: 8 0 1 8 10 2 8
94: 8 0 1 9 18 8 12
95: 8 0 1 10 30 12 15
96: 8 0 1 11 45 15 18
97: 8 0 1 12 63 18 17
98: 8 0 1 13 80 17 18
99: 8 0 1 14 98 18 18
100: 8 0 1 15 116 18 20
101: 9 1 0 1 6 6 8
102: 9 1 0 2 14 8 3
103: 9 1 0 3 17 3 6
104: 9 1 0 4 23 6 6
105: 9 1 0 5 29 6 5
106: 9 1 0 6 34 5 18
107: 9 1 0 7 42 18 18
108: 9 1 0 8 61 08 17
109: 9 1 0 9 77 17 19
110: 9 1 0 10 96 19 19
111: 9 1 0 11 115 19 18
112: 9 1 0 12 133 18 20
113: 10 1 0 1 1 1 6
114: 10 1 0 2 7 6 16
115: 10 1 0 3 23 16 17
116: 10 1 0 4 40 17 17
117: 10 1 0 5 57 17 8
118: 10 1 0 6 65 8 18
119: 10 1 0 7 83 18 17
120: 10 1 0 8 100 17 18

```
121:   10 1 0 9 118 18 18
122:   10 1 0 10 136 18 19
123:   10 1 0 11 155 19 20
124:   11 1 0 1 2 2 7
125:   11 1 0 2 9 7 5
126:   11 1 0 3 14 5 13
127:   11 1 0 4 27 13 19
128:   11 1 0 5 46 19 19
129:   11 1 0 6 65 19 18
130:   11 1 0 7 83 18 19
131:   11 1 0 8 102 19 17
132:   11 1 0 9 119 17 20
133:   12 1 0 1 1 1 6
134:   12 1 0 2 7 6 15
135:   12 1 0 3 22 15 19
136:   12 1 0 4 41 19 17
137:   12 1 0 5 58 17 19
138:   12 1 0 6 77 19 17
139:   12 1 0 7 94 17 19
140:   12 1 0 8 103 19 19
141:   12 1 0 9 121 19 20
142:   13 0 0 1 1 1 0
143:   13 0 0 2 1 0 2
144:   13 0 0 3 3 2 3
145:   13 0 0 4 6 3 16
146:   13 0 0 5 22 16 12
147:   13 0 0 6 44 12 17
148:   13 0 0 7 61 17 18
149:   13 0 0 8 79 18 15
150:   13 0 0 9 94 15 16
151:   13 0 0 10 110 16 17
152:   13 0 0 11 127 17 20
153:   14 0 0 1 0 0 0
154:   14 0 0 2 0 0 0
155:   14 0 0 3 0 0 0
156:   14 0 0 4 0 0 0
157:   14 0 0 5 0 0 0
158:   14 0 0 6 0 0 0
159:   14 0 0 7 0 0 0
160:   14 0 0 8 0 0 14
161:   14 0 0 9 14 14 15
162:   14 0 0 10 29 15 18
163:   14 0 0 11 47 18 16
164:   14 0 0 12 63 16 18
165:   14 0 0 13 81 18 18
166:   14 0 0 14 99 18 17
167:   14 0 0 15 116 17 17
168:   14 0 0 16 133 17 19
169:   14 0 0 17 152 19 19
170:   14 0 0 18 161 19 19
171:   14 0 0 19 180 19 20
172:   15 0 0 1 2 2 2
173:   15 0 0 2 4 2 8
174:   15 0 0 3 12 8 4
175:   15 0 0 4 16 4 1
176:   15 0 0 5 17 1 9
177:   15 0 0 6 26 9 15
178:   15 0 0 7 41 15 16
179:   15 0 0 8 57 16 17
180:   15 0 0 9 74 17 17
```

```
181:  15 0 0 10 91 17 19
182:  15 0 0 11 110 19 19
183:  15 0 0 12 129 19 20
184:  16 0 0 1 1 1 1
185:  16 0 0 2 2 1 0
186:  16 0 0 3 2 0 0
187:  16 0 0 4 2 0 4
188:  16 0 0 5 6 4 4
189:  16 0 0 6 10 4 5
190:  16 0 0 7 15 5 7
191:  16 0 0 8 22 1 13
192:  16 0 0 9 35 13 17
193:  16 0 0 10 52 17 15
194:  16 0 0 11 77 15 15
195:  16 0 0 12 92 15 19
196:  16 0 0 13 111 19 18
197:  16 0 0 14 129 18 20
198:  12SEL0 2 3 4 5 6 7 0 0 0 0 0
199:   7LIN 7  1 .5
200:   8LIN 7 -1 21
201:   7LOG 7
202:   8LOG 8
203:   9MPY 2 4
204:  10MPY 3 4
205:  11MPY 2 3
206:  12MPY 11 4
207:     C*C
208:          --- VAR2 = BELL
209:          --- VAR3 = EXPERIMENTAL
210:          --- VAR4 = TRANSITION NUMBER
211:          --- VAR5 = TOTAL SUCCESSES
212:          --- VAR6 = PREVIOUS SUCCESSES
213:          --- VAR7 = LOG(CURRENT SUCCESSES + .5)
214:          --- VAR8 = LOG(20.5 - CURRENT SUCCESSES)
215:         0 MODEL 0 2 3 2*3 6
216:   7SEL 7 8 1 2 3 11 6
217:   2LXM
218:    END
```

APPENDIX D

LAMDA Dictionary

*x*IOP*yyyyy*	Name	Type*	Description [Modifiers]	Page
*x*ADD *yy*, *zz*	add	*T*	Make variable *xx* the sum of variables *yy* and *zz*	271
BDY *xx*, *yy*, *zz*	boundary	*T*	Discard all cases except those for which $yy \leqq$ (variable xx) $\leqq zz$	—
CHG *xx*, *yy*, *zz*	change	*A*	Change the entry in row *xx* and column *yy* to the value *zz*	270
COL *xx*	column	*A*	Following line(s) replace column *xx*	270
*x*CRT	create	*A*	Create *C*-file with *xx* variables assuming values designated by the next *xx* lines	278
C*C		*A*	Read the *C*-file into the *C*-file [SEL, transformations]	272
C*R*yyyyy*		*A*	Read the *C*-file into the *R*-file according to the format *yyyyy* (default: *yyyyy* = --- for (8F10.4)) [INT]	274
DIM *xx*, *yy*	dimension	*A*	Specify that *C*-file has *xx* rows, *yy* columns	267
*x*DIS*yyyyy*	display	*P*	Display the *xx* variables in *yyyyy* [PLT, PRT, PRS]	276
*x*DIV *yy*, *zz*	divide	*T*	Make variable *xx* the result of dividing variable *yy* by variable *zz*	—
END	end	*A*	Return to operating system	267
ENT	enter	*P*	Following line(s) are data for *C*-file	267
*x*EXP *yy*	exponential	*T*	Make variable *xx* equal to *e* raised to the power of variable *yy*	—
*x*FIT*yyyyy*, *zz*	fit	*A*	Standardize the $xx - 1$ variables in *yyyyy*, and replace variable *zz* by its least-squares predictor [WGT]	282
*x*GEN	generating dist.	*P*	Following line(s) contain the *xx* values of the generating distribution	96
*x*G=V		*P*	Replace the generating distribution by variable *xx*	97
HOR *xx*, *yy*, *zz*	horizontal	*P*	*xx* spaces for horizontal scale, beginning at *yy* and going to *zz*	277
*x*IND *yy*, *zz*, *ww*	indicate	*T*	Make variable *xx* equal to 1 if $zz <$ (variable yy) $\leqq ww$; 0 otherwise	—
INT*yyyyy*	integer	*P*	Declare integer output, with decimal shifts in *yyyyy*	274

*xx*IOP*yyyyy*	Name	Type*	Description [Modifiers]	Page
*xx*LEX*yyyyy*	lex. order	*A*	Lexicographically order the *C*-file, according to the *xx* variables in *yyyyy*	280
*xx*LIN *yy*, *zz*, *ww*	linear	*T*	Make the variable *xx* equal to the value *ww* plus the value *zz* times the variable *yy*	—
*xx*LOG *yy*	log	*T*	Make variable *xx* the natural log of variable *yy*	—
LXM	linear exponential	*A*	Fit linear exponential model [SEL, GEN, G = V]	96, 197, 292
LXX	model	*A*	Extended LXM [SEL, GEN, G = V]	233
*xx*MPY *yy*, *zz*	multiply	*T*	Make variable *xx* the product of variables *yy* and *zz*	272
NOP	no operator	—	Erase all pending passive commands	275
*xx*NRM*yyyyy*	norm	*A*	Normalize the *xx* variables in *yyyyy* [WGT]	281
*xx*N*C*yyyyy*		*A*	Real the *N*-file into the *C*-file, format *yyyyy* [SEL, transformations]	268
*xx*N+C*yyyyy*		*A*	Add the *N*-file to the *C*-file format *yyyyy* [SEL, transformations]	269
PAK	pack	*A*	Pack the rows of the *C*-file, with frequencies in the WGT variable	279
PLT	plot	*A*	Plot variables [VER, HOR, SYM, DIS]	277
*xx*POW *yy zz*	power	*T*	Make variable *xx* equal to variable *yy* raised to the power *zz*	—
PRS *xx*	print short	*A*	Print the first *xx* rows of the *C*-file [DIS]	270
PRT	print	*A*	Print the *C*-file [DIS]	269
*xx*RFQ	relative frequency	*A*	Divide the positive values of variable *xx* by their sum	—
ROW *xx*	row	*A*	Following line(s) replace row *xx*	270
RPT	repeat	—	Prevent preceding active command from erasing passive commands	276
*xx*R*C*yyyyy*		*A*	Read the *xx* variables in the *R*-file into the *C*-file, format *yyyyy* (default: *yyyyy* = --- for (8F10.4))	274
*xx*SEL*yyyyy*	select	*P*	Select the *xx* variables in *yyyyy* [transformations]	273
*xx*STD*yyyyy*	standardize	*A*	Standardize the *xx* variables in *yyyyy* [WGT]	281
STS	status	—	Report number of rows and columns	269
*xx*SUB*yy*, *zz*	subtract	*T*	Make variable *xx* equal to variable *yy* minus variable *zz*	272
*xx*SUM*yyyyy*	summary	*A*	Print summary statistics [WGT]	276
*xx*SYM*yyyyy*	symbols	*P*	Use the *xx* symbols in *yyyyy* for plotting	278
VER *xx*, *yy*, *zz*	vertical	*P*	*xx* lines for vertical scale, beginning at *yy* and going to *zz*	277
*xx*V=G		*P*	Replace variable *xx* by the generating distribution—	
*xx*WGT	weight	*P*	Declare variable *xx* as the weight variable	279

*A = active, *P* = passive, *T* = transformation.

References

GENERAL BOOKS

Andersen, E. B. (1980). *Discrete Statistical Models with Social Science Applications*. Amsterdam, North-Holland Publishing Co.

Bishop, Y. M. M., Stephen E. Fienberg, and Paul W. Holland (1975). *Discrete Multivariate Analysis: Theory and Practice*. Cambridge, Mass., The MIT Press.

Cox, D. R. (1970). *The Analysis of Binary Data*. London, Methuen and Co., Ltd.

Everitt, B. S. (1977). *The Analysis of Contingency Tables*. London, Halsted Press.

Fienberg, Stephen E. (1980). *The Analysis of Cross-Classified Categorical Data*. Second Edition, Cambridge, Mass., The MIT Press.

Gokhale, D. V., and Solomon Kullback (1978). *The Information in Contingency Tables*. New York, Marcel Dekker, Inc.

Haberman, Shelby J. (1974). *The Analysis of Frequency Data*. Chicago, University of Chicago Press.

Haberman, Shelby J. (1978). *Analysis of Qualitative Data*. New York, Academic Press.

Knoke, David, and Peter J. Burke (1980). *Log-Linear Models*. Beverly Hills, Sage Publications.

Plackett, R. L. (1974). *The Analysis of Categorical Data*. London, Griffin.

Reynolds, H. T. (1977). *The Analysis of Cross-Classifications: An Introduction for Social Scientists*. New York, Free Press.

Upton, Graham, J. G. (1978). *The Analysis of Cross-Tabulated Data*. New York, John Wiley and Sons.

DATA AND PROBLEM REFERENCES

Chapter I

Table I.1—Murphy, Raymond J., and Richard T. Morris (1961). Occupational situs, subjective class identification, and political affiliation. *American Sociology Review* **26**:383–392.

Table I.2—MacMahon, B., P. Cole, M. Lin, C. R. Lowe, A. P. Mirra, B. Ravnihar, E. J. Salber, V. G. Valaoras, and S. Yuasa (1970). Age at first birth and breast cancer risk. *Bulletin of the World Health Organization* **43**:209–221.

Table I.3—Marcuse, Edgar K., and M. Gilbert Grand (1973). Epidemiology of diphtheria in San Antonio, Tex., 1970. *Journal of the American Medical Association* **224**:305–310.

Table I.4—Same as Table I.1.

Table I.5—Hodge, Robert W. (1966). Occupational mobility as a probability process. *Demography* **3**:19–34.

Table I.6—Mantel and Brown (1973).

Table I.7—Brogden, W. J. (1965). Sensory pre-conditioning. In Henry Goldstein et al. (eds.), *Controversial Issues in Learning*. New York, Appleton-Century-Crofts, pp. 189–196.

Table I.8—Young, Christabel M. (1979). The role of medical factors in the failure to achieve desired family size. *Journal of Biosocial Science* **11**:159–171.

Chapter II

Table II.2—Hanes, Richard C. (1977). Lithic tools of the Dirty Shame rock shelter: typography and distribution. *Tebiwa* (Miscellaneous papers of the Idaho State Museum of Natural History), No. 6, p. 16.

Table II.3—(No author) (1979). 23rd Annual AMS Survey. *Notices of the American Mathematical Society* **26**:382–386.

Table II.5—Bergman, Abraham B., C. George Ray, Margaret A. Pomeroy, Patricia W. Wahl, and J. Bruce Beckwith (1972). Studies in the sudden infant death syndrome in King County, Washington, III. Epidemiology. *Pediatrics* **49**:860–870.

Problem II.18—Same as Table II.5.

Problems II.19, II.20—Author's data.

Chapter III

Table III.11—Peart, A. F. W. (1949). An outbreak of poliomyelitis in Canadian Eskimos in wintertime. Epidemiological features. *Canadian Journal of Public Health* **40**:410.

Table III.14—Same as Table I.3.

Table III.17—Author's data.

Table III.19—Same as Table II.3.

Table III.21—Elder, Glen H., Jr., (1974). *Children of the Great Depression*. Chicago, University of Chicago Press, p. 317.

Table III.24—Schumann, Howard, and Stanley Presser (1977). Attitude measurement and the gun control paradox. *Public Opinion Quarterly* **41**:427–438.

Table III.26—Cisin, Ira H., and Dean I. Mannheimer (1971). Marijuana use among adults in a large city and suburb. In A. J. Singer (ed.), Marijuana: chemistry, pharmacology, and patterns of social use. *Annals of the New York Academy of Science* **191**:222–234.

Tables III.28, III.30—Author's data.

Table III.33—Weiss, N. S. (1972). Cigarette smoking and arteriosclerosis obliterans: an epidemiological approach. *American Journal of Epidemiology* **95**:17–25.

Table III.35—Cochran, William G. (1955). A test of a linear function of the deviations between observed and expected numbers. *Journal of the American Statistical Association* **50**:377–397.

Problem III.12—Hildebrand, Milton, Robert C. Wilson, and Evelyn R. Dienst (1971). *Evaluating University Teaching*. Center for Research and Development in Higher Education, University of California, Berkeley.

Problem III.17—Simon (1974).

Problem III.18—Bahadur (1961).

Problem III.19—Fienberg and Larntz (1976).

Chapter IV

Table IV.10—Hedlund, Ronald D. (1977). Crossover voting in a 1976 open presidential primary. *Public Opinion Quarterly* **41**:498–515.

Table IV.19—Katz, Elihu, and Paul F. Lazarsfeld (1966). *Personal Influence*. New York, The Free Press, p. 292.

Table IV.20—Fuchs, Camil (1979). Possible biased inferences in tests for average partial association. *The American Statistician* **33**:120–126. [See also, Mantel, Nathan (1980), Letter, *The American Statistician* **34**:190–191.]

Problem IV.1—Same as Table I.1.

Problem IV.2—Sarason, Seymour B., and George Mandler (1952). Some correlates of test anxiety. *The Journal of Abnormal and Social Psychology* **47**:810–817.

Problem IV.3—Bhapkar (1979).

Problem IV.5—Same as Table III.35.

Problem IV.6—Leatherwood, Stephen, James R. Gilbert, and Douglas G. Chapman (1978). An evaluation of some techniques for aerial censuses of bottlenosed dolphins. *Journal of Wildlife Management* **42**:239–250.

Problem IV.8—Tuchman, G., and N. Fortin (1980). Edging women out: some suggestions about the structure of opportunities and the Victorian novel. *Signs* **6**:308–325.

Problem IV.9—Modified from Ray, R. M. (1976). A new C(alpha) test for 2×2 tables. *Communications in Statistics* **A6**:545–563. Data modified from Brown, Jr., B. W. (1980). The crossover experiment for clinical trials. *Biometrics* **36**:69–79.

Chapter V

Table V.2—Cornfield, Jerome, Tavia Gordon, and Willie W. Smith (1961). Quantal response curves for experimentally uncontrolled variables. *Bulletin of the International Statistical Institute* **38**:97–115.

Table V.4—Bell, C. H. (1974). The efficiency of phosphine against diapausing larvae of *Ephestia elutella* (Lepidoptera) over a wide range of concentrations and exposure times. *Journal of Stored Product Research* **15**:53–58.

Table V.6—Rehberg, Richard A., and Walter E. Schafer (1968). Participation in interscholastic athletics and college expectation. *American Journal of Sociology* **73**:732–740.

Table V.9—Hackney, Peter A., and Thomas E. Linkous (1978). Striking behavior of the largemouth bass and use of the binomial distribution for its analysis. *Transactions of the American Fisheries Society* **107**:682–688.

Table V.12—Same as Table III.33.

Problem V.1—DeVore, Phillip W., and Ray J. White (1978). Daytime responses of brown trout (*Salmo trutta*) to cover stimuli in stream channels. *Transactions of the American Fisheries Society* **107**:763–771.

Problem V.2—Seegert, Gregory L., Arthur S. Brooks, John R. Vande Castle, and Kenneth Gradall (1979). The effects of monochloramine on selected riverine fishes. *Transactions of the American Fisheries Society* **108**:88–96.

Problem V.3—Sullins, G. Lamont, and B. J. Verts (1978). Baits and baiting techniques for control of Belding's ground squirrels. *Journal of Wildlife Management* **42**:890–896.

Problem V.4—Same as Table V.9.

Problem V.5—March, James G. (1966). Purely legislative representation as a function of election results. In Paul F. Lazarsfeld (ed.), *Readings in Mathematical Social Science*. Cambridge, Mass., The MIT Press (reprinted from *Public Opinion Quarterly*, 1957, p. 21). Also (1966) *The U.S. Book of Facts, Statistics and Information*. New York, Pocket Books, Inc., p. 381.

Problem V.6—Breslow (1976).

Problem V.9—Dasen, Pierre R., and Robert D. Christie (1972). A regression phenomenon in the conservation of weight. *Archives de Psychologie* **41**:145–152.

Problem I.11—Marks, H. L., and R. D. Wyatt (1979). Genetic resistance to aflatoxin in Japanese quail. *Science* **206**:1329–1330.

Chapter VI

Table VI.1—Same as Table I.7.

Table VI.3—Vitaliano, Peter Paul (1978). The use of logistic regression for modeling risk factors with application to non-melanoma skin cancer. *American Journal of Epidemiology* **108**:402–414 (corrected by communication from the author).

Problem VI.1—Mantel (1966).

Problem VI.3—Anderson (1972).

Problem VI.12—Holford, T. R. (1980). The analysis of rates and survivorship using log-linear models. *Biometrics* **36**:299–305.

GROUP I REFERENCES*

Aickin, Mikel (1983). Serial *P*-values. *Journal of Statistical Planning and Inference* 7:243–256.

Bahadur, R. R. (1961). A representation of the joint distribution of responses to *n* dichotomous items. In Herbert Solomon (ed.), *Studies in Item Analysis and Prediction*. Stanford, Stanford University Press, pp. 158–168.

Bartlett, M. A. (1935). Contingency table interactions. *Journal of the Royal Statistical Society*, Supplement **2**:248–252.

Billingsley, Patrick (1961). *Statistical Inference for Markov Processes*. Chicago, University of Chicago Press.

Birch, M. W. (1963). Maximum likelihood in three-way contingency tables. *Journal of the Royal Statistical Society, Series B* **25**:220–233.

Birch, M. W. (1964a). A new proof of the Pearson–Fisher theorem. *Annals of Mathematical Statistics* **35**:817–824 (also **36**:344).

Bock, R. Darrell (1969). Estimating multinomial response relations. In R. C. Bose (ed.) *Contributions to Statistics and Probability: Essays in Memory of S. N. Roy*. Chapel Hill, University of North Carolina Press.

Brown, Charles C. (1975). On the use of indicator variables for studying the time-dependence of parameters in a response-time model. *Biometrics* **31**:863–872.

Brown, Morton B. (1976). Screening effects in multidimensional contingency tables. *Applied Statistics* **25**:37–46.

Cochran, W. G. (1952). The chi-squared test of goodness of fit. *Annals of Mathematical Statistics* **23**:315–345.

*See the Preface for an explanation regarding the following groups of references.

Cox, D. R. (1958). The regression analysis of binary sequences. *Journal of the Royal Statistical Society, Series B* **20**:215–242.

Darroch, J. N., and S. D. Silvey (1963). On testing more than one hypothesis. *Annals of Mathematical Statistics* **34**:555–567.

Dempster, A. P. (1971). An overview of multivariate data analysis. *Journal of Multivariate Analysis* **1**:316–346.

Fienberg, Stephen E. and Kinley Larntz (1976). Log linear representation for paired and multiple comparison models. *Biometrika* **63**:245–254.

Fix, Evelyn, J. L. Hodges, and E. L. Lehmann (1959). The restricted chi-square test. In Ulf Grenander (ed.), *Probability and Statistics: The Harald Cramer Volume*. Stockholm, Almqvist and Wiksell, pp. 92–107.

Goodman, L. A. (1969). On partitioning chi-squared and detecting partial association in three-way contingency tables. *Journal of the Royal Statistical Society, Series B* **31**:486–498.

Haberman, Shelby J. (1973). The analysis of residuals in cross-classified tables. *Biometrics* **29**:205–220.

Hutchinson, T. P. (1979). The validity of the chi-square test when expected frequencies are small: a list of recent research references. *Communications in Statistics A* **8**:327–335.

Ireland, C. T., H. H. Ku, and S. Kullback (1969). Symmetry and marginal homogeneity of an $r \times r$ contingency table. *Journal of the American Statistical Association* **64**:1323–1341.

Ireland, C. T., and S. Kullback (1968). Contingency tables with given marginals. *Biometrika* **55**:179–188.

Ku, Harry H., and Solomon Kullback (1974). Loglinear models in contingency table analysis. *American Statistician* **28**:115–122.

Kullback, S., M. Kupperman, and H. H. Ku (1962). Tests for contingency tables and Markov chains. *Technometrics* **4**:573–608.

Lauritzen, Steffen L. (1975). General exponential models for discrete observations. *Scandinavian Journal of Statistics* **2**:23–33.

Mantel, Nathan (1966). Models for complex contingency tables and polychotomous dosage response curves. *Biometrics* **22**:83–95.

Mantel, Nathan (1973). Synthetic retrospective studies and related topics. *Biometrics* **29**:479–486.

Mantel, Nathan, and Charles Brown (1973). A logistic reanalysis of Ashford and Sowden's data on respiratory symptoms in British coal miners. *Biometrics* **29**:649–665.

Manten, Nathan, and David P. Byar (1978). Marginal homogeneity, symmetry and independence. *Communications in Statistics A* **7**:953–976.

Nelder, J. A. (1974). Log linear models for contingency tables: a generalization of classical least squares. *Applied Statistics* **23**:323–329.

Nerlove, Marc, and S. James Press (1973). *Univariate and Multivariate Log-linear and Logistic Models*. Santa Monica, Rand Corporation.

Nerlove, Marc, and S. James Press (1976). *Multivariate Log-linear Probability Models for the Analysis of Qualitative Data*. Discussion Paper 1, Center for Statistics and Probability, Northwestern University.

Neyman, J. (1949). Contribution to the theory of the chi-square test. In J. Neyman (ed.) *Proceedings of the (First) Berkeley Symposium on Mathematical Statistics and Probability*. Berkeley, University of California Press, pp. 239–274.

Prentice, R. L., and R. Pyke (1979). Logistic disease incidence models and case-control studies. *Biometrika* **66**:403–411.

Simon, Gary (1973). Additivity of information in exponential family probability laws. *Journal of the American Statistical Association* **68**:472–482.

Walker, Strother H., and David B. Duncan (1967). Estimation of the probability of an event as a function of several independent variables. *Biometrika* **54**:167–179.

GROUP II REFERENCES

Aickin, Mikel (1979). Existence of MLEs for discrete linear exponential models. *Annals of the Institute of Statistical Mathematics A* **31**:103–113.

Altham, Patricia M. E. (1975). Quasi-independent triangular contingency tables. *Biometrics* **31**:233–238.

Andersen, A. Holst (1974). Multidimensional contingency tables. *Scandinavian Journal of Statistics* **1**:115–127.

Andersen, E. B. (1970). Asymptotic properties of conditional maximum likelihood estimators. *Journal of the Royal Statistical Society, Series B* **32**:283–301.

Anderson, J. A. (1972). Separate sample logistic discrimination. *Biometrika* **59**:19–35.

Berkson, J. (1955). Maximum likelihood and minimum chi-square estimates of the logistic function. *Journal of the American Statistical Association* **50**:130–162.

Bhapkar, Vasant P. (1979). On tests of marginal symmetry and quasi-symmetry in two and three-dimensional contingency tables. *Biometrics* **35**:417–426.

Bhapkar, V. P., and Grant W. Somes (1976). Multiple comparisons of matched proportions. *Communications in Statistics A* **5**:17–25.

Billingsley, Patrick (1968). *Convergence of Probability Measures*. New York, John Wiley and Sons.

Bradley, Ralph A. (1976). Science, statistics, and paired comparisons. *Biometrics* **32**:213–232.

Breslow, N. (1976). Regression analysis of the log odds ratio: a method for retrospective studies. *Biometrics* **32**:409–416.

Breslow, N., and W. Powers (1978). Are there two logistic regressions for retrospective studies? *Biometrics* **34**:100–105.

Cochran, W. G. (1950). The comparison of percentages in matched samples. *Biometrika* **37**:256–266.

Cox, D. R. (1966). Some procedures connected with the logistic qualitative response curve. In F. N. David (ed.), *Research Papers in Statistics: Festschrift for J. Neyman*. New York, John Wiley and Sons.

Cox, D. R. (1972). The analysis of multivariate binary data. *Applied Statistics* **21**:113–120.

Darroch, J. N. (1962). Interactions in multifactor contingency tables. *Journal of the Royal Statistical Society, Series B* **24**:251–263.

Darroch, J. N. (1974). Multiplicative and additive interaction in contingency tables. *Biometrika* **61**:207–214.

Day, N. E., and D. P. Byar (1979). Testing hypotheses in case-control studies—equivalence of Mantel–Haenszel statistics and logit score tests. *Biometrics* **35**:623–630.

Day, N. E., and D. F. Kerridge (1967). A general maximum likelihood discriminant. *Biometrics* **23**:313–323.

Diamond, Earl L. (1963). The limiting power of categorical chi-square tests analogous to normal analysis of variance. *Annals of Mathematical Statistics* **34**:1432–1441.

Dwass, Meyer (1976). *Exact Tests in Log-Linear Models*. Discussion Paper No. 6, Center for Statistics and Probability, Northwestern University.

Feller, William (1966). *An Introduction to Probability Theory and Its Applications*, Vols. I and II. New York, John Wiley and Sons.

Gabriel, K. R. (1966). Simultaneous test procedures for multiple comparison on categorical data. *Journal of the American Statistical Association* **61**:1081–1096.

Goodman, Leo A. (1965a). On simultaneous confidence intervals for multinomial proportions. *Technometrics* **7**:247–254.

Goodman, Leo A. (1978). *Analyzing Qualitative/Categorical Data*. Cambridge, Abt Books.

Haberman, Shelby J. (1974a). Log-linear models for frequency tables with ordered classifications. *Biometrics* **30**:589–600.

Holm, Sture (1979). A simple sequentially rejective multiple test procedure. *Scandinavian Journal of Statistics* **6**:65–70.

Johnson, N. L. and S. Kotz (1969). *Discrete Distributions*. Boston, Houghton Mifflin Co.

Killion, Ruth A., and Douglas A. Zahn (1976). A bibliography of contingency table literature. *International Review of Statistics* **44**:71–112.

Kullback, Solomon (1968). *Information Theory and Statistics*. New York, Dover Publications, Inc. [Reprinted from (1959) *Information Theory and Statistics*. New York, John Wiley and Sons.]

Kullback, S., and M. A. Khairat (1966). A note on minimum discrimination information. *Annals of Mathematical Statistics* **37**:279–280.

Lancaster, H. O. (1949). The derivation and partition of chi-square in certain discrete distributions. *Biometrika* **36**:117–129.

Lancaster, H. O. (1969). *The Chi-squared Distribution*. New York, John Wiley and Sons.

Lewis, Don, and C. J. Burke (1949). The use and misuse of the chi-square test. *Psychological Bulletin* **46**:433–489. [Reprinted along with comments and rejoinders in (1971) Joseph A. Steger (ed.) *Readings in Statistics for the Behavioral Scientist*. New York, Holt, Rinehart and Winston, Inc., pp. 55–131.]

Mann, H. B., and A. Wald (1943). On stochastic limit and order relationships. *Annals of Mathematical Statistics* **14**:217–226.

Moore, R. H., and R. K. Zeigler (1967). The use of non-linear regression methods for analyzing sensitivity and quantal response data. *Biometrics* **23**:563–566.

Murthy, V. K., and A. V. Gafarian (1970). Limiting distributions of some variations of the chi-square statistics. *Annals of Mathematical Statistics* **41**:188–194.

O'Neill, M. E. (1980). The distribution of higher-order interactions in contingency tables. *Journal of the Royal Statistical Society* **42**:357–365.

Page, E. (1977). Approximations to the cumulative normal function and its inverse for use on a pocket calculator. *Applied Statistics* **26**:75–76.

Patil, G. P. (1965). *Classical and Contagious Discrete Distributions*. Oxford, Pergamon Press.

Pearson, K. (1900). On a criterion that a given system of deviations from the probable, in the case of a correlated system of variables, is such that it can be reasonably supposed to have arisen from random sampling. *Philosophical Magazine Series 5* **50**:157–175.

Ponnapalli, R. (1976). Deficiencies of minimum discrepancy estimators. *Canadian Journal of Statistics* **4**:33–50.

Quade, Dana, and Ibrahim A. Salama (1975). A note on minimum chi-square statistics in contingency tables. *Biometrics* **31**:953–956.

Rao, C. R. (1973). *Linear Statistical Inference and Its Applications*. Second Edition. New York, John Wiley and Sons.

Read, Campbell B. (1977). Partitioning chi-square in contingency tables: a teaching approach. *Communications in Statistics A* **6**:553–562.

Roy, S. N., and M. A. Kastenbaum (1956). On the hypothesis of no "interaction" in a multiway contingency table. *Annals of Mathematical Statistics* **27**:749–757.

Simon, Gary (1974). Alternative analysis of the singly-ordered contingency table. *Journal of the American Statistical Association* **69**:971–976.

Sundberg, Rolf (1975). Some results about decomposable (or Markov-type) models for multidimensional contingency tables: distribution of marginals and partitioning. *Scandinavian Journal of Statistics* **2**:71–79.

Vajda, Igor (1971). Chi-alpha-divergence and generalized Fisher's Information. In *Information Theory, Statistical Decision Functions, Random Processes. 6th Prague Conference*. Prague, Academia.

Wackerly, D. D., and F. H. Dietrich (1976). Pairwise comparison of matched proportions. *Communications in Statistics A* **5**:1455–1467.

Wilks, S. S. (1938). The large-sample distribution of the likelihood ratio for testing composite hypotheses. *Annals of Mathematical Statistics* **9**:60–62.

Wilson, E. B., and M. M. Hilferty (1931). The distribution of chi-square. *Proceedings of the National Academy of Science* **17**:694.

Yates, F. (1934). Contingency tables involving small numbers and the chi-square test. *Journal of the Royal Statistical Society Supplement* **1**:217–235.

Zar, Jerrold, H. (1978). Approximations for the percentage points of the chi-squared distribution. *Applied Statistics* **27**:280–290.

GROUP III REFERENCES

Abramowitz, Milton, and Irene A. Stegun (1965). *Handbook of Mathematical Functions*. New York, Dover Publications, Inc.

Aitkin, M. A. (1979). A simultaneous test procedure for contingency table models. *Applied Statistics* **28**:233–242.

Altham, P. M. E. (1970a). The measurement of association of rows and columns for an $r \times s$ contingency table. *Journal of the Royal Statistical Society, Series B* **32**:63–73.

Altham, P. M. E. (1970b). The measurement of association in a contingency table: three extensions of the cross-ratios and metric methods. *Journal of the Royal Statistical Society, Series B* **32**:395–407.

Altham, P. M. E. (1971). The analysis of matched proportions. *Biometrika* **58**:561–576.

Altham, P. M. E. (1979). Detecting relationships between categorical variables observed over time: a problem of deflating a chi-squared statistic. *Applied Statistics* **28**:115–125.

Altschul, R. E., and R. Marcuson (1979). A generalized Scheffe method of multiple comparisons. *Communications in Statistics* **A8**:271–281.

Anderson, J. A., and S. C. Richardson (1979). Logistic discrimination and bias correction in maximum likelihood estimation. *Technometrics* **21**:71–78.

Anderson, T. W., and L. A. Goodman (1957). Statistical inference about Markov chains. *Annals of Mathematical Statistics* **28**:89–110.

Andrich, David (1979). A model for contingency tables having an ordered response classification. *Biometrics* **35**:403–415.

Anscombe, F. J. (1956). On estimating binomial response relations. *Biometrika* **43**:461–464.

Armitage, P. (1955). Tests for linear trends in proportions and frequencies. *Biometrics* **11**:375–386.

Armitage, P. (1966). The chi-square test for heterogeneity of proportions after adjustment for stratification. *Journal of the Royal Statistical Society, Series B* **28**:150–163.

Baker, R. J. (1977). Exact distributions derived from two-way tables. *Applied Statistics* **26**:199–206.

Belle, G. V., and R. G. Cornell (1971). Strengthening tests of symmetry in contingency tables. *Biometrics* **27**:1074–1078.

Benedetti, J., and M. B. Brown (1978). Strategies for the selection of log-linear models. *Biometrics* **34**:680–686.

Bennett, B. M., and C. Kaneshiro (1974). On the small-sample properties of the Mantel–Haenszel test for relative risk. *Biometrika* **61**:233–236.

Berkson, J. (1944). Application of the logistic function to bio-assay. *Journal of the American Statistical Association* **39**:357–365.

Berkson, J. (1949). Minimum chi-square and maximum likelihood solution in terms of a linear transform, with particular reference to bioassay. *Journal of the American Statistical Association* **44**:273–278.

Berkson, J. (1951). Relative precision of minimum chi-square and maximum likelihood estimates of regression coefficients. In J. Neyman (ed.), *Proceedings of the Second Berkeley Symposium on Mathematical Statistics and Probability*. Los Angeles, University of California Press, pp. 471–479.

Berkson, J. (1953). A statistically precise and relatively simple method of estimating the bio-assay with quantal response, based on the logistic function. *Journal of the American Statistical Association* **48**:565–599.

Berkson, J. (1968). Application of minimum logit chi-square estimate to a problem of Grizzle with a notation on the problem of no interaction. *Biometrics* **24**:75–96.

Berkson, J. (1972). Minimum discrimination information, the no interaction problem, and the logistic function. *Biometrics* **28**:443–468.

Berkson, J. (1978). In dispraise of the exact test. *Journal of Statistical Planning and Inference* **2**:27–42.

Berkson, J., and B. N. Nagnur (1974). A note on the minimum chi-square(1) estimate and a LAMST chi-square in the no interaction problem. *Journal of the American Statistical Association* **69**:1038–1040.

Bhapkar, V. P. (1961). Some tests for categorical data. *Annals of Mathematical Statistics* **32**:72–83.

Bhapkar, V. P. (1966). A note on the equivalence of two test criteria for hypotheses in categorical data. *Journal of the American Statistical Association* **61**:228–235.

Bhapkar, V. P., and G. G. Koch (1968a). Hypotheses of no interaction in multidimensional contingency tables. *Technometrics* **10**:107–123.

Bhapkar, V. P., and G. G. Koch (1968b). On the hypotheses of no interaction in contingency tables. *Biometrics* **24**:567–594.

Bhat, B. R., and S. R. Kulharni (1966). LAMP test of linear and loglinear hypotheses in multinomial experiments. *Journal of the American Statistical Association* **61**:236–245.

Bhat, B. R., and B. N. Nagnur (1965). Locally asymptotically most stringent tests and Lagrangian multiplier tests of linear hypotheses. *Biometrika* **52**:459–468.

Birch, M. W. (1964b). The detection of partial association, I: the 2 × 2 case. *Journal of the Royal Statistical Society, Series B* **26**:313–324.

Birch, M. W. (1965). The detection of partial association II: the general case. *Journal of the Royal Statistical Society, Series B* **27**:111–124.

Bishop, Y. M. M. (1969). Full contingency tables, logits, and split contingency tables. *Biometrics* **25**:383–399.

Bishop, Y. M. M. (1971). Effects of collapsing multidimensional contingency tables. *Biometrics* **27**:545–562.

Bishop, Y. M. M., and S. E. Fienberg (1969). Incomplete two-dimensional contingency tables. *Biometrics* **22**:119–128.

Bock, R. D. (1972). Estimating item parameters and latent ability when responses are scored in two or more nominal categories. *Psychometrika* **37**:29–51.

Bowker, A. H. (1948). A test for symmetry in contingency tables. *Journal of the American Statistical Association* **43**:572–574.

Brown, D. T. (1959). A note on approximations to discrete probability distributions. *Information and Control* **2**:386–392.

Brown, M. B. (1974). Identification of the sources of significance in two-way contingency tables. *Applied Statistics* **23**:405–413.

Chambers, E. A., and D. R. Cox (1967). Discrimination between alternative binary response models. *Biometrika* **54**:573–578.

Chapman, D. G., and R. C. Meng (1966). The power of chi-square tests for contingency tables. *Journal of the American Statistical Association* **61**:965–975.

Chapman, J-A. W. (1976). A comparison of the χ^2, $-2\log R$, and multinomial probability criteria for significance tests when expected frequencies are small. *Journal of the American Statistical Association* **71**:854–863.

Chen, T. T., and S. E. Fienberg (1974). Two-dimensional contingency tables with both completely and partially cross-classified data. *Biometrics* **30**:629–642.

Chen, T. T., and S. E. Fienberg (1976). The analysis of contingency tables with incompletely classified data. *Biometrics* **32**:133–144.

Clayton, D. G. (1974). Some odds ratio statistics for the analysis of ordered categorical data. *Biometrika* **61**:525–531.

Cochran, W. G. (1954). Some methods for strengthening the common chi-square tests. *Biometrics* **10**:417–451.

Cohen, J. E. (1971). Estimation and interaction in a censored 2 × 2 × 2 contingency table. *Biometrics* **27**:379–386.

Conover, W. J. (1974). Some reasons for not using the Yates continuity correction on a 2 × 2 contingency tables (with comments). *Journal of the American Statistical Association* **69**:374–382.

Cornfield, J. (1962). A statistical problem arising from retrospective studies. In J. Neyman (ed.), *Proceedings of the Third Berkeley Symposium on Mathematical Statistics and Probability*. Los Angeles, University of California Press, Vol. 4, pp. 135–148.

Cox, D. R. (1958). Two further applications of a model for binary regression. *Biometrika* **45**: 562–565.

Cox, D. R. (1966). A simple example of a comparison involving quantal data. *Biometrika* **53**:215–220.

Cox, D R. (1970b). The continuity correction. *Biometrika* **57**:217–219.

Cox, D. R., and E. Lauh (1967). A note on the graphical analysis of multidimensional contingency tables. *Technometrics* **9**:481–488.

Craddock, J. M., and C. R. Flood (1970). The distribution of the chi-square statistic in small contingency tables. *Applied Statistics* **19**:173–181.

Daly, C. (1962). A simple test for trends in a contingency table. *Biometrics* **18**:114–119.

Darroch, J. N., and D. Ratcliff (1972). Generalized iterative scaling for loglinear models. *Annals of Mathematical Statistics* **43**:1470–1480.

Davidson, Roger R., and Peter H. Farquhar (1976). A bibliography on the method of paired comparisons. *Biometrics* **32**:241–252.

Duncan, O. D. (1975). Partitioning polytomous variables in multiway contingency analysis. *Social Science Research* **4**:167–182.

Dyke, G. V., and H. D. Patterson (1952). Analysis of factorial arrangements when the data are proportions. *Biometrics* **8**:1–12.

Edwards, A. W. F. (1963). The measurement of association in a 2×2 table. *Journal of the Royal Statistical Society, Series A* **126**:109–114.

Farewell, V. T. (1979). Some results on the estimation of logistic models based on retrospective data. *Biometrika* **66**:27–32.

Fienberg, S. E. (1968). The geometry of an $r \times c$ contingency table. *Annals of Mathematical Statistics* **39**:1186–1190.

Fienberg, S. E. (1969). Preliminary graphical analysis and quasi-independence for two-way contingency tables. *Applied Statistics* **18**:153–168.

Fienberg, S. E. (1970a). Quasi-independence and maximum likelihood estimation in incomplete contingency tables. *Journal of the American Statistical Association* **65**:1610–1616.

Fienberg, S. E. (1970b). An iterative procedure for estimation in contingency tables. *Annals of Mathematical Statistics* **41**:907–917.

Fienberg, S. E. (1972a). The multiple-recapture census for closed populations and incomplete 2^k contingency tables. *Biometrika* **59**:591–603.

Fienberg, S. E. (1972b). The analysis of incomplete multiway contingency tables. *Biometrics* **28**:177–202.

Fienberg, S. E. (1979a). Log-linear representation for paired comparison models with ties and within-pair order effects. *Biometrics* **35**:479–481.

Fienberg, S. E. (1979b). The use of chi-squared statistics for categorical data problems. *Journal of the Royal Statistical Society, Series B* **44**:54–64.

Fienberg, S. E., and P. W. Holland (1970). Methods for eliminating zero counts in contingency tables. In G. P. Patil (ed.), *Random Counts in Scientific Work*, The Pennsylvania State University Press, pp. 233–260.

Fienberg, S. E., and P. W. Holland (1973). Simultaneous estimation of multinomial cell probabilities. *Journal of the American Statistical Association* **68**:638–691.

Fischer, P. (1972). On the inequality $\Sigma p(i)f(p(i)) > \Sigma p(i)f(q(i))$. *Metrika* **18**:199–208.

Fleiss, J. L., and B. S. Everitt (1971). Comparing the marginal totals of square contingency tables. *British Journal of Mathematical and Statistical Psychology* **24**:117–123.

Forthofer, R. N., and G. G. Koch (1973). An analysis of compounded functions of categorical data. *Biometrics* **29**:43–157.

Freeman, Jr., D. H., and J. F. Jekel (1980). Table selection and log-linear models. *Journal of Chronic Disease* **33**:512–524.

Fryer, J. G. (1971). On the homogeneity of the marginal distributions of a multidimensional contingency table. *Journal of the Royal Statistical Society, Series A* **134**:368–371.

Fuchs, C., and R. Kenett (1980). A test for detecting outlying cells in the multinomial distribution and two-way contingency tables. *Journal of the American Statistical Association* **75**:395–398.

Gail, M. H. (1972). Mixed quasi-independence models for categorical data. *Biometrics* **28**:703–712.

Garside, G. R., and C. Mack (1976). Actual Type I error probabilities for various tests in the homogeneity case of the 2 × 2 contingency table. *American Statistician* **30**:18–21.

Gart, J. J. (1966). Alternative analyses of contingency tables. *Journal of the Royal Statistical Society, Series B* **28**:168–179.

Gart, J. J. (1969). An exact test for comparing matched proportions in crossover designs. *Biometrika* **56**:75–80.

Gart, J. J. (1970). Some simple graphically oriented statistical methods for discrete data. In G. P. Patil (ed.), *Random Counts in Scientific Work*. University Park, Pennsylvania State University Press, Vol. 1, pp. 171–191.

Gart, J. J. (1970). Point and interval estimation of the common odds ratio in the combination of 2 × 2 tables with fixed marginals. *Biometrika* **57**:471–475.

Gart, J. J. (1971a). On the ordering of contingency tables for significance tests. *Technometrics* **13**:910–911.

Gart, J. J. (1971b). The comparison of proportions; a review of significance tests, confidence intervals, and adjustments for stratification. *Review of the International Statistical Institute* **29**:148–169.

Gart, J. J. (1972). Interaction tests for 2 × s × t contingency tables. *Biometrika* **59**:309–316.

Gart, J. J. (1978). The analysis of ratios and cross-product ratios of Poisson variates with applications to incidence rates. *Communications in Statistics* **A7**:917–938.

Gart, J. J., and D. G. Thomas (1972). Numerical results on approximate confidence limits for the odds ratio. *Journal of the Royal Statistical Society, Series B* **34**:441–447.

Gart, J. J., and J. R. Zweifel (1967). On the bias of various estimators of the logit and its variance with application to quantal bioassay. *Biometrika* **54**:181–187.

Godambe, A. B., and W. L. Harness (1975). Normal approximation to the distribution of a cell entry in a 2 × 2 × 2 contingency table. *Communications in Statistics* **4**:699–709.

Gokhale, D. V. (1971). An iterative procedure for analysing log-linear models. *Biometrics* **27**:681–687.

Gokhale, D. V. (1972). Analysis of log-linear models. *Journal of the Royal Statistical Society, Series B* **34**:371–376.

Gokhale, D. V. (1973). Approximating discrete distributions with applications. *Journal of the American Statistical Association* **68**:1009–1012.

Gokhale, D. V., and N. S. Johnson (1978). A class of alternatives to independence in contingency tables. *Journal of the American Statistical Association* **73**:800–804.

Gokhale, D. V., and S. Kullback (1978). The minimum discrimination information approach in analyzing categorical data. *Communications in Statistics* **A7**:987–1005.

Good, I. J. (1963). Maximum entropy for hypothesis formulation, especially for multidimensional contingency tables. *Annals of Mathematical Statistics* **34**:911–934.

Good, I. J., T. N. Gover, and G. J. Mitchell (1970). Exact distributions for chi-square and for the likelihood-ratio statistics for the equiprobable multinomial distribution. *Journal of the American Statistical Association* **65**:267–283.

Goodman, L. A. (1963a). On methods for comparing contingency tables. *Journal of the Royal Statistical Society, Series A* **126**:94–108.

Goodman, L. A. (1963b). On Plackett's test for contingency table interactions. *Journal of the Royal Statistical Society, Series B* **25**:179–188.

Goodman, L. A. (1964a). Simultaneous confidence intervals for cross-product ratios in contingency tables. *Journal of the Royal Statistical Society, Series B* **26**:86–102.

Goodman, L. A. (1964b). Simple methods for analyzing three-factor interaction in contingency tables. *Journal of the American Statistical Association* **59**:319–352.

Goodman, L. A. (1964c). Interactions in multidimensional contingency tables. *Annals of Mathematical Statistics* **35**:632–646.

Goodman, L. A. (1964d). A short computer program for the analysis of transaction flows. *Behavioral Science* **9**:176–186.

Goodman, L. A. (1965b). On the multivariate analysis of three dichotomous variables. *American Journal of Sociology* **71**:290–301.

Goodman, L. A. (1965c). On the statistical analysis of mobility tables. *American Journal of Sociology* **70**:564–585.

Goodman, L. A. (1968). The analysis of cross-classified data: independence, quasi-independence, and interactions in contingency tables with or without missing entries. *Journal of the American Statistical Association* **63**:1091–1131.

Goodman, L. A. (1971a). Partitioning of chi-square, analysis of marginal contingency tables, and estimation of expected frequencies in multi-dimensional contingency tables. *Journal of the American Statistical Association* **66**:339–344.

Goodman, L. A. (1971b). Some multiplicative models for the analysis of cross-classified data. In L. LeCam et al. (ed.,) *Proceedings of the Sixth Berkeley Symposium*. Los Angeles, University of California Press, pp. 649–696.

Goodman, L. A. (1971c). The analysis of multidimensional contingency tables: stepwise procedures and direct estimation methods for building models for multiple classifications. *Technometrics* **13**:33–61.

Goodman, L. A. (1973). Guided and unguided methods for selecting models for a set of T multidimensional contingency tables. *Journal of the American Statistical Association* **68**:165–175.

Goodman, L. A. (1975). On the relationship between two statistics pertaining to tests of three-factor interaction in contingency tables. *Journal of the American Statistical Association* **70**:624–625.

Goodman, L. A. (1979). On quasi-independence in triangular contingency tables. *Biometrics* **35**:651–655.

Grizzle, J. E. (1961). A new method of testing hypotheses and estimating parameters for the logistic model. *Biometrics* **17**:372–385.

Grizzle, J. E. (1967). Continuity correction in the chi-square test for 2 × 2 tables. *American Statistician* **21**:28–33.

Grizzle, J. E. (1971). Multivariate logit analysis. *Biometrics* **27**:1057–1062.

Grizzle, J. E., C. F. Starmer, and G. G. Koch (1969). Analysis of categorical data by linear models. *Biometrics* **25**:489–504.

Grizzle, J. E., and O. D. Williams (1972a). Log-linear models and tests of independence for contingency tables. *Biometrics* **28**:137–156.

Grizzle, J. E., and O. D. Williams (1972b). Contingency tables having ordered response categories. *Journal of the American Statistical Association* **67**:55–63.

Haberman, S. J. (1972). Log-linear fit for contingency tables. *Applied Statistics* **21**:218–225.

Haberman, S. J. (1973a). Log-liner models for frequency data: sufficient statistics and likelihood equations. *Annals of Statistics* **1**:617–632.

Haberman, S. J. (1975). Direct products and linear models for complete factorial tables. *Annals of Statistics* **3**:314–333.

Haberman, S. J. (1976). Generalized residuals for log-linear models. In *Proceedings of the Ninth International Biometrics Conference*. Boston, The Biometric Society, Vol. 1, pp. 104–122.

Haberman, S. J. (1977). Log-linear models and frequency tables with small expected cell counts. *Annals of Statistics* **5**:1148–1169.

Haldane, J. B. S. (1937). The exact value of the moments of the distribution of chi-square used as a test of goodness-of-fit, when expectations are small. *Biometrika* **29**:113–143.

Haldane, J. B. S. (1955). Substitutes for chi-square. *Biometrika* **42**:265–266.

Hannan, J. (1960). Consistency of maximum likelihood estimation in discrete distributions. In I. Olkin (ed.) *Contributions to Probability and Statistics: Essays in Honor of Harold Hotelling*, Stanford University Press, pp. 249–257.

Healy, M. J. R. (1969). Exact tests of significance in contingency tables. *Technometrics* **11**:393–395.

Hitchcock, S. E. (1962). A note on the estimation of the parameters of the logistic function using the minimum logit chi-square method. *Biometrika* **49**:250–252.

Hitchcock, S. E. (1966). Tests of hypotheses about the parameters of the logistic function. *Biometrika* **53**:535–544.

Hocking, R. R., and H. H. Oxspring (1974). The analysis of partially categorized contingency data. *Biometrics* **30**:469–483.

Holford, T. R., C. White, and J. L. Kelsey (1978). Multivariate analysis for matched case-control studies. *American Journal of Epidemiology* **107**:245–256.

Imrey, P. B., W. D. Johnson, and G. G. Koch (1976). An incomplete contingency table approach to paired comparison experiments. *Journal of the American Statistical Association* **71**:614–623.

Imrey, P. B., E. Sobel, and M. E. Francis (1979). Analysis of categorical data obtained by stratified random sampling. *Communications in Statistics* **A8**:653–670.

Ireland, C. T., and S. Kullback (1968). Minimum discrimination information estimation. *Biometrics* **24**:707–713.

Irwin, J. O. (1949). A note on the subdivision of chi-square into components. *Biometrika* **36**:130–134.

Kastenbaum, M. A. (1960). A note on the additive partitioning of chi-square in contingency tables. *Biometrics* **16**:416–422.

Kastenbaum, M. A. (1970). A review of contingency tables. In R. C. Bose (ed.) *Essays in Probability and Statistics*. University of North Carolina Press, pp. 407–438.

Kastenbaum, M. A. (1974). Analysis of categorical data: some well known analogues and some new concepts. *Communications in Statistics* **3**:401–417.

Kastenbaum, M. A., and D. E. Lamphiear (1959). Calculation of chi-square to test the no three-factor interaction hypothesis. *Biometrics* **15**:107–115.

Katti, S. J. (1973). Exact distribution for the chi-square test in the one-way table. *Communications in Statistics* **2**:435–447.

Kerridge, D. F. (1961). Inaccuracy and inference. *Journal of the Royal Statistical Society, Series B* **23**:184–194.

Kincaid, W. M. (1962). The combination of $2 \times m$ contingency tables. *Biometrics* **18**:224–228.

Koch, G. G., and D. W. Reinfurt (1971). The analysis of categorical data from mixed models. *Biometrics* **27**:157–173.

Koch, G. G., W. D. Johnson, and H. D. Tolley (1972). A linear models approach to the analysis of survival and the extent of disease in multinomial contingency tables. *Journal of the American Statistical Association* **67**:783–796.

Koehler, K. J., and K. Larntz (1980). An empirical investigation of goodness-of-fit statistics for sparse multinomials. *Journal of the American Statistical Association* **75**:336–344.

Ku, H. H. (1963). A note on contingency tables involving zero frequencies and the $2I$ test. *Technometrics* **5**:398–400.

Ku, H. H., R. Varner, and S. Kullback (1971). On the analysis of multidimensional contingency tables. *Journal of the American Statistical Association* **66**:55–64.

Kullback, S. (1968). Probability densities with given marginals. *Annals of Mathematical Statistics* **39**:1236–1243.

Kullback, S. (1971). Marginal homogeneity of multidimensional contingency tables. *Annals of Mathematical Statistics* **42**:594–606.

Kullback, S. (1973). Estimating and testing interaction parameters in the log-linear model. *Biometrische Zeitschrift* **15**:371–388.

Kullback, S., and M. Fisher (1973). Partitioning second-order interaction in three-way contingency tables. *Applied Statistics* **22**:172–184.

Kullback, S., and M. Fisher (1975). Multivariate logit analysis. *Biometrische Zeitschrift* **17**:139–146.

Kullback, S., and R. Leibler (1951). On information and sufficiency. *Annals of Mathematical Statistics* **22**:79–86.

Kullback, S., and P. N. Reeves (1975). Analysis of interactions between categorical variables. *Biometrische Zeitschrift* **17**:3–12.

Lancaster, H. O. (1951). Complex contingency tables treated by the partition of chi-square. *Journal of the Royal Statistical Society, Series B* **13**:242–249.

Lancaster, H. O. (1950). The exact partition of chi-square and its application to the problem of the pooling of small expectations. *Biometrika* **37**:267–270.

Lancaster, H. O. (1969). Contingency tables of higher dimensions. *Bulletin of the International Statistical Institute* **43**:143–151.

Lancaster, H. O. (1971). The multiplicative definition of interaction. *Australian Journal of Statistics* **13**:36–44.

Larntz, K. (1978). Small sample comparisons of exact levels for chi-square goodness-of-fit statistics. *Journal of the American Statistical Association* **73**:253–263.

Larntz, K., and S. Weisberg (1976). Multiplicative models for dyad formation. *Journal of the American Statistical Association* **71**:455–461.

Lawal, H. B. (1980). Tables of percentage points of Pearson's goodness-of-fit statistic for use with small expectations. *Applied Statistics* **29**:292–298.

Lazarsfeld, Paul F., and Neil W. Henry (1968). *Latent Structure Analysis*. Boston, Houghton Mifflin Co.

Lee, S. K. (1977). On the asymptotic variances of u terms in loglinear models of multidimensional contingency tables. *Journal of the American Statistical Association* **72**:412–419.

Lewis, B. N. (1962). On the analysis of interaction in multi-dimensional contingency tables. *Journal of the Royal Statistical Society, Series A* **125**:88–117.

Light, R. J., and B. H. Margolin (1971). An analysis of variance for categorical data. *Journal of the American Statistical Association* **66**:534–544.

Little, R. E. (1968). A note on estimation for quantal response data. *Biometrika* **55**:578–579.

Madsen, M. (1976). Statistical analysis of multiple contingency tables. Two examples. *Scandinavian Journal of Statistics* **3**:97–106.

Mantel, Nathan (1963). Chi-square tests with one degree of freedom; extensions of the Mantel–Haenszel procedure. *Journal of the American Statistical Association* **58**:690–700.

Mantel, Nathan (1970). Incomplete contingency tables. *Biometrics* **26**:291–304.

Mantel, N. (1973). Synthetic retrospective studies and related topics. *Biometrics* **29**:479–486.

Mantel, N., and J. L. Fleiss (1975). The equivalence of the generalized McNemar tests for marginal homogeneity in 2^3 and 3^3 tables. *Biometrics* **31**:727–729.

Mantel, N., and J. L. Fleiss (1980). Minimum expected cell size requirements for the Mantel–Haenszel one-degree-of-freedom chi-square test and a related rapid procedure. *American Journal of Epidemiology* **112**:129–134.

Mantel, N., and S. W. Greenhouse (1968). What is the continuity correction. *American Statistician* **22**:27–30.

Mantel, Nathan, and W. Haenszel (1959). Statistical aspects of the analysis of data from retrospective studies. *Journal of the National Cancer Institute* **22**:719–748.

Mantel, N., and B. F. Hankey (1975). The odds ratios of a 2×2 contingency table. *American Statistician* **29**:143–145.

Margolin, B. H., and R. J. Light (1974). An analysis of variance for categorical data, II; small sample comparisons with chi-square and other competitors. *Journal of the American Statistical Association* **69**:755–764.

McNemar, Q. (1947). Note on the sampling error of the difference between correlated proportions or percentages. *Psychometrika* **12**:153–157.

Mitra, S. K. (1958). On the limiting power function of the frequency chi-square test. *Annals of Mathematical Statistics* **29**:1221–1233.

Mosteller, F. (1968). Association and estimation in contingency tables. *Journal of the American Statistical Association* **63**:1–28.

Nagnur, B. N. (1969). LAMST and the hypothesis of no three factor interaction in contingency tables. *Journal of the American Statistical Association* **64**:207–215.

Nam, J. (1971). On two tests for comparing matched proportions. *Biometrics* **27**:945–959.

Odoroff, C. L. (1970). A comparison of minimum logit chi-square estimation and maximum likelihood estimation in $2 \times 2 \times 2$ and $3 \times 2 \times 2$ contingency tables: tests for interaction. *Journal of the America Statistical Association* **65**:1617–1631.

Patil, K. D. (1974). Interaction test for three-dimensional contingency tables. *Journal of the American Statistical Association* **69**:164–168.

Patil, G. P., and S. W. Joshi (1968). *A Dictionary and Bibliography of Discrete Distributions.* London, Oliver and Boyd.

Pearl, Raymond, and Lowell J. Reed (1920). On the rate of growth of the population of the United States since 1790 and its mathematical representation. *Proceedings of the National Academy of Sciences* **6**:275–288.

Pike, M. C., A. P. Hill, and P. G. Smith (1980). Bias and efficiency in logistic analysis of stratified case-control studies. *International Journal of Epidemiology* **9**:89–95.

Plackett, R. L. (1962). A note on interactions in contingency tables. *Journal of the Royal Statistical Society, Series B* **24**:162–166.

Plackett, R. L. (1964). The continuity correction in 2×2 tables. *Biometrika* **51**:327–337.

Plackett, R. L. (1969). Multidimensional contingency tables. A survey of models and methods. *Bulletin of the International Statistical Institute* **43**:133–142.

Potthoff, R. F., and M. Whittinghill (1966). Testing for homogeneity I. The binomial and multinomial distributions. *Biometrika* **53**:167–182.

Prentice, R. L. (1976). Use of the logistic model in retrospective studies. *Biometrics* **32**:599–606.

Radhakrishna, S. (1965). Combination of results from several 2×2 contingency tables. *Biometrics* **21**:86–98.

Read, C. B. (1977). Partitioning chi-square in contingency tables: a teaching approach. *Communications in Statistics* **A6**:553–562.

Robbins, Herbert, and E. J. G. Pitman (1949). Applications of the method of mixtures to quadratic forms in normal variates. *Annals of Mathematical Statistics* **20**:552–560.

Savage, I. R. (1973). Incomplete contingency tables: condition for the existence of unique MLE. In P. Jagars and L. Rade (eds.), *Mathematics and Statistics, Essays in Honour of Harold Bergstrom*. Goteborg, Chalmers Institute of Technology, pp. 87–90.

Simpson, E. H. (1951). The interpretation of interaction in contingency tables. *Journal of the Royal Statistical Society, Series B* **13**:238–241.

Slakter, M. J. (1968). Accuracy of an approximation to the power of the chi-square goodness of fit test with small but equal expected frequencies. *Journal of the American Statistical Association* **63**:912–924.

Stuart, A. (1955). A test for homogenity of the marginal distributions in a two-way classification. *Biometrika* **42**:412–416.

Sugiura, N. (1974). Maximum likelihood estimates for the logit model and the iterative scaling method. *Communications in Statistics* **3**:985–993.

Sugiura, N., and M. Otake (1968). Numerical comparison of improved methods of testing in contingency tables with small frequencies. *Annals of the Institute of Statistical Mathematics* **20**:507–517.

Sugiura, N., and M. Otake (1973). Approximate distribution of the maximum of $c - 1$ chi-square statistics (2×2) derived from $2 \times c$ contingency tables. *Communications in Statistics* **1**:9–16.

Sugiura, N., and M. Otake (1974). An extension of the Mantel–Haenszel procedure to k $2 \times c$ contingency tables and the relation to the logit model. *Communications in Statistics* **3**:829–842.

Tarone, R. E., and J. J. Gart (1980). On the robustness of combined tests for trends in proportions. *Journal of the American Statistical Association* **75**:110–116.

Thomas, D. G. (1971). Exact confidence limits for the odds ratio in a 2 × 2 table. *Applied Statistics* **20**:105–110.

Thomas, D. G., and J. J. Gart (1977). A table of exact confidence limits for differences and ratios of two proportions and their odds ratios. *Journal of the American Statistical Association* **72**:73–76.

Watson, G. S. (1959). Some recent results in chi-square goodness of fit tests. *Biometrics* **15**:440–468.

Wermuth, N. (1976). Model search among multiplicative models. *Biometrics* **32**:253–264.

Whittaker, J., and M. Aitkin (1970). A flexible strategy for fitting complex log-linear models. *Biometrics* **34**:487–495.

Whittemore, A. S. (1978). Collapsibility of multidimensional contingency tables. *Journal of the Royal Statistical Society, Series B* **40**:328–340.

Wilks, S. S. (1935). The likelihood test of independence in contingency tables. *Annals of Mathematical Statistics* **6**:190–196.

Williams, O. D., and J. E. Grizzle (1972). Analysis of contingency tables having ordered response categories. *Journal of the American Statistical Association* **67**:55–63.

Wilson, E. B. (1925). The logistic or autocatalytic grid. *Proceedings of the National Academy of Sciences* **11**:451–456.

Yarnold, J. K. (1970). The minimum expectation in chi-square goodness of fit tests and the accuracy of approximations for the null distribution. *Journal of the American Statistical Association* **65**:864–886.

Yates, F. (1961). Marginal percentages in multiway tables of quantal data with disproportionate frequencies. *Biometrics* **17**:1–9.

Zelen, M. (1971). The analysis of several 2 × 2 contingency tables. *Biometrika* **58**:129–137.

Index

Aickin, M., 132, 181
Aitkin, M. A., 181
Anderson, E. B., 225, 264
Anderson, T. W., 264
Approximation, local, 256
Association, 147ff
  marginal, 147, 172
  partial, 147, 172
  see also Ln odds ratio

Bahadur, R. R., 65
Bartlett, M. A., 181
Benedetti, J., 181
Berkson, J., 225
Bhapkar, V. P., 182
Billingsley, P., 66, 264
Binomial, *see* Model
Birch, M. W., 181
Bishop, Y. M. M., 181, 182, 263
BMDP, 225
Bock, R. D., 225
Bonferroni, 153, 182
Breslow, N., 264
Brown, M. B., 181
Buffon's needle problem, 61
Burke, C. J., 23

Category, 2
Cell, 2
  fixed empty, 18, 130, 176
  random empty, 18, 38
C-file, 266
Chi-square:
  information statistic, 52
  likelihood statistic, 12, 51
  Matusita's statistic, 51
  Neyman's statistic, 51
  Pearson's statistic, 12, 31, 37, 50, 58
  see also Distribution
Complement, 69, 82
Conditional independence, *see* Model
Conditionalization, 148
Confidence intervals, *see* Intervals, confidence
Contingency table, 2
Convergence:
  in distribution, 32
  in probability, 31
  problems, 232
  rapid, 76ff, 86, 151
Covariance, 26, 46, 57, 198
Covariates, 19, 184
Cox, D. R., 225
Cross-classification, 3
Crossover design, 178
Cube law, 221

Darroch, J. N., 181, 264
DELTA, 232
Delta method, 35
df, 37
Dietrich, F. H., 182
Discrepancy:
  convex, 49
  definition of, 47, 55
  equivalence of, 48

Discrepancy *(Continued)*
smooth, 47
see also Chi-square
Discrete, 2
Distribution:
chi-square, 12, 37
conditional, 4, 6, 22
empirical, 7, 14
fitted, 14
generating, 69, 72
marginal, 4, 6, 28
multinomial, 32
noncentral chi-square, 52, 55
normal, 33
true, 7
Dyke, G. V., 225

ECTA, 226

Factors, 3
nested, 17
Fienberg, S. E., 181
Fix, E., 132
Forthofer, R. N., 227
F-RES, 152

Gafarian, A. V., 264
GEN, 96, 197
GLIM, 227
Gokhale, D. V., 180, 263
Goodman, L. A., 180, 181, 182, 226, 264
Gram-Schmidt, 28
G-RES, 152
Grizzle, J. E., 227

Haberman, S. J., 132, 181, 182
Hilferty, M. M., 67
Hodges, J. L., 132
Holland, P. W., 181
Holm, S., 182
Hutchinson, T. P., 66

Identifiability, *see* Parameter
Independence, affine linear, 27, 127. *See also* Model
Indicators, 27, 60, 65, 105
Inference:
conditional, 190, 240ff
reversed, 240ff
see also Test
Interference, 180
Intervals:
confidence, 8, 10
multiple confidence, 10, 153
IPF, 263
Ireland, C. T., 180
Irwin, J. O., 66
ITERATIONS, 232

Johnson, N. S., 180

Kastenbaum, M. A., 181
Kerridge, D. F., 66
Khairat, M. A., 131
Koch, G. G., 227
Ku, H. H., 180, 181, 264
Kullback, S., 67, 131, 132, 180, 181, 264
Kupperman, M., 180, 264

LAMDA, 16, 265ff
commands, 335
Lancaster, H. O., 65, 66
Lauritzen, S. L., 132
Least squares, 46
Lehmann, E. L., 132
Level:
definition of, 3
serial, 137ff
significance, 38
summation over, 3
Lewis, D., 23
Ln odds, 8, 245
ratio, 12, 21, 59, 114, 130, 186, 210, 252
Log ratios, 36
LXM, 96, 197, 292
LXX, 233

Mann, H. B., 66
Mantel, N., 225
Marginalization, 149
MCLT, *see* Theorem
MDE:
estimator, 83
principle, 81
MDI, 84
Mean, 26, 57, 153

Möbius function, 142
Model:
  backward, 241
  binomial, 72
  chain of, 134
  of conditional independence, 6, 15, 21, 113, 148
  definition of, 5
  exponential, 71
  of flow, 166
  forward, 241
  hierarchical, 140, 146, 148ff, 155
  of independence, 4-6, 11, 21, 74, 110, 112, 194
  joint, 241
  linear, 69
  linear exponential, 184
  of local independence, 75
  logistic regression, 185, 198, 200, 204, 208
  of marginal homogeneity, 70, 115, 126, 175, 193
  Markov, 234ff, 259, 261
  maximal chain of, 139
  maximal rejected, 141, 156
  minimal confirmed, 141, 156
  paired comparisons, 117, 130, 131, 167
  Poisson, 223
  Poisson regression, 212
  of quasi-independence, 130
  saturated, 105, 110
  of symmetry, 70, 108, 126, 175
  transition, 14
  truncated gamma, 74, 106
  truncated geometric, 123
  truncated inverse binomial, 73
  truncated Poisson, 74
Multinomial, *see* Distribution
Murthy, V. K., 264
MWT, *see* Theorem

Nelder, J. A., 227
Newton-Raphson, 230ff
Neyman, J., 132
N-file, 268
Normal, *see* Distribution
Normalization, 281
Normalizing factor, 71

Page, E., 67
Parameter:
  identifiability, 3, 69, 127, 145
  in independence model, 5
  log-linear, 6, 9, 15, 141ff
  noncentrality, 52
  nuisance, 196
Patterson, H. D., 225
Piaget, 222
Pitman, E. J. G., 264
Plateau, 98
Plot:
  command PLT, 277
  uniform, 153, 160-161, 163
Power, asymptotic, 54, 56, 78, 93
Powers, W., 264
Prediction, 245
Probability distribution, 3
Projection, 58
P-values:
  chi-square, 39
  normal, 39
  programs for, 47
  serial, 94, 137

Rao, C. R., 66, 132
Ratcliff, D., 264
RES-F, 152
RES-G, 152
Residual:
  definition of, 151
  standardized, 9
  use in model fitting, 161
R-file, 274
Robbins, H., 264
Roy, S. N., 181

SAS, 227
SEL, 46, 197
Sequences, contiguous, 52, 77
Silvey, S. D., 181
Simpson, E. H., 24
S-method, 154
Somes, G. W., 182
Space, function, 26
Standardization, 28, 30, 46, 57, 280

Starmer, C. F., 227
STD, 46

Table:
analysis, 92, 136ff, 156
collapsed, 4, 146, 169, 251-252, 254
four-way, 17, 169
one-dimensional, 3, 100ff
ordered, 11, 120, 128
periodic, 43, 101
three-dimensional, 6
triangular, 16, 214
two-dimensional, 3
type 2, 7
type 7, 9
type 2^2, 11, 40, 110
type 2^3, 113, 155
type 2^4, 114
type 2x3x4, 158
type $2x3^2$, 162
type $2^2/2^2$, 210
type 3x4, 112
type 3^6, 120
type $3x2^2$, 112
type 4^2, 13, 115
type 2^2 x 9, 14
type 5^2, 117

Test:
exact conditional, 192
Fisher-Irwin, 194
homogeneity, 246ff.
hypothesis, 38, 56, 93
McNemar's, 193
serial, 93
t-function, 187, 198
Theorem:
central limit (MCLT), 33, 66
Mann-Wald (MWT), 35, 66
T-value, 101

u-functions, 187, 198
Unlikelihood, 188

Vajda, I., 66
Variance, 27
Varner, R., 180, 181

Wackerly, D. D., 182
Wald, A., 66
WGT, 97
Whittmore, A. S., 182
Wilson, E. B., 67

Yule's Q, 59

Zar, J., 67

Here is a unified treatment of the theory, application, and computation required for the statistical analysis of discrete data. Bringing together the most important findings of the past two decades, the text offers statisticians complete, in-depth coverage of the procedures appropriate for log-linear analysis of multifactor contingency tables, as well as logistic covariate analysis of response data falling into a contingency table. In addition to the classical large-sample theory for discrete data, it presents some newer results, including the use of serial inference in situations where there are many hypotheses to be tested.

Throughout the volume, numerous examples from actual experiments and surveys help link theory to a broad range of practical applications and help to resolve many of the difficulties and ambiguities prevalent in the statistical literature. Greatly facilitating analysis as well is a computer program, LAMDA, which is specifically tailored to the methods and techniques presented. Summaries of all theoretical results provide easy reference for readers not requiring detailed proofs.

The text also features a unique classification system which distinguishes the relative importance of each section and whether it is primarily of interest for theory, application, or computation. In addition, to help readers wade through the vast literature in this rapidly expanding field, bibliographical references are categorized according to their importance and readability.

LINEAR STATISTICAL ANALYSIS OF DISCRETE DATA represents a valuable working tool for statisticians in both industry and academia. It will also aid epidemiologists in the analysis of complex studies of the relationships between risk factors and disease, and actuaries in the prediction of mortality and accident rates.

MIKEL AICKIN has done extensive research into the statistical analysis of discrete data by log-linear models, simultaneous inference, and statistical aspects of clinical trials. A former assistant professor of statistics at Arizona State University and the University of Kansas, Dr. Aickin is a member of the Institute of Mathematical Statistics, the Biometric Society, and the Society for Clinical Trials. He received his Ph.D. in biomathematics from the University of Washington.

Applied Probability and Statistics (Continued)

DEMING • Sample Design in Business Research
DODGE and ROMIG • Sampling Inspection Tables, *Second Edition*
DOWDY and WEARDEN • Statistics for Research
DRAPER and SMITH • Applied Regression Analysis, *Second Edition*
DUNN • Basic Statistics: A Primer for the Biomedical Sciences, *Second Edition*
DUNN and CLARK • Applied Statistics: Analysis of Variance and Regression
ELANDT-JOHNSON and JOHNSON • Survival Models and Data Analysis
FLEISS • Statistical Methods for Rates and Proportions, *Second Edition*
FRANKEN, KÖNIG, ARNDT, and SCHMIDT • Queues and Point Processes
GALAMBOS • The Asymptotic Theory of Extreme Order Statistics
GIBBONS, OLKIN, and SOBEL • Selecting and Ordering Populations: A New Statistical Methodology
GNANADESIKAN • Methods for Statistical Data Analysis of Multivariate Observations
GOLDBERGER • Econometric Theory
GOLDSTEIN and DILLON • Discrete Discriminant Analysis
GREENBERG and WEBSTER • Advanced Econometrics: A Bridge to the Literature
GROSS and CLARK • Survival Distributions: Reliability Applications in the Biomedical Sciences
GROSS and HARRIS • Fundamentals of Queueing Theory
GUPTA and PANCHAPAKESAN • Multiple Decision Procedures: Theory and Methodology of Selecting and Ranking Populations
GUTTMAN, WILKS, and HUNTER • Introductory Engineering Statistics, *Third Edition*
HAHN and SHAPIRO • Statistical Models in Engineering
HALD • Statistical Tables and Formulas
HALD • Statistical Theory with Engineering Applications
HAND • Discrimination and Classification
HILDEBRAND, LAING, and ROSENTHAL • Prediction Analysis of Cross Classifications
HOAGLIN, MOSTELLER, and TUKEY • Understanding Robust and Exploratory Data Analysis
HOEL • Elementary Statistics, *Fourth Edition*
HOEL and JESSEN • Basic Statistics for Business and Economics, *Third Edition*
HOLLANDER and WOLFE • Nonparametric Statistical Methods
IMAN and CONOVER • Modern Business Statistics
JAGERS • Branching Processes with Biological Applications
JESSEN • Statistical Survey Techniques
JOHNSON and KOTZ • Distributions in Statistics
Discrete Distributions
Continuous Univariate Distributions—1
Continuous Univariate Distributions—2
Continuous Multivariate Distributions
JOHNSON and KOTZ • Urn Models and Their Application: An Approach to Modern Discrete Probability Theory
JOHNSON and LEONE • Statistics and Experimental Design in Engineering and the Physical Sciences, Volumes I and II, *Second Edition*
JUDGE, HILL, GRIFFITHS, LÜTKEPOHL and LEE • Introduction to the Theory and Practice of Econometrics
JUDGE, GRIFFITHS, HILL and LEE • The Theory and Practice of Econometrics
KALBFLEISCH and PRENTICE • The Statistical Analysis of Failure Time Data